Technologies to Advance Automation in Forensic Science and Criminal Investigation

Chung-Hao Chen
Old Dominion University, USA

Wen-Chao Yang
National Central Police University, Taiwan

Lijia Chen
Henan University, China

A volume in the Advances in Digital Crime, Forensics, and Cyber Terrorism (ADCFCT) Book Series

IGI Global
PUBLISHER of TIMELY KNOWLEDGE

Published in the United States of America by
IGI Global
Information Science Reference (an imprint of IGI Global)
701 E. Chocolate Avenue
Hershey PA, USA 17033
Tel: 717-533-8845
Fax: 717-533-8661
E-mail: cust@igi-global.com
Web site: http://www.igi-global.com

Library of Congress Cataloging-in-Publication Data

Names: Chen, Chung-Hao, 1974- editor. | Yang, Wen-Chao, 1975- editor.
Title: Technologies to advance automation in forensic science and criminal investigation / Chung-Hao Chen, Wen-Chao Yang, and Lijia Chen, editors.
Description: Hershey, PA : Information Science Reference, 2021. | Includes bibliographical references and index. | Summary: "This book is focused on technologies to advance automation in forensic science and criminal investigation, including digital evidence and computer technology, photoelectronics, jurisprudence, behavioral science, linguistics, Forensic digital multimedia authentication and origin, cyber security, video surveillance system and technologies used in modern forensics"-- Provided by publisher.
Identifiers: LCCN 2021026862 (print) | LCCN 2021026863 (ebook) | ISBN 9781799883869 (hardcover) | ISBN 9781799883876 (paperback) | ISBN 9781799883883 (ebook)
Subjects: LCSH: Criminal investigation--Technological aspects. | Electronics in criminal investigation. | Electronic evidence. | Forensic sciences--Technological aspects. | Digital forensic science.
Classification: LCC HV8073 .T2266 2021 (print) | LCC HV8073 (ebook) | DDC 363.25--dc23
LC record available at https://lccn.loc.gov/2021026862
LC ebook record available at https://lccn.loc.gov/2021026863

This book is published in the IGI Global book series Advances in Digital Crime, Forensics, and Cyber Terrorism (ADCFCT) (ISSN: 2327-0381; eISSN: 2327-0373)

British Cataloguing in Publication Data
A Cataloguing in Publication record for this book is available from the British Library.

All work contributed to this book is new, previously-unpublished material.
The views expressed in this book are those of the authors, but not necessarily of the publisher.

For electronic access to this publication, please contact: eresources@igi-global.com.

Advances in Digital Crime, Forensics, and Cyber Terrorism (ADCFCT) Book Series

ISSN:2327-0381
EISSN:2327-0373

Editor-in-Chief: Bryan Christiansen Global Research Society, LLC, USA Agnieszka Piekarz Independent Researcher, Poland

MISSION

The digital revolution has allowed for greater global connectivity and has improved the way we share and present information. With this new ease of communication and access also come many new challenges and threats as cyber crime and digital perpetrators are constantly developing new ways to attack systems and gain access to private information.

The **Advances in Digital Crime, Forensics, and Cyber Terrorism (ADCFCT) Book Series** seeks to publish the latest research in diverse fields pertaining to crime, warfare, terrorism and forensics in the digital sphere. By advancing research available in these fields, the **ADCFCT** aims to present researchers, academicians, and students with the most current available knowledge and assist security and law enforcement professionals with a better understanding of the current tools, applications, and methodologies being implemented and discussed in the field.

COVERAGE

- Digital Crime
- Criminology
- Computer Virology
- Cyber Terrorism
- Data Protection
- Database Forensics
- Network Forensics
- Watermarking
- Global Threat Intelligence
- Encryption

IGI Global is currently accepting manuscripts for publication within this series. To submit a proposal for a volume in this series, please contact our Acquisition Editors at Acquisitions@igi-global.com or visit: http://www.igi-global.com/publish/.

The Advances in Digital Crime, Forensics, and Cyber Terrorism (ADCFCT) Book Series (ISSN 2327-0381) is published by IGI Global, 701 E. Chocolate Avenue, Hershey, PA 17033-1240, USA, www.igi-global.com. This series is composed of titles available for purchase individually; each title is edited to be contextually exclusive from any other title within the series. For pricing and ordering information please visit http://www.igi-global.com/book-series/advances-digital-crime-forensics-cyber/73676. Postmaster: Send all address changes to above address.

Titles in this Series

For a list of additional titles in this series, please visit:
www.igi-global.com/book-series/advances-digital-crime-forensics-cyber/73676

Intelligence and Law Enforcement in the 21st Century
Eugene de Silva (Virginia Research Institute, USA) and Asanga Abeyagoonesekera (Parliament of Sri Lanka, Sri Lanka)
Information Science Reference • © 2021 • 253pp • H/C (ISBN: 9781799879046) • US $225.00

Social Engineering and Information Warfare Operations Emerging Research and Opportunities
Rhonda L. Johnson (Upper Iowa University, USA)
Information Science Reference • © 2021 • 150pp • H/C (ISBN: 9781799842705) • US $145.00

Evaluating Emerging Threats and New Research Opportunities in Digital Crime and Forensics
Rhonda Johnson (Upper Iowa University, USA)
Information Science Reference • © 2021 • 350pp • H/C (ISBN: 9781799822288) • US $195.00

Confluence of AI, Machine, and Deep Learning in Cyber Forensics
Sanjay Misra (Covenant University, Nigeria) Chamundeswari Arumugam (Sri Sivasubramaniya Nadar College of Engineering, India) Suresh Jaganathan (Sri Sivasubramaniya Nadar College of Engineering, India) and Saraswathi S. (Sri Sivasubramaniya Nadar College of Engineering, India)
Information Science Reference • © 2021 • 248pp • H/C (ISBN: 9781799849001) • US $225.00

Cyber Security Auditing, Assurance, and Awareness Through CSAM and CATRAM
Regner Sabillon (Universitat Oberta de Catalunya, Spain)
Information Science Reference • © 2021 • 260pp • H/C (ISBN: 9781799841623) • US $195.00

For an entire list of titles in this series, please visit:
www.igi-global.com/book-series/advances-digital-crime-forensics-cyber/73676

IGI Global
PUBLISHER of TIMELY KNOWLEDGE

701 East Chocolate Avenue, Hershey, PA 17033, USA
Tel: 717-533-8845 x100 • Fax: 717-533-8661
E-Mail: cust@igi-global.com • www.igi-global.com

To My Parents, Wife (Chiung-Ju), and Children (Ryan and Rachel)

To My Parents, Wife (Yi-Chen) and Children (Sheng-Wei, and Tzu-Ching)

To My Parents, Wife (Lijuan Zou) and Children (Mufan Chen and Yifan)

Table of Contents

Preface xiii

Acknowledgment xvii

Chapter 1
Advances in Forensic Geochemistry 1
Elhoucine Essefi, University of Gabes, Tunisia

Chapter 2
Advances in Forensic Geophysics: Magnetic Susceptibility as a Tool for Environmental Forensic Geophysics 15
Elhoucine Essefi, University of Gabes, Tunisia

Chapter 3
Advances in Forensic Sedimentology 37
Elhoucine Essefi, University of Gabes, Tunisia

Chapter 4
Advances of Forensic Remote Sensing Applications in the Face of Transnational Organized Crime and Terrorism 48
Elhoucine Essefi, University of Gabes, Tunisia

Chapter 5
Cross-Layer Learning: A Deep Learning-Based Forensic Framework for IoT Systems 62
Tushar Mane, Symbiosis Institute of Technology, Symbiosis International University (Deemed), India
Ambika Pawar, Symbiosis Institute of Technology, Symbiosis International University (Deemed), India

Chapter 6
Fire Investigation and Ignitable Liquid Residue Analysis 91
Sachil Kumar, Naif Arab University for Security Sciences, Saudi Arabia
Anu Singla, Bundelkhand University, India
Ruddhida R. Vidwans, Jain University, India

Chapter 7
Large Feature Mining With Ensemble Learning for Image Forgery Detection. 119
Qingzhong Liu, Sam Houston State University, USA
Tze-Li Hsu, Sam Houston State University, USA

Chapter 8
Forensic Camera Identification in Social Networks via Camera Fingerprint 148
Tzuhuan Lin, New Taipei City Police Department, Taiwan
Yu-Ru Wang, Yilan County Government Police Bureau, Taiwan

Chapter 9
Vehicle License Plate Recognition With Deep Learning 161
Chi-Hsuan Huang, National Taiwan University, Taiwan
Yu Sun, National Taiwan University, Taiwan
Chiou-Shana Fuh, National Taiwan University, Taiwan

Chapter 10
Leveraging Machine Learning in Financial Fraud Forensics in the Age of Cybersecurity 220
Md Ariful Haque, Old Dominion University, USA
Sachin Shetty, Old Dominion University, USA

Conclusion 250

Compilation of References 255

About the Contributors 283

Index 288

Detailed Table of Contents

Preface xiii

Acknowledgment xvii

Chapter 1
Advances in Forensic Geochemistry 1
Elhoucine Essefi, University of Gabes, Tunisia

This chapter is meant to give the state of the art of forensic geochemistry and recent advances. In terms of forensic organic geochemistry, detecting mature organic matter including polluting hydrocarbons follows an experimental procedure by using recent experimental analytical techniques. However, the interpretation of these results needs an understanding of the geochemical context to make a distinction between the natural and the human made origin of oil. Infrared data coupled with statistical analyses would have an important relevance for the detection of the pollution during the Anthropocene, which is marked an increasing human pollution reaching the level of environmental crimes. In terms of nuclear and isotopic forensic geochemistry, recent studies provided that nuclear forensics considers the fact that some measurable parameters or signatures are distinctive.

Chapter 2
Advances in Forensic Geophysics: Magnetic Susceptibility as a Tool for Environmental Forensic Geophysics 15
Elhoucine Essefi, University of Gabes, Tunisia

Traditionally, forensic geophysics involves the study, search, localization, and mapping of buried objects or elements within soil, buildings, or water using geophysics tools for legal purposes. Recently, with the evolution of environmental crimes, forensic geophysics gave special care to detection, location, and quantification of polluting products. New techniques including the magnetic susceptibility have emerged to investigate this type of crimes. After discussing the state of the art of forensic geophysics, this chapter proposed the magnetic susceptibility as an efficient tool of environmental crimes detection. A case study of pollution detection was proposed from Tunisia. Being a fast and cheap technique, magnetic surveys represent a real promise for environmental forensic geophysics.

Chapter 3
Advances in Forensic Sedimentology 37
Elhoucine Essefi, University of Gabes, Tunisia

Forensic sedimentology is a relatively recently realized field. Sedimentological methods used to solve cases have evolved as the field has developed, beginning with simple identification of minerals and progressing to the examination of individual grains using highly advanced scanning electron microscopes. More simple methods, such as color analysis, are still used today, but in addition, forensic sedimentologists look at surface textures and grain size distribution. For instance, quartz grains were used in a forensic technique as sediment fingerprint. The particle size distribution is one of the important tests when analysing sediments and soils in geological studies. For forensic work, the particle size distribution of sometimes very small samples requires precise determination using a rapid and reliable method with a high resolution. FRITSCH laser granulometer offers rapid and accurate sizing of particles in the range 0.04–2000 μm for a variety of sample types, including soils, unconsolidated sediments, dusts, powders, and other particulate materials.

Chapter 4
Advances of Forensic Remote Sensing Applications in the Face of Transnational Organized Crime and Terrorism 48
Elhoucine Essefi, University of Gabes, Tunisia

This chapter aims to investigate advance and relevance of remote sensing in detecting the increasing transnational terrorist and crimes acts. This work should take into the widest definition of transnational crimes and terrorist activities and the link between. Geopolitics has created a favor climate for the setting of transnational crimes and terrorism at the Tunisian-Libyan borders. A possible future scenario is the fall of a military base with high technology arms in the hand of terrorist groups. Remote would be relevant by monitoring of terrorist mobility and their number evolution, arms quality and quantity within the base and the region, linked illegal activities funding terrorist groups (human trafficking from Africa to Europe, arms trade towards Mali, and smuggling).

Chapter 5
Cross-Layer Learning: A Deep Learning-Based Forensic Framework for IoT Systems 62
Tushar Mane, Symbiosis Institute of Technology, Symbiosis International University (Deemed), India
Ambika Pawar, Symbiosis Institute of Technology, Symbiosis International University (Deemed), India

Deep learning-based investigation mechanisms are available for conventional forensics, but not for IoT forensics. Dividing the system into different layers according to their functionalities, collecting data from each layer, finding the correlating factor, and using it for pattern detection is the fundamental concept behind the proposed intelligent system. The authors utilize this notion for embedding intelligence in forensics and speed up the investigation process by providing hints to the examiner. They propose a novel cross-layer learning architecture (CCLA) for IoT forensics. To the best of their knowledge, this is the first attempt to incorporate deep learning into the forensics of the IoT ecosystem.

Chapter 6
Fire Investigation and Ignitable Liquid Residue Analysis 91
Sachil Kumar, Naif Arab University for Security Sciences, Saudi Arabia
Anu Singla, Bundelkhand University, India
Ruddhida R. Vidwans, Jain University, India

A fire investigation is a difficult and challenging task. An investigator's basic task at a fire scene is two-fold: first, to ascertain the origin of the fire and, second, to closely investigate the site of origin and try to determine what triggered a fire to start at or near that spot. Usually, an investigation would begin by attempting to obtain a general view of the site and the fire damage; this may be achieved at ground level or from an elevated location. Following this, one may examine the materials available, the fuel load, and the condition of the debris at different locations. Surprisingly, the science of fire investigation is not stagnant, and each year, more information to assist investigators in determining the location and cause of a fire by diligent observation of the scene and laboratory study of fire debris is released. This chapter is split into two sections. The first section discusses the general procedures to be used during a fire investigation, and the second section discusses laboratory analysis of ignitable liquid residue analysis.

Chapter 7
Large Feature Mining With Ensemble Learning for Image Forgery Detection.119
Qingzhong Liu, Sam Houston State University, USA
Tze-Li Hsu, Sam Houston State University, USA

The detection of different types of forgery manipulation including seam-carving in JPEG images is a hot spot in image forensics. Seam carving was originally designed for content-aware image resizing. It is also being used for forgery manipulation. It is still very challenging to effectively identify the seam carving forgery under recompression. To address the highly challenging detection problems, this chapter introduces an effective approach with large feature mining. Ensemble learning is used to deal with the high dimensionality and to avoid overfitting that may occur with some traditional learning classifier for the detection. The experimental results validate the efficacy of proposed approach to detecting JPEG double compression and exposing the seam-carving forgery while the JPEG recompression is proceeded at the same quality and a lower quality, which is generally much harder for traditional detection methods. The methodology introduced in this chapter provides a strategy and realistic approach to resolve the highly challenging problems in image forensics.

Chapter 8
Forensic Camera Identification in Social Networks via Camera Fingerprint 148
Tzuhuan Lin, New Taipei City Police Department, Taiwan
Yu-Ru Wang, Yilan County Government Police Bureau, Taiwan

Image-related crimes cause the urgent demand for tracing the origin of digital images. The breakthrough is a passive detection method via photo response non-uniformity (PRNU) analysis proposed by Lukáš et al. Recently, digital images are often shot with handheld devices (such as smartphones) and transmitted using social media (such as LINE). Most of the images are distorted (such as compressed and resized) during transmission. Previous studies are less focused on the impact of transmission compression through social networks. Thirty-one different Apple mobile phones were used to capture digital images in the experiment. Images were uploaded to the photo album via LINE software and then downloaded. The modified signed peak correlation energy (MSPCE) statistics is used to evaluate the correlation between the PRNU values of the disputed images and the pattern noise of the experimental devices. Experimental results show that the PRNU analysis method can effectively trace the source of the shot device using the distorted images which are compressed and resized during the transmission in LINE.

Chapter 9
Vehicle License Plate Recognition With Deep Learning 161
Chi-Hsuan Huang, National Taiwan University, Taiwan
Yu Sun, National Taiwan University, Taiwan
Chiou-Shana Fuh, National Taiwan University, Taiwan

In this chapter, an AI (artificial intelligence) solution for LPR (license plate recognition) on moving vehicles is proposed. The license plates in images captured with cameras on moving vehicles have unpredictable distortion and various illumination which make traditional machine vision algorithms unable to recognize the numbers correctly. Therefore, deep learning is leveraged to recognize license plate in such challenging conditions for better recognition accuracy. Additionally, lightweight neural networks are chosen since the power supply of scooter is quite limited. A two-stage method is presented to recognize license plate. First, the license plates in captured images are detected using CNN (convolutional neural network) model and the rotation of the detected license plates are corrected. Subsequently, the characters are recognized as upper-case format (A-Z) and digits (0-9) with second CNN model. Experimental results show that the system achieves 95.7% precision and 95% recall at high speed during the daytime.

Chapter 10
Leveraging Machine Learning in Financial Fraud Forensics in the Age of Cybersecurity 220
Md Ariful Haque, Old Dominion University, USA
Sachin Shetty, Old Dominion University, USA

Financial sectors are lucrative cyber-attack targets because of their immediate financial gain. As a result, financial institutions face challenges in developing systems that can automatically identify security breaches and separate fraudulent transactions from legitimate transactions. Today, organizations widely use machine learning techniques to identify any fraudulent behavior in customers' transactions. However, machine learning techniques are often challenging because of financial institutions' confidentiality policy, leading to not sharing the customer transaction data. This chapter discusses some crucial challenges of handling cybersecurity and fraud in the financial industry and building machine learning-based models to address those challenges. The authors utilize an open-source e-commerce transaction dataset to illustrate the forensic processes by creating a machine learning model to classify fraudulent transactions. Overall, the chapter focuses on how the machine learning models can help detect and prevent fraudulent activities in the financial sector in the age of cybersecurity.

Conclusion 250

Compilation of References 255

About the Contributors 283

Index 288

Preface

In the current scenario of Forensic Science and Criminal Investigation, experts are facing many challenges, including huge amount of data, tiny pieces of evidences in the chaotic and complex environment, traditional laboratory structures and sometimes insufficient knowledge, which may lead to failure of investigation or miscarriage of justice. The scope of this book is focused on technologies to advance automation in forensic science and criminal investigation, including digital evidence and computer technology, photoelectronics, jurisprudence, behavioral science, linguistics, Forensic digital multimedia authentication and origin, cyber security, video surveillance system and technologies used in modern forensics. Topics are covered in this book are as follows:

Chapter 1 is meant to give the state of the art of forensic geochemistry and recent advances. In terms of forensic organic geochemistry, detecting mature organic matter including polluting hydrocarbons follows an experimental procedure by using recent experimental analytical techniques. However, the interpretation of these results need an understanding of the geochemical context to make a distinction between the natural and the human made origin of oil. Infrared data coupled with statistical analyses would have an important relevance for the detection of the pollution during the Anthropocene, which is marked an increasing human pollution reaching the level of environmental crimes. In terms of nuclear and isotopic forensic geochemistry, recent studies provided that nuclear forensics considers the fact that some measurable parameters or signatures are distinctive.

In Chapter 2, traditionally, forensic geophysics involves the study, search, localization and mapping of buried objects or elements within soil, buildings, or water using geophysics tools for legal purposes. Recently, with the evolution of environmental crimes, forensic geophysics gave special care to detection, location and quantification of polluting products. New techniques including the magnetic susceptibility have emerged to investigate this type of crimes. After discussing the state of the art of forensic geophysics, this chapter proposed the magnetic susceptibility as efficient tool of environmental crimes detection. A case study of pollution detection was proposed from Tunisia. Being a fast and cheap technique, magnetic surveys represent a real promise for environmental forensic geophysics.

In Chapter 3, forensic sedimentology is a relatively recently realized field. Sedimentological methods used to solve cases have evolved as the field has developed, beginning with simple identification of minerals and progressing to the examination of individual grains using highly advanced scanning electron microscopes. More simple methods, such as color analysis, are still used today, but in addition, forensic sedimentologists look at surface textures and grain size distribution. For instance, quartz grains were used in a forensic technique as sediment fingerprint. The particle size distribution is one of the important tests when analysing sediments and soils in geological studies. For forensic work, the particle size distribution of sometimes very small samples requires precise determination using a rapid and reliable method with a high resolution. FRITSCH laser granulometer offers rapid and accurate sizing of particles in the range 0.04–2000 µm for a variety of sample types, including soils, unconsolidated sediments, dusts, powders and other particulate materials. Reliable results are possible for sample weights of just 50 mg. Discrimination between samples is performed on the basis of the shape of the particle size curves and statistical measures of the size distributions.

Chapter 4 aims to investigate advance and relevance of remote sensing in detecting the increasing transnational terrorist and crimes acts. This work should take into the widest definition of transnational crimes and terrorist activities and the link between. Geopolitics has created a favor climate for the setting of transnational crimes and terrorism at the Tunisian-Libyan borders. A possible future scenario is the fall of a military base with high technology arms in the hand of terrorist groups. Remote would be relevant such situation by monitoring of terrorist mobility and their number evolution, arms quality and quantity within the base and the region, linked illegal activities funding terrorist groups (human trafficking from Africa to Europe, arms trade towards Mali, and smuggling).

In Chapter 5, deep learning is proving to be more robust and accurate in comparison with the rest of the machine learning techniques due to the support of larger datasets. Deep linking in Internet of Things (IoT) environments is achievable with the help of deep learning. Deep Linking Protocol enables IoT objects and applications to interact involuntarily with each other. Smart home example gives an idea of how things in a smart home can collaborate to form a system. As of now, deep learning has been applied in Network Forensics only. We believe Deep learning assisted investigation of end-to-end IoT applications will certainly help speed up the investigation process with fewer errors and leftovers.

In Chapter 6, fire investigation is the study of fire-related incidents using fire dynamics. When a tragic fire happens, experts must decide where and how the fire began, as well as whether it was accidental or deliberate. If the fire was started on intentionally, the investigator's results could lead to felony charges. If there were deaths as a result of the fire, an individual may be charged with murder. If they suspect that a fire was lit deliberately, they want to see some signs of an ignitable liquid residue (ILR). Spotting the source of the fire allows the investigator to gather residue that could contain substances suggestive of, say, fuel or hydrocarbons that are trademarks of other ignitable liquids which can be used to cause fires.

In Chapter 7, the detection of different types of forgery manipulation including seam-carving in JPEG images is a hot spot in image forensics. Seam carving was originally designed for content-aware image resizing. It is also being used for forgery manipulation. It is still very challenging to effectively identify the seam carving forgery under recompression. To address the highly challenging detection problems, this chapter introduces an effective approach with large feature mining. Ensemble learning is used to deal with the high dimensionality and to avoid overfitting that may occur with some traditional learning classifier for the detection. The experimental results validate the efficacy of proposed approach to detecting JPEG double compression and exposing the seam-carving forgery while the JPEG recompression is proceeded at the same quality and a lower quality, which is generally much harder for traditional detection methods. The methodology introduced in this chapter provides a strategy and realistic approach to resolve the highly challenging problems in image forensics.

In Chapter 8, image-related crimes cause the urgent demand for tracing the origin of digital images. The breakthrough is a passive detection method via Photo Response Non-uniformity (PRNU) analysis proposed by Lukáš et al. Traditionally, digital images are often shot with handheld devices (such as smartphones) and transmitted using social media (such as LINE). Most of the images are distorted (such as compressed and resized) during transmission. Previous studies are less focused on the impact of transmission compression through social networks. In this research, 31 different Apple mobile phones were used to capture digital images, which were uploaded to the photo album via LINE software, and then downloaded. The Modified Signed Peak Correlation Energy (MSPCE) statistics is used to evaluate the correlation between the PRNU of the disputed image and the pattern noise of the experimental devices. Experimental results show that the PRNU analysis method could effectively trace the source of the imaging device using the distorted images which are compressed and resized during the transmission in LINE.

In Chapter 9, an AI (Artificial Intelligence) solution for LPR (License Plate Recognition) on moving vehicles is proposed. The license plates in images captured with cameras on moving vehicles have unpredictable distortion and various illumination which make traditional machine vision algorithms unable to recognize the numbers correctly. Therefore, deep learning is leveraged to recognize license plate in such challenging conditions for better recognition accuracy. Additionally, lightweight neural networks are chosen since the power supply of scooter is quite limited. A two-stage method is presented to recognize license plate. First, the license plates in captured images are detected using CNN (Convolutional Neural Network) model and the rotation of the detected license plates are corrected. Subsequently, the characters are recognized as upper-case format (A-Z) and digits (0-9) with second CNN model. Experimental results show that our system achieves 95.7% precision and 95% recall at high speed during the daytime.

In Chapter 10, financial sectors are lucrative cyber-attack targets because of their immediate financial gain. As a result, financial institutions face challenges in developing systems that can automatically identify security breaches and separate fraudulent transactions from legitimate transactions. Today, organizations widely use machine learning techniques to identify any fraudulent behavior in customers' transactions. However, machine learning techniques are often challenging because of financial institutions' confidentiality policy, leading to not sharing the customer transaction data. This chapter discusses some crucial challenges of handling cybersecurity and fraud in the financial industry and building machine learning-based models to address those challenges. We utilize an open-source e-commerce transaction dataset to illustrate the forensic processes by creating a machine learning model to classify fraudulent transactions. Overall, the chapter focus on how the machine learning models can help detect and prevent fraudulent activities in the financial sector in the age of cybersecurity.

Acknowledgment

We are indebted to all IGI publisher, reviewers, and chapter authors to the book.

Chapter 1
Advances in Forensic Geochemistry

Elhoucine Essefi
University of Gabes, Tunisia

ABSTRACT

This chapter is meant to give the state of the art of forensic geochemistry and recent advances. In terms of forensic organic geochemistry, detecting mature organic matter including polluting hydrocarbons follows an experimental procedure by using recent experimental analytical techniques. However, the interpretation of these results needs an understanding of the geochemical context to make a distinction between the natural and the human made origin of oil. Infrared data coupled with statistical analyses would have an important relevance for the detection of the pollution during the Anthropocene, which is marked an increasing human pollution reaching the level of environmental crimes. In terms of nuclear and isotopic forensic geochemistry, recent studies provided that nuclear forensics considers the fact that some measurable parameters or signatures are distinctive.

1. INTRODUCTION

Forensic geochemistry is study of organic (de Oliveira et al., 2020; Poulson and Drever, 2020) and inorganic elements (John et al., 2021) occurring when environmental crimes related to pollution of soil, surface and/or deep water, atmosphere and rocks had taken place. The geochemical analysis of traces in the forensic field is able to determine the origin and type of contaminants and the processes of interaction between chemical contaminants and contaminated (Whyte et al., 2021). Geochemical surveys can also provide important support for the investigation of events such as murder,

DOI: 10.4018/978-1-7998-8386-9.ch001

rape, theft (Dawson et al., 2021). The geochemical analyzes mainly applied in the forensic field are the analysis of trace elements (Sutliff-Johansson, 2021) and the analysis of radioactive and stable isotopes (Ubelaker and Francescutti, 2020; Migeon et al., 2020). The isotopic composition of geological material can provide relevant information to delineate crime trends. Moreover, thanks to this type of analysis, it is possible to date the material and sometimes the processes suffered from the same material (Kristo, 2020). This chapter is meant to review recent advances in forensic geochemistry dealing with organic pollutants as well as inorganic contaminants such as trace elements, radioactive and stable isotopes.

2. FORENSIC ORGANIC GEOCHEMISTRY

Biomarkers are one of the most important hydrocarbon classes in petroleum (Malmborg et al., 2020). They are one of the most resistant groups of hydrocarbons to degradation in the environment (Thienpont et al., 2021). Using gas chromatography/mass spectrometry (GC/MS) we can detect biomarkers in small quantities (ppm and sub ppm) in the presence of a wide variety of other petroleum hydrocarbon types. Biomarkers are one of the most degradation resistant groups of hydrocarbons in the environment. They have formed under a variety of conditions and geological ages resulting in several biomarker fingerprints. Chemical analysis of biomarkers gives very important information to environmental legal investigations from the point of view of determining the source of spilled oil, differentiating and correlating oils and monitoring the degradation process and eroding condition of oils under a wide variety of conditions. Biomarkers have for many years been widely used in petroleum exploration and reservoir geochemistry to obtain information of value to geochemists such as the thermal maturity of the oil, the type of source material, the depositional environment of the source rock, the approximate geological age of the source rock and the degree of biodegradation of the oil (Zhou et al., 2021). This chapter provided a brief overview and discussion of biomarker chemistry, biomarker characterization and quantification, biomarker distributions, eroding effects on biomarker composition.

2.1. Methods of Forensic Organic Geochemistry

Organic Matter Extraction and Analysis

The use of biomarker data is a long-standing tool in petroleum system analysis, but it is also widely used in forensic geochemistry to identify oil spills (Nuzzo and Gehlen, 2021). Biomarkers are complex molecular fossils derived from once-existing organisms and can be measured in crude oil and its host rocks. They provide powerful tools for associated oil wells (Chua et al., 2021). The amount of biomarkers that are trace components of organic matter/oil can be measured very accurately on the selected ion monitoring mode (GC/MS-SIM) by gas chromatography-mass spectrometry (Fig.1). For qualitative and quantitative geochemistry in the petroleum industry and forensic geochemistry of oil spills, it is important to reliably prove legal decisions about the source of pollution because the source of crude oil or refined products released into the environment is usually unknown, so effective identification based on the correlation with the suspected source is an important part of assessing oil spills (Zacharias et al., 2021). Two main processes make it difficult to correlate spilled oil with its source: weathering of the oil in the environment and mixing with hydrocarbons from different sources. Weathering includes processes that occur in the early stages of oil spills, namely proliferation, evaporation, dispersion, emulsification and dissolution, as well as long-term processes that determine the final destination of the oil, such as photooxidation, precipitation and biodegradation. The quantitative and qualitative analyses of hydrocarbons proceeded in four stages in Essefi, (2013): the elimination of elementary sulfur, the extraction of the organic matter, the purification and eventually the detection of different components. (1) Desulphurization: The sample of sediment was first placed in the steamroom at 40 C° to be dehydrated; then, it was crushed manually by a mortar. Only the fraction that was smaller than 63um was analyzed. Due to its high content the sample, sulfur bothers the separation of the saturated and aromatic hydrocarbons. Thus, it should be fixed within a column on an amalgam of copper and zinc. Amalgam preparation consisted in: (1) dissolving 200 mg of crystallized sulfate of copper within 2 liters of distilled water under agitation; (2) dissolving 30 g of powder of zing within 50 ml of distilled water; (3) shedding drop by drop the mixture of zinc within the solution of copper in order to avoid pulverizing of amalgam. The obtained pasty mixture with a cupric color permitted the fixation of the elementary sulfur. The operation of sulfur elimination was carried out on a glass column (Pasteur Pipette), having 10 cm of length and 5 mm of diameter. The column was washed many times with acetone in order to eliminate water; then, it was rinsed with $CHCL_3$. The sample was placed on the amalgam and washed many times with $CHCL_3$. During this operation, the amalgam maintained the sulfur and became black. The de-sulfured extract was recuperated within a pipe with calibrated haemolysis in order to be weighted after total evaporation of the solvent. (2) Extraction of the organic matter: An amount of 10 g of the de-sulfured fine fraction was placed in a cellulose cartridge of a soxlet (Fig. 1a). The organic solvent composed of 100 ml of methanol (MeOH) and 200 ml of dichloromethane (CH_2Cl_2) evaporate under the heat effect. Hence, it joined the cooling agent while passing by the tube supply main; thus, it condensed to fall within the cartridge of extraction. The sample of sediment macerated with organic solvent.

Once the cartridge was full, the liquid accompanied by the extracted organic matter turned over in the balloon; the cycle continued making the balloon progressively enriched by the organic matter. It was necessary to count approximately 7 hours to extract by this method all the organic matter of a sample. Thereafter, the balloon was placed in a Marie bath to evaporate solvent and recuperate the organic matter. Then, the concentration contained in a sample was obtained by weighing the dry residual. (3) Purification of hydrocarbons: The purification of different components of hydrocarbons was done by fractionation. To be able to dose with precision only hydrocarbons, other components should be eliminated from the organic matter; the residual suspended with dichloromethane passed through a burette containing of the florisil, which is a gel that can be crossed only by hydrocarbons and others negligible lipidic compounds. The total hydrocarbons were obtained after the passage in the burette. The concentration of these total hydrocarbons is determined also by weighting the dry residual obtained after evaporation. To facilitate the detection of hydrocarbons by chromatography, the saturated are separated from non-polar compounds. The total residual hydrocarbon was included in hexane then passed in a silica gel column; the fraction saturated with hydrocarbons may be hence recovered. In the same silica column non-polar hydrocarbons were fairly elected by a hexane-dichloromethane mixture with the proportions of 2/3-1/3. This recovered fraction corresponds to Polycyclic Aromatic Hydrocarbons. The percentage of the polar fraction may be obtained by simple subtraction of Alkanes and Aromatic Polycyclic Hydrocarbons from Total Hydrocarbons. (4) Detection of hydrocarbons by chromatography: The detection of hydrocarbons of the saturated fraction was carried out by Gas Chromatography (GC) coupled with Mass Spectroscopy (MS) (Fig.1b). Thus, these aliphatic hydrocarbons were analyzed by gas chromatography–mass spectrometry HP 6890-HP 5973 MSD combination (Agilent Technologies, Wilmington, DE, USA). The GC was used with a 30m fused-silica column (0.25 mm i.d.) coated with 5% phenyl methyl siloxane. Helium was used as the carrier gas at a flow rate of 1.4 ml/min. The temperature program was as follows: temperature held at 40 °C for 1 min, then increased from 40 to 120 °C at 30 °C min^{-1}, 120 to 300 °C at 5 °C min^{-1} with the final isothermal hold at 300 °C for 20 min. The mass spectrometer operated in the electron impact (EI) mode at 70 eV (ionization energy) and scanned from 50 to 650 Dalton. Compounds were identified by comparison with published mass spectra and relative retention times.

Infrared Spectroscopy

Fourier transform infrared (FT-IR) spectroscopy technology has attracted the attention of forensics due to its high sensitivity and selectivity and near real-time detection of the analyzed samples. However, the amount of information obtained is caused by the

Figure 1. Materials to extract the organic matter and detect hydrocarbons: (a) soxhlet and (b) CGP

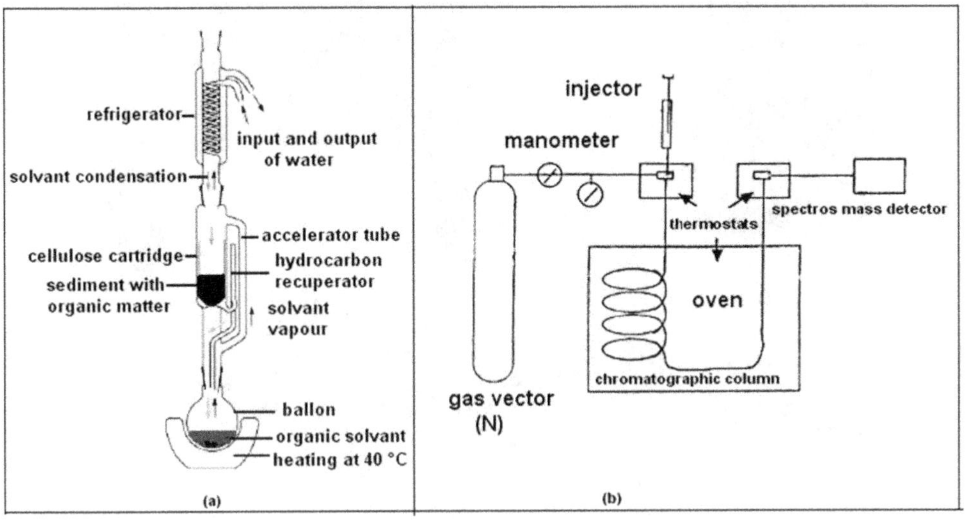

complexity of the spectral power to be measured. Use additional data processing. The use of multivariate statistical methods to analyze FT-IR data seems to be necessary for extracting features and making appropriate estimates. And determine the spectrum obtained. Possible methods to characterize the spectral characteristics of explosives in explosive residues. All spectra were obtained after analyzing the fragment samples after the explosion. These explosions were carried out using three explosives: C-4, 2,4, 6-trinitrotoluene (TNT) and pentaerythritol tetranitrate (PETN). Prepared and executed. Tested two known multivariate statistical methods, namely hierarchical cluster analysis (HCA) and principal component analysis (PCA), in order to use a wide range of wavelength data (4000-600 cm-1) in the collected spectra. Divided into different categories. Tests seem to indicate that PCA is the best choice for accomplishing the above tasks. It was found that only three main components carried more than 99.6% of the variance in the sample. The results show that even in relatively large data sets, the combination of FT-IR and multivariate methods is very suitable for identification and differentiation. Therefore, forensic laboratories can use these methods for rapid detection and analysis. Infrared spectroscopy with Fourier transformation or FTIR: Fourier Transformed Infra-Red Spectroscopy (Griffiths, P. R. and de Haseth, J. A. 1986) is based on the absorption of infrared radiation by the analyzed material.It allowsvia the detection of characteristic vibrations of the chemical bonds present in the material. Also,it is relatively easy to obtain spectra from samples in solution or in the liquid, solid or gaseousstates. This technique is

Figure 2. Infrared spectroscopy apparatus

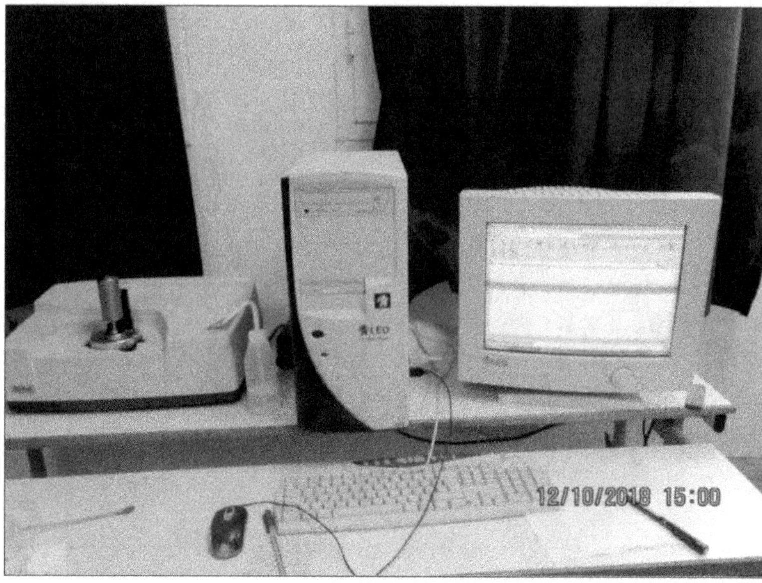

widely used in analytical laboratories for the structural characterization of organic - inorganic materials.

2.2. Cases of Environmental Investigations

a. Organic Geochemistry in Sidi El Hani Discharge Playa, Eastern Tunisia: Distinction Between Pollution and Natural Migration

The oil spill could have a significant impact on the environment and ecosystems. However, the source of the oil spill is not always obvious because it is natural oil leakage or man-made, such as pipeline leakage or large tank leakage (Essefi et al., 2014). Therefore, the source of hydrocarbons is an important part of environmental remediation. Detecting hydrocarbon at the surface of saline systems is In order to avoid localized assumption of hydrocarbons migration and to generalize this phenomenon on the discharge playa, 10 samples were collected from different sites along the sebkha; they were mixed in order to obtain a representative sample. All analyses were carried out on this mixture. The elimination of the elementary sulphur shows that the sample form Sidi El Hani discharge playa is rich in sulphur (5g/kg). This strong presence may be explained by the activity of anaerobic bacteria in a reduced environment; as it may be explained, by sulphur migration from a petroleum reservoir. This high amount of sulphur is noticed in many saline environments

Figure 3. Concentrations of the organic matter (OM) and Total Hydrocarbons (THC) in the mixed sample of sediment from Sidi El Hani discharge playa

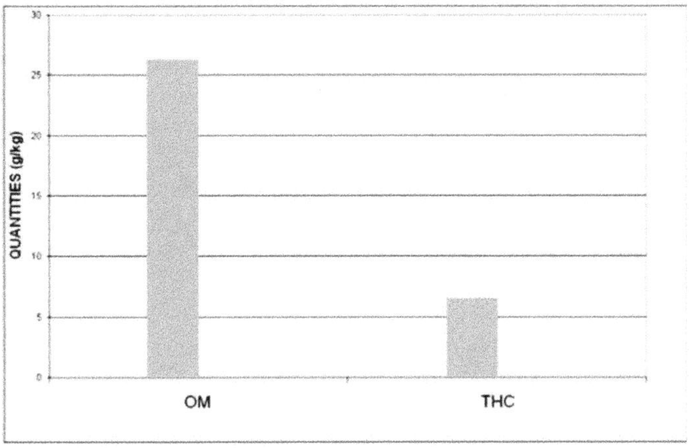

such as the sebkha of Moknine. It is also detected in Sidi Kilani petroleum field. After weighting the dry residues, the different fractions are obtained. Fig. 3. shows the relative percentages of organic matter (OM) and Total Hydrocarbons (THC). Comparably to results of the organic matter in sebkha of Moknine, these amounts are relatively high and may have an allochthonous origin.

On the other hand, Fig. 4 shows the distribution of different components of the total hydrocarbons such as Alkanes ALK Aromatic Polycyclic Hydrocarbons APHs and Polar Fraction PF. The predominance of the Polar Fraction (75%) indicates a primarily biological origin of hydrocarbons. Nevertheless, the presence of relatively high amount of APHs (15%) represents actually a geochemical anomaly in such environments. The hydrocarbons thus have a double origin; though the principal origin remains biological and autochthonous, the allochthonous origin is more than likely to have a considerable contribution. Added to the hypothesis of a human induced pollution previously advocated, the hypothesis of migrating hydrocarbons could by no means be excluded in this region especially with the presence of many factors enhancing such a phenomenon.

Interpretation of the Chromatogram

The chromatogram of saturated hydrocarbons (Fig. 5) may be divided in two parts. First, alkanes with low molecular weight range from C13 to C20. Second, alkanes with strong molecular weight range between C21 to C31. The absence alkanes with short chain (C1 to C10) on the chromatogram may be explained by their volatile aspect. The bimodal distribution is also noticed. The first mode is characterized by

Figure 4. Concentrations of the Total Hydrocarbons (THC), Polar Fraction (PF), Alkane (ALK) and Aromatic Polycyclic Hydrocarbons (APH) in the mixed sample of sediment from Sidi El Hani discharge playa

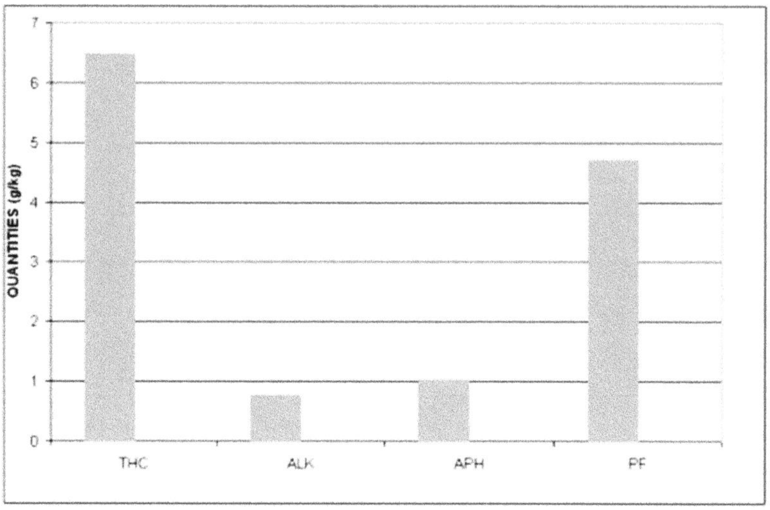

a predominance of C15, C17 and C1 alkanes. This disparity is characteristic of an algal origin; the hydrocarbons would have been formed after the maturation of the organic matter rejected by the algae. These short n-alkanes are generally attributed to autochthonous origin, since they are ubiquitously observed in aquatic organisms, such as bacteria and algae. This specific distribution of n-alkanes has previously been attributed to bacterial or algal origin in similar contexts. The second mode includes alkanes with strong molecular weight; it is also characterized by a predominance of alkanes made up of odd carbon atom. This disparity testifies a continental natural origin of hydrocarbons. They result primarily from the remains of plants. These long chain n-alkanes indicate a higher plant input since these compounds are abundant in epicuticular waxes of vascular plants. It is worth noting that C21 and C23 prove an allochthonous origin. The pristane and the phytane are products of the diagenesis of chlorophyl. The pristine corresponds to the evolution of chlorophyl in a reduced environment and the phystane in an oxidized environment. These two products appear on the chromatograms the ratio of pristane and phytane can thus inform us on the conditions of oxidation and reduction of the environment. In the studied case, without calculation one notes that this ratio is lower than 1, it thus states that the hydrocarbons develop in a reduced environment.

Figure 5. Chromatogram of saturated hydrocarbons in the mixed sample of sediment from Sidi El Hani discharge playa

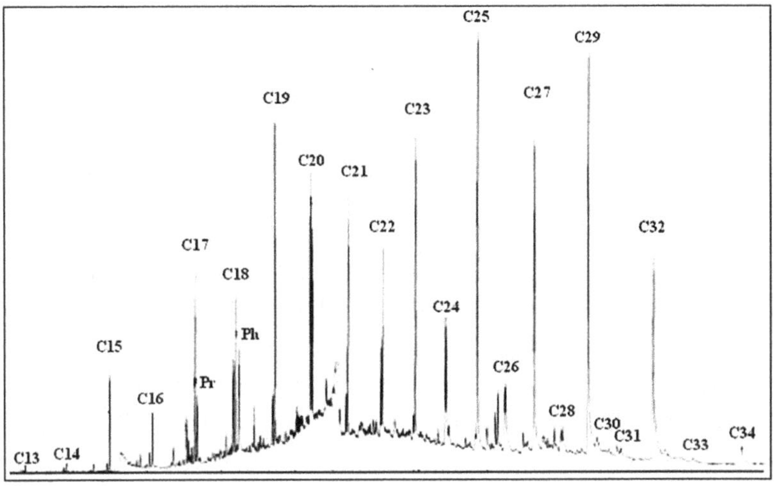

b. Pollution in Chott Djerid

During the Anthropocene human pollution has worldwide increased (Essefi, 2020; 2021) to reach the level of environmental crimes. For instance, wetlands in Tunisia collected the byproducts of many environmental crimes taken place in their hydrological and hydrogeological drainage basins. Infrared investigation of the core from Chott El Jerid shows the individualization of the top of core with obvious signature related to the Anthropocene (Fig. 6).

3. NUCLEAR AND ISOTOPIC FORENSIC GEOCHEMISTRY

The information inherent to a nuclear material is revealed by the application of nuclear forensics associated in nuclear security is an important element (John et al., 2021). Nuclear forensics is the detailed investigation performed on a nuclear and other radioactive material in order to identify the possible origin, intended use and hazards related to the material, following its seizure from illicit trafficking or diversion (Varga et al., 2020). Nuclear forensics considers the fact that some measurable parameters or signatures are distinctive primarily based on the geological supply of the material under investigation (Eppich, 2020). A large number of samples of nuclear and different radioactive materials from around the world have been successfully studied and characterized while reminiscent signatures of uranium

Figure 6. Signature widespread human impact

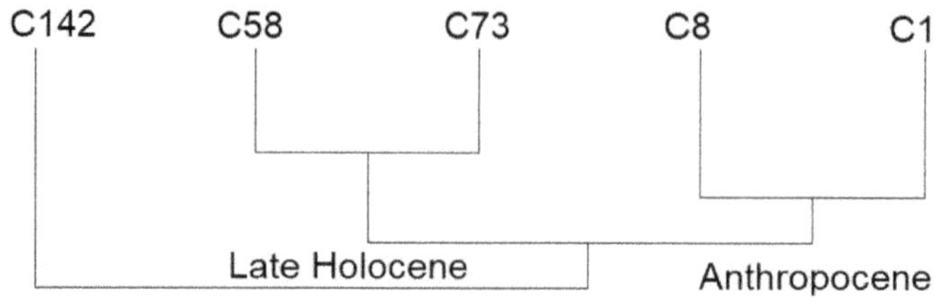

isotopic compositions and ratios, elemental impurities, rare-earth element elements (REE) patterns (Carter, 2021) Neodymium (Nd) (Reinhard et al., 2021), Strontium (Sr) (Pastoor et al., 2020) and Lead (Pb) (Kupi et al., 2020) have been determined. Rare earth elements (REE) or lanthanides composition of uranium ores is a powerful tool to understand metal deposits formation (Smith and Dietz, 2021). The ability to use various analytical methods to identify the source of nuclear materials has become an important goal of nuclear forensics (Nitrean, 2020). Although new tools have been developed in the past few years, it is clear that multiple tools are still needed to accurately identify the source (Mclain, 2020). A new isotope instrument based on molybdenum isotopes, which can be used in nuclear forensics. Among uranium minerals, the largest fractionation of Mo isotopes is in sedimentary deposits, while some magmatic minerals have limited fractionation of Mo isotopes. Ultimately, it can be used to distinguish high-temperature uranium from low-temperature uranium (Yu et al.,2020). When a sufficient understanding of the ore processing flow chart is available, it can be used to restore the Mo isotopic composition of the ore, or to provide information about the properties of the uranium ore. If the mineral composition can be estimated, it can be combined with other instruments to make a specific UOC method. Our research shows the potential of Mo isotopes as useful tools in nuclear forensics. Geochemistry and isotope tracers have been used in forensic state research. In the past four decades, the methods used have changed significantly. More and more geochemical indicators (including stable isotopes) are used for geomorphological and stratigraphic techniques. Lead isotopes have been well studied and have been used in a variety of environmental media. Since the early 1990s, advances in analytical chemistry have made it possible to accurately characterize other unconventional stable isotope systems in geological materials. The potential of these unconventional isotopes as indicators in ecological systems is not well understood, but some of these isotopes such as Cd, Cu, Cr, Hg, Sb, and Zn show great promise. In addition, some of these isotopes (Cu, Cr and Hg), it can provide important information about the biogeochemical process of water circulation. The resulting hydrological data may be subject to judicial review.

4. CONCLUSION

One century ago, Georg Popp was the first scientist presenting in court a case in which the geological makeup of soils was used to secure a criminal conviction. Ever since, significant advances in the theory and practice of forensic geosciences have taken place. As branch of forensic geosciences, forensic geochemistry has been used in standard contamination characterization with methodology developed in organic geochemistry, a combination we refer to as Forensic Environmental Geochemistry. Other geochemical proxies were also used for forensic purposes including Nuclear and isotopic forensic geochemistry.

REFERENCES

Carter, R. (2020). *Investigation of Rare Earth Element Signatures and Surrogate Nuclear Melt Glass for Nuclear Forensic Analysis* (Doctoral dissertation). University Of Notre Dame.

Chua, C. C., Kwok, H., Yan, J., Cuthbertson, D., Aggelen, G. V., Brunswick, P., & Shang, D. (2021). Development of a tiered analytical method for forensic investigation of mixed lubricating oil samples. *Environmental Forensics*, 1–13. doi:10.1080/15275922.2021.1907821

Dawson, L. A., Di Maggio, R. M., McKinley, J., Di Capua, G., Peppoloni, S., & Pringle, J. (2021). A proposal for a White Paper on Geoethics in Forensic Geology. *Geological Society, 508.*

de Oliveira, O. M., Queiroz, A. F. D. S., Cerqueira, J. R., Soares, S. A., Garcia, K. S., Pavani Filho, A., ... Moreira, Í. T. (2020). Environmental disaster in the northeast coast of Brazil: Forensic geochemistry in the identification of the source of the oily material. *Marine Pollution Bulletin*, *160*, 111597. doi:10.1016/j.marpolbul.2020.111597 PMID:32896715

Eppich, G. R. (2020). Nuclear forensics: How science helps stop the trafficking of nuclear materials. *Bulletin of the Atomic Scientists*, *76*(5), 263–270. doi:10.1080/00963402.2020.1806587

Essefi, E. (2013). *Wet Aeolian Sedimentology and Sequence Stratigraphy within the Terrestrial Analogues in Eastern Tunisia: Implications for Wet Aeolian Sedimentology and Sequence Stratigraphy on Mars* (Doctoral dissertation). National Engineering School of Sfax, Sfax, Tunisia.

Essefi, E. (2020). On the Geochemistry and Mineralogy of the Anthropocene. *Int J Water Wastewater Treat*, *6*(2).

Essefi, E. (2021). Record of the Anthropocene-Great Acceleration along a core from the coast of Sfax, southeastern Tunisia. *Turkish Journal of Earth Sciences*.

Essefi, E., Tagorti, M. A., Touir, J., & Yaich, C. (2013). Hydrocarbons migration through groundwater convergence toward saline depressions: A case study, Sidi El Hani discharge playa, Tunisian Sahel. *International Scholarly Research Notices*.

Gharsalli, N., Essefi, E., Baydoun, R., & Yaich, C. (2020). The Anthropocene and great acceleration as controversial epoch of human-induced activities: Case study of the Halk el Menjel Wetland, Eastern Tunisia. *Applied Ecology and Environmental Research*, *18*(3), 4137–4166. doi:10.15666/aeer/1803_41374166

John, S. O. O., Usman, I. T., Akpa, T. C., & Ibrahim, U. (2021, February). Rare earth elements in Uranium ore for nuclear forensic application. *IOP Conference Series. Earth and Environmental Science*, *655*(1), 012075. doi:10.1088/1755-1315/655/1/012075

Kristo, M. J. (2020). Nuclear forensics. In *Handbook of Radioactivity Analysis* (Vol. 2, pp. 921–951). Academic Press. doi:10.1016/B978-0-12-814395-7.00013-1

Kupi, T. G., Uushona, V., Mathuthu, M., Coetzee, M., & van Tonder, D. (2020). Using lead isotope ratios to distinguish between samples of diffrent uranium mines. *Journal of Radioanalytical and Nuclear Chemistry*, *324*(1), 1–5. doi:10.100710967-020-07048-1

Malmborg, J., Kooistra, K., Kraus, U. R., & Kienhuis, P. (2020). Evaluation of light petroleum biomarkers for the 3rd edition of the European Committee for Standardization methodology for oil spill identification (EN15522-2). Environmental Forensics, 1-15.

Mclain, D., Canaday, J., Chamberlain, D., Deangeles, K., Naik, S., Quigley, N., ... El-Jaby, A. (2020). *Approaches for Developing and Sustaining the Sealed Radioactive Source and Material Component of a National Nuclear Forensics Library* (No. IAEA-TECDOC--1896 (SUPPLEMENTARY FILES)). Academic Press.

Migeon, V., Fitoussi, C., Pili, E., & Bourdon, B. (2020). Molybdenum isotope fractionation in uranium oxides and during key processes of the nuclear fuel cycle: Towards a new nuclear forensic tool. *Geochimica et Cosmochimica Acta*, *279*, 238–257. doi:10.1016/j.gca.2020.03.046

Nitrean, A. (2020). *Development of interagency cooperation for nuclear forensics examinations in Moldova* (No. IAEA-TECDOC--1896 (SUPPLEMENTARY FILES)). Academic Press.

Nuzzo, M., & Gehlen, M. (2021). *Forensic use of diagnostic biomarker ratios in environmental oil spill geochemistry*. Academic Press.

Pastoor, K. J., Kemp, R. S., Jensen, M. P., & Shafer, J. C. (2021). Progress in Uranium Chemistry: Driving Advances in Front-End Nuclear Fuel Cycle Forensics. *Inorganic Chemistry*, *60*(12), 8347–8367. doi:10.1021/acs.inorgchem.0c03390 PMID:33619961

Poulson, S. R., & Drever, J. I. (2020). Organic contaminant stable isotope geochemistry, and application to contaminant characterization and remediation. In *Groundwater 2000* (pp. 335–336). CRC Press. doi:10.1201/9781003078593-165

Reinhard, A., Inglis, J., Steiner, R., LaMont, S., & Palacz, Z. (2021). Isotopic analysis of sub-nanogram neodymium loads using new ATONA™ amplifiers. *Rapid Communications in Mass Spectrometry*, *35*(7), e9032. doi:10.1002/rcm.9032 PMID:33340159

Smith, C. D., & Dietz, M. L. (2021). Support loading effects on the performance of an extraction chromatographic resin: Toward improved separation of trivalent lanthanides. *Talanta*, *222*, 121541. doi:10.1016/j.talanta.2020.121541 PMID:33167249

Sutliff-Johansson, S. (2021). *Tracing anthropogenic Technology Critical Elements using Environmental Forensics: Case studies at Kallholmsfjärden and the Bothnian Bay, Northern Sweden* (Doctoral dissertation). Luleå University of Technology.

Thienpont, J. R., Yang, Z., Hall, R. I., Wolfe, B. B., Hollebone, B. P., & Blais, J. M. (2021). Tracking petrogenic hydrocarbons in lakes of the Peace-Athabasca Delta in Alberta, Canada using petroleum biomarkers. *Environmental Pollution*, *286*, 117286. doi:10.1016/j.envpol.2021.117286 PMID:33990046

Ubelaker, D. H., & Francescutti, C. (2020). The role of stable isotope analysis in forensic anthropology. *Forensic Science and Humanitarian Action: Interacting with the Dead and the Living*, 273-284.

Varga, Z., Wallenius, M., Nicholl, A., & Mayer, K. (2020). Assessment of uranium inhomogeneity and isotope imaging for nuclear forensics. *Spectrochimica Acta. Part B, Atomic Spectroscopy*, *171*, 105920. doi:10.1016/j.sab.2020.105920

Whyte, C. J., Vengosh, A., Warner, N. R., Jackson, R. B., Muehlenbachs, K., Schwartz, F. W., & Darrah, T. H. (2021). Geochemical evidence for fugitive gas contamination and associated water quality changes in drinking-water wells from Parker County, Texas. *The Science of the Total Environment*, *780*, 146555. doi:10.1016/j.scitotenv.2021.146555 PMID:34030322

Yu, C. D., Wang, K. X., Liu, X. D., Cuney, M., Pan, J. Y., Wang, G., Zhang, L., & Zhang, J. (2020). Uranium mineralogical and chemical features of the Na-metasomatic type uranium deposit in the Longshoushan metallogenic belt, northwestern China. *Minerals (Basel)*, *10*(4), 335. doi:10.3390/min10040335

Zacharias, D. C., Gama, C. M., & Fornaro, A. (2021). Mysterious oil spill on Brazilian coast: Analysis and estimates. *Marine Pollution Bulletin*, *165*, 112125. doi:10.1016/j.marpolbul.2021.112125 PMID:33582423

Zhou, C., Yu, S., Huang, W., Zhang, H., Xiao, Z., Zeng, L., Huang, W., & Pan, C. (2021). Oil maturities, mixing and charging episodes in the cratonic regions of the Tarim Basin, NW China: Insight from biomarker and diamondoid concentrations and oil bulk properties. *Marine and Petroleum Geology*, *126*, 104903. doi:10.1016/j.marpetgeo.2021.104903

Chapter 2
Advances in Forensic Geophysics:
Magnetic Susceptibility as a Tool for Environmental Forensic Geophysics

Elhoucine Essefi
University of Gabes, Tunisia

ABSTRACT

Traditionally, forensic geophysics involves the study, search, localization, and mapping of buried objects or elements within soil, buildings, or water using geophysics tools for legal purposes. Recently, with the evolution of environmental crimes, forensic geophysics gave special care to detection, location, and quantification of polluting products. New techniques including the magnetic susceptibility have emerged to investigate this type of crimes. After discussing the state of the art of forensic geophysics, this chapter proposed the magnetic susceptibility as an efficient tool of environmental crimes detection. A case study of pollution detection was proposed from Tunisia. Being a fast and cheap technique, magnetic surveys represent a real promise for environmental forensic geophysics.

1. INTRODUCTION

Traditional crimes may damage a person or a small group while environmental crimes threaten the whole humanity. These newly emerging crimes were not given the worthy care. Instead, governors remain tolerant with wrongdoers. In this vein, forensic investigations would pave the way for the applications of laws in face of

DOI: 10.4018/978-1-7998-8386-9.ch002

these new emergent crimes. In terms of criminal issues, to elaborate a successful criminal conviction, locating forensically important evidence is of great importance (Alajmi et al., 2020; De Giorgi et al., 2020; Jakubec et al., 2021; Malejko et al., 2020; Pringle et al., 2021; Wallace et al., 2021). Searched objects for forensic investigation are variable including landmines (Madzunya et al., 2021) and improvised explosive devices (Baldaino et al., 2021; Vanderheyden et al., 2020) or improvised explosive devices (IEDs) (Franceschetti et al., 2021), illegally buried explosives (Shvedchikova et al., 2021) and weapons (Harvey and Sekulla, 2021), mass genocide graves (Ekštajn et al., 2021), drugs and weapons caches (Cadwell, 2020) and clandestine graves of murder victims (Colledge, 2020). With less than 3 m under the ground level (Koşaroğlu et al., 2021; Różycki et al., 2020), burials are generally not quite deep in these cases. In terms of environmental crimes, toxic waste present in illegal dumps has become a serious growing issue (Davis et al., 2021; Qasim et al., 2020; Morrison, 2020). Along water bodies with poor visibility or large search areas, water-based forensic geoscience surveys prove vital to assist police and environmental divers (Burnier et al., 2020; Kootker et al., 2020). Variable forensic techniques are currently in use; in many countries; a search strategist is by obligation involved in a case at an early stage to decide upon the highest probability of search success (Swain, 2020). The use of geoscientific methods has been increasingly adopted and reported to locate of clandestinely buried material (Mansegosa et al., 2021; Molina et al., 2020). Investigations start from the large-scale remote sensing methods (Sharma et al., 2020; Pensieri et al., 2020), surface geomorphology changes (de Bruin and Schmitz, 2021), aerial (Rocke et al., 2021) and ultraviolet photography (De Angelis et al., 2020; Mia et al., 2021), to ground-based observations of vegetation changes (Finkelstein et al., 2020), thermal imaging (Listos et al., 2021; Deveci et al., 2020), soil type and depositional environment (Xu et al., 2020; Hachem et al., 2020; Sangwan et al., 2020), near-surface geophysics (Weiss and Roy, 2020), diggability surveys (Lockwood, and Masters, 2021) and probing of anomalous areas before topsoil removal and finally controlled excavation and recovery. Near-surface geophysical methods are based on the contrast detection between the signal of the target and the and the noise of the host materials. Although the use of ground penetrating radar or GPR (Amran et al., 2020; Damiata et al., 2020; Elis et al., 2020; Schneider et al., 2020; Kelly et al., 2021; Berezowski et al., 2021) as the dominant geophysical method, recent studies suggested multi-methods approaches (Osbourne et al., 2020). For example, both Electro-Magnetics or EM (Mankoff et al., 2020) and its reciprocal electrical resistivity techniques are relatively fast to acquire and resulting anomalous areas can then be further investigated by higher resolution methods. Concerning the limitations of the GPR method, other studies showed that the GPR is not relevant in certain search environments including clayey material (Canata et al., 2020) and heterogeneous soil types (Dong et al., 2020), which are significantly attenuating radar signal amplitudes. Also, in water-logged, saline-rich and wooded environments (Ferrell et al., 2020), GPR method showed its limitation. On the other hand, since EM methods relies on metallic objects having good conductivity and transmissivity, metal detectors are actually active. Magnetic methods are passive measurements measuring slight changes in the Earth's magnetic field and its derivative (magnetic susceptibility) due to nearby objects. Previous studies (Pringle et al., 2015) argued that magnetic approach, due to the possibility of interference, is not the best tool for forensic investigation and control search studies. All substances do have magnetic properties and, when a magnetic field is applied to soil and rock, the degree of magnetization can be measured as the magnetic susceptibility or MS in SI dimensionless units. MS causes are complex and are a combination of mineralogical dia-, para- and ferro/ferri-magnetism and geochemical magnetization related to trace elements (Essefi, 2021). Nonetheless, recent studies (Leucci, 2020) proposed that magnetic surveys are quite relevant after removing of artefacts created during data collection. There are wide variations in measured MS reported between different rock and soil types with the largest values being partly attributable to the relative proportions of magnetic minerals present in the material. In soils, the presence of the ferrimagnetic mineral maghemite (Fe_2O3, γ-Fe_2O_3) has a dominant effect on the magnetic susceptibility and is a low-temperature, oxidation weathering product of the strongly magnetic minerals magnetite and titanomagnetite. The magnetic susceptibility has therefore been used for site soil characterization, forensic trace evidence and environmental forensic pollution studies (de Caritat et al., 2021) namely during the polluted era of the Anthropocene (Essefi, 2021). In urban depositional environments, it has a poorly resolved a simulated clandestine grave (Colledge, 2020). Magnetic surveys commonly used in archaeological searches (Mahaney, 2020) have been shown to successfully locate areas of historic surface burning as the weakly magnetic iron oxide minerals in the soil are transformed into the highly magnetic minerals magnetite and maghemite. Environmental crimes are basically recorded at the pedological covers. In a geoforensic investigation, the demonstration the relation between variations in the soil properties and dissimilarity between samples is an important issue. Magnetic parameters such as magnetic susceptibility are important in characterizing materials, and detectable quantities of magnetic minerals are almost always found in soils. In geoforensic investigations, it is important to not only obtain samples that are representative and comparable, but also develop standardized protocols and frameworks of best practice to better evaluate soil properties in a representative manner. Several experiments on the magnetic susceptibility of soils have been conducted recently, mostly related to pollution monitoring and mapping. Although there are only a few reported applications of magnetic susceptibility in forensic investigations ((Shina et al., 2021), the importance of the use of this technique for forensic applications has been reported in work relating to forensic archaeology, detecting clandestine graves and forensic searches for buried targets. Most of the research and applications deal with the survey application of portable magnetometers, and the advantages and limitations of the technique are reported (Shina et al., 2021). The magnetic susceptibility measurement of soils recovered from a crime scene, at room temperature, is non-destructive; therefore, the same material is available for further analysis with any other technique. Additionally, it does not require sample preparation, and can be used as a simple and rapid method that is operable with small sample sizes; thus, it is an advantageous method for the study of soils being used as trace evidence in forensic investigations.

This chapter aims to validate the potential usefulness of magnetic susceptibility surface surveys as a forensic search technique. It will also be briefly compared to other, more commonly utilized forensic geophysics techniques and finally suggest best practice for such MS surveys. This study is based on research and experimentation based on soil samples for forensic applications. The magnetic susceptibility measurement methodology, reproducibility and accuracy in analysis were tested on soil samples collected from a polluted location in Tunisia.

2. THEORETICAL BACKGROUND OF MAGNETISM

2.1. Magnetic Force

The magnetic force could be defined by the application of the Law of Coulomb for a magnetic pole:

$$\vec{F} = \frac{m_1 m_2}{\mu r^2}\vec{r} \quad (1)$$

Where

F = force in dynes (cm.g/s2 = 105 N)
m1; m2 are magnetic masses or poles (emu) R is the distance between the two poles
r is the unit vector according to the line joining m1; m2
µ = permeability of the domain around the poles = 1 in the vacuum and the air

The concept of pole is artificial because it cannot exist by itself; it needs a pair. If two poles of 1 emu are distanced of 1 cm in the vacuum, the force between them will be 1 dyne. The force is gravitational if the two poles are opposite signs; it is repulsive if they are of the same sign. By convention, a pole is positive if it is attracted by the magnetic north of the ground and negative if it is attracted by the southern pole.

2.2. Magnetic Field

A magnetic mass m1 brought to the point M of space is subjected to the attraction emanating of another magnetic mass m located at a distance r of the point M. The intensity of the magnetic field H is defined as the force exerted on a unit pole:

$$\vec{H} = \frac{\vec{F}}{m_1} = \frac{m}{\mu r^2}\vec{r} \tag{2}$$

It is supposed that m1 is not enough large to assign the field H to the point of measurement, i.e. m1 is largely lower than M. In CGS units, the unit of the field is the oersted. In geophysics, we measure variations of about 10-4 times the amplitude of the terrestrial magnetic field, which is approximately 0.5 Oe.

2.3. Magnetic Moment

There is no free magnetic mass. Only the dipole, association of two poles - m and +m separated by a distance l, have a physical significance. The magnetic moment M of the dipole is a directed vector following the line joining - m with +m, directed - m to +m and of intensity:

$$\vec{M} = ml\vec{r} \tag{3}$$

2.4. Intensity of Magnetization

A magnetic body placed in an external magnetic field will be magnetized by induction. The intensity of magnetizing is proportional to the force of the field and its direction of the field. It is defined like the magnetic moment per unit of volume:

$$\vec{I} = \frac{\vec{M}}{v} \tag{4}$$

It often bears the magnetic name of polarization because induction tends to align the dipoles of the magnetic body.

2.5. Magnetic Susceptibility

The magnetic susceptibility is expressed by the facility of the body to be magnetized in a weak applied magnetic field. It characterizes the ratio of the induced magnetization when applying a low intensity magnetic field and the field which creates it. It can be volume (K) or mass (χ). the degree of magnetization will be determined by its magnetic susceptibility K.

$$k = \frac{\vec{I}}{\vec{H}} \quad \vec{I} = k\vec{H} \tag{5}$$

In magnetic prospection, the magnetic susceptibility is a fundamental parameter, since the magnetic response of the rocks is a function of the contents in magnetic material, which will have a susceptibility much larger than that of the rock itself.

2.6. Magnetic Induction

A magnetic body placed in an external magnetic field H, will have its magnetic poles more or less aligned under the effect H, producing a H0 field related to the intensity of magnetization I. The magnetic induction will be then the total field within the body

$$\begin{aligned} &\vec{B} = \vec{H} + \vec{H}' \\ &CGS\ unit = \vec{H} + 4\pi\vec{I} \\ &= \left(1 + 4\pi k\right)\vec{H} \end{aligned} \tag{6}$$

By definition, the ratio between the induction on the principal magnetic field is the permeability μ

$$\begin{aligned} &\vec{B} = \left(1 + 4\pi k\right)\vec{H} \\ &= \mu\vec{H} \end{aligned} \tag{7}$$

In system CGS,

$$\begin{aligned} &\vec{B} : gauss \\ &\vec{H} : Oersted \end{aligned}$$

Thus, the magnetic permeability should be written as μ = (1 + 4ïk). In general, one can write μ = μ 0 (1 + 4ïk) with μ 0 = 1 for the vacuum (CGS unit). In the SI units, B = μ 0 H in the vacuum (or in the air) if the material is polarizable.

$$\begin{aligned}\vec{B} &= \mu_o\left(\vec{H}+\vec{H}'\right) \quad \vec{H}' = \vec{I} \\ &= \mu_o\vec{H} + \mu_o k\vec{H}' \\ &= \mu_o\left(1+k\right)\vec{H}\end{aligned} \tag{8}$$

Where

μ = permeability of material
μ 0 = permeability of the vacuum (of the air) = 4ï10 -7 (IF)
B = in Tesla or Weber/m2
H = in A/m

2.7. Magnetostatic Potential

As in gravimetry, the magnetic field H is a conservative field. It can thus be expressed like the gradient of a scalar potential. The potential is the necessary work to move a unit pole between two points in the magnetic field.

$$H\left(\vec{r}\right) = -\nabla A\left(\vec{r}\right) \tag{9}$$

Where r is given by:

$$\nabla = \frac{\partial}{\partial x}\vec{i} + \frac{\partial}{\partial y}\vec{j} + \frac{\partial}{\partial z}\vec{k} \tag{10}$$

And, A, the potential,

$$A\left(\vec{r}\right) = -\int_{\infty}^{r} H\left(\vec{r}\right)dr = \frac{m}{\mu r} \tag{11}$$

The method of prospection is based on the existence of magnetic contrast of susceptibility in the Earth's crust. Any body placed in an external magnetic field, such that of the ground, acquires an induced magnetization proportional to the induction field and in the same direction. The total field is then in unit CGS: When one makes a quantitative interpretation of magnetic data, it is thus necessary to take account of Mr. Otherwise, serious mistakes in interpretation can result from.

$$\vec{H}_T = \vec{H} + 4\pi k\vec{H} + \vec{M}_R \quad (12)$$

3. DIFFERENT TYPES OF MAGNETISM

The soil consists of many minerals having marked magnetic properties. The magnetic behavior of materials is considered as their initial reaction when they are subjected to a weak field. Acquired magnetization is proportional to the field applied by a factor called magnetic susceptibility.

3.1. Diamagnetism

It is the more known behavior. It corresponds to the acquisition of a very weak negative induced magnetization when a magnetic field is applied. In the absence of field, the magnetic moments of the atoms and molecules are null. As diamagnetic examples of minerals we quote water, calcite, quartz and feldspar. For forensic investigation, in most soils, quartz, calcium carbonate, orthoclase, organic matter and water are diamagnetic. In some cases, such as pure silica sands and pure limestone, the diamagnetic component will have a significant effect on the magnetic susceptibility of the studied soil (Fitzpatrick and Baker, 2006).

3.2. Paramagnetism

This behavior appears when certain atoms or molecules constituting a mineral have non nil magnetic moment (spin), related to an incomplete filling of the electron layers of the atoms. Acquired induced magnetization corresponds then to an alignment of the elementary moments on the field applied and thus has the same direction. It remains weak, but of intensity definitely higher than that observed in the diamagnetic behavior. This alignment is opposed by thermal agitation. Magnetic susceptibility is thus inversely proportional to the absolute temperature and null residual magnetization. Olivine, pyroxene or biotite are minerals presenting this type of behavior. The paramagnetic susceptibility appears to be a very specific probe that could be used in forensic science analyses.

3.3. Ferromagnetism

Below a temperature threshold called Curie temperature (TC), the elementary moments solidify and their orientations become coherent. The material as a whole shows a very strong magnetization. Iron presents this type of behavior at the ambient temperatures, having hence the name ferromagnetism. The ferromagnetic bodies can also carry residual magnetizations. In the usual vocabulary, the word ferromagnetism may bebe used to speak about ferromagnetic materials in a strict sense as we have just defined, as well as the antiferromagnetic or ferrimagnetic materials. The magnetic susceptibility of human body reflects the total composition of the tissue, and it depends on the contribution of diamagnetic, paramagnetic and ferromagnetic behaviors of all components (Sant'Ovaia et al., 2015). In an application of the Anhysteretic Remanent Magnetization (ARM) for environmental magnetism on crime detection in a highway traffic accident (Manrong et al., 2009), it has been proven that ARM is especially sensitive to fine-grained single domain (SD) ferromagnetic minerals.

3.4. Ferrimagnetism

This behavior is very close to the ferromagnetism. The difference holds in the fact that there exists in the crystal two types of atoms carrying different magnetic moments. The moments are antiparallel but do not compensate themselves. The ferrimagnetic materials can acquire a strong residual or induced magnetization. Magnetite and the greigite are examples of ferrimagnetic. The antiferromagnetism is a form of ferromagnetism where the orientation of the spins is alternate. There is thus a compensation and the resulting macroscopic moment is null. In certain materials, the shape of the crystals as well as the thermal agitation make that the antiparallelism is slightly deviated; then the compensation is not any more and one macroscopic moment appears. Hematite is an example of deviated antiferromagnetic. In many soil samples, the magnetic susceptibility is largely determined by the ferrimagnetic minerals. magnetic signal of these sediments is controlled by ferrimagnetic minerals magnetite-like minerals and a minor contribution of antiferromagnetic carriers (such as hematite minerals)

4. MATERIALS

The Research Unit Sedimentary Dynamics and Environment is equipped with an MS2B Bartington magnetometer (Fig. 1). This sensor is designed for use with 1" (25.4 mm) and 7/8" (23 mm) cubic boxes, 20 cc and 10 cc cylindrical bottles and 1" cylindrical cores. Sample insertion and removal is facilitated by a hand operated platen. The sensor is normally fitted with a moving platen with a 27.5 mm stem. All details of experiment may recuperated from the technical manual provided with the machine. This apparatus was used in many environmental, climatic and environmental studies (Essefi, 2020, 2021a,b; Essefi et al., 2013, 2014, 2015a,b,c,d), but without special focus of the forensic aspects leading to legal acts of the pollution origins . Fabian et al. (2016) used MS2B Bartington for forensic aspects. Ever since, many forensic studies dealt on the magnetic susceptibility as tool for detecting traces used as arguments in courts (Nde et al., 2019; Zawadzki et al., 2020).

Figure 1. MS2B Bartington magnetometer at the Research Unit Sedimentary Dynamics and Environment

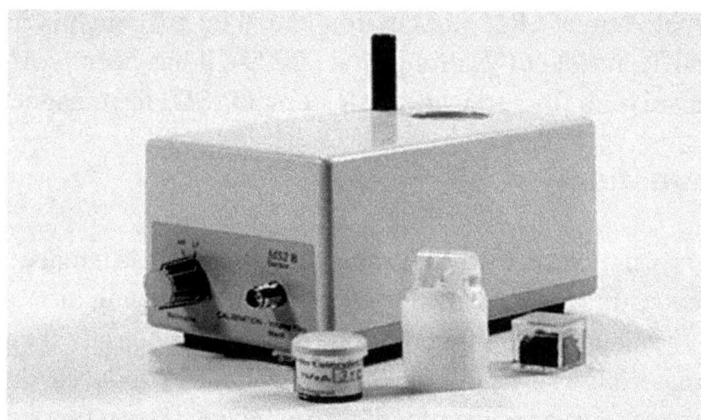

The sensor is specifically calibrated for use with a 10 cc sample container with internal dimensions 24 mm diameter x 23 mm height and a base external diameter of 26 mm maximum. Suitable polystyrene sample containers produced from virgin plastic and sealed with a polyethylene lid are available from Bartington Instruments. The sensor characteristics are optimized to produce no measurable response to sample conductivity e.g. even in 25% saline solution. Even metallic conducting material can be measured provided it is divided into a granular form to reduce the apparent diamagnetic contribution that might otherwise result due to eddy currents. Some technical specifications of the MS2B are shown in Tab.1.

5. SOIL FORMING FACTORS AND MAGNETISM

As outlined by Jenny (1941), factors of soils formation are used conventionally to provide a framework within which we analyze the variability of soil properties.

Table 1. Technical specifications of the MS2B (Operation Manual for MS2 Magnetic Susceptibility System)

Calibration accuracy	1% (10ml calibration sample provided)
Measurement period: x1 range	1.5s SI (1.2s CGS)
x0.1 range	1.2s SI (12s CGS)
Operating frequencies: LF	0.465kHz ±1%
HF	4.65kHz ±1%
Amplitude of applied field	250μT peak ±10% (LF and HF)
Maximum resolution	2 x 10^{-6} SI (vol), 2 x 10^{-7} CGS (LF and HF)
HF/LF cross calibration	0.1% worst case (can be adjusted using calibration sample)
Temperature induced drift: sample to sensor differential	±0.05 x 10^{-5} SI/°C/minute (±0.05 x 10^{-6} CGS/°C/minute) LF and HF
Enclosure material	High-impact ABS
Weight	0.7kg
Dimensions	210 x 145 x 110mm
Sample cavity internal diameter	36mm

5.1. Parent Material

Any soil forensic study should take into account the parental material of the studied soil (Fitzpatrick, 2009). There are at least four aspects in which parent material operates as a factor for soil magnetism. First, a supply of Fe from parent material is essential for Secondary Ferrimagnetic Mineral SFM production. Second, high concentrations of ferrimagnetic minerals in surface soils may result through the accumulation of resistant PFMs through weathering of igneous and sedimentary rocks. Third, resistant metamorphic and sedimentary rocks are often correlated with elevated topography and as a result, they are associated with lower temperatures and elevated orographic rainfall affecting SFM production. Fourth, parent material influences many of the local physical and chemical soil conditions, such as texture, drainage, and pH.

5.2. Climate

Identifying human remains by forensic search teams is difficult in varying terrain and climates (Molina et al., 2015). Many studies have attempted to identify the climate controls on soil and palaeosol magnetism. The effects of rainfall have been widely reported as an important causal factor in the magnetic enhancement of paleosols. These observations are consistent with other arguments arguing that the major role of rainfall lies in driving hydrolysis reactions and releasing Fe from primary minerals. On the other hand, the effects of temperature are generally less well studied and would seem less pronounced, although laboratory studies suggest that temperature may be a key factor in magnetic enhancement through the conversion of ferrihydrite to maghemite.

5.3. Relief and Drainage

Relief is important with respect to magnetic enhancement because it is intimately linked with soil drainage and translocation of soil particles. Commonly, magnetic susceptibility increases down slope due to increasing soil fines.

5.4. Soil Organisms and Vegetation

Within the soil profile, organisms may have a variety of effects on magnetic mineral formation. The efficiency of a fermentation mechanism is expected to depend on the population and activity of DIRB. These heterotrophic bacteria require moderate temperature (N10 °C required for significant activity), adequate moisture, organic carbon (acetate and other short-chain fatty acids) as electron donors, and intermediate to alkaline pH. These are conditions that are often best met within the rhizosphere, rather than the bulk soil. Low levels of organic matter might therefore represent limiting conditions.

5.5. Time

Like all pedogenic processes, magnetic enhancement is time dependent but few studies provide conclusive data for rates of mineral formation. Singer et al. (1992) reported that susceptibility had shown continued enhancement for over 240,000 years. These changes were attributed to the greater cumulative effects of the weathering of parent material and release of Fe-bearing minerals for potential conversion to SFMs. magnetic susceptibility may attain 'saturation' status rapidly within a few hundred to a few thousand years, other views emphasize the interactions between climate and time. Previous comprehensive analyses of soil factors and three magnetic susceptibility parameters in surface soils shows that the importance of factors is broadly: 1) parent material and drainage; 2) mean annual rainfall, slope and altitude, land use, organic carbon and pH. Mean annual temperature and time do not appear to be significant factors at these spatial and temporal scales.

6. APPLICATION OF THE MAGNETIC SUSCEPTIBILITY FOR FORENSIC PURPOSES: POLLUTION OF THE COAST OF SFAX

The use of the magnetic susceptibility for environmental purposes has gained a noticeable popularity. The use of this tool allows inferring the natural conditions as well as the human induced activity. On the one hand, it follows the contamination of surface sediments of the coast of Sfax (Fig.2) by polluting materials such as heavy metals. On the other hand, it records the variation of depositional environment.

Figure 2. Contamination of the coast of Sfax by zinc
(Ghannem et al., 2011)

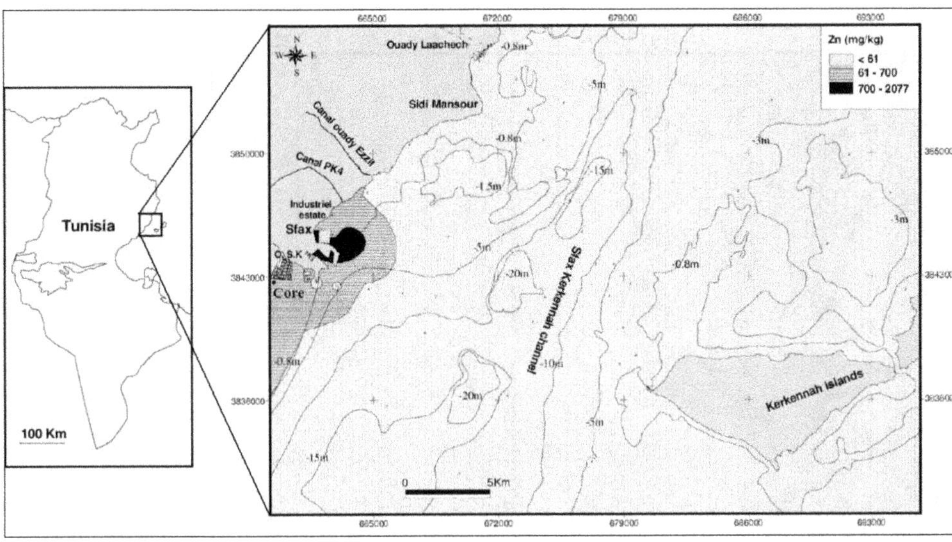

The coast of Sfax has also witnessed ever increasing urbanising activities. Many studies (e.g., Ghannem et al., 2011) have noticed the damage. The southern coast contains, in particular, a site of collecting wastes of the town of Sfax, the liquid and solid rejections of SIAPE, the basins of storage of the "margines" and the station of purification. The Tunisian government has taken serious measurements that heavily coasted on the national budget to clean the polluted northern coast of Sfax by funding the Taparura Project and the controlled zone of phospho-gypsum waste. But results of these remedial solutions are not guaranteed for sure. The first ten centimeters of the core are characterized by high values of magnetic susceptibility. These values are due the pollution of surface sediment with heavy metals. This contamination was recently discussed by Ghannem et al. (2011). Such kind of studies is really a forensic support to point out the responsibilities of all the surrounding vicinities.

Figure 3. Record of the mean of the grain size, low frequency magnetic susceptibility, high frequency magnetic susceptibility and independent frequency magnetic susceptibility along a core collected from the coast of Sfax

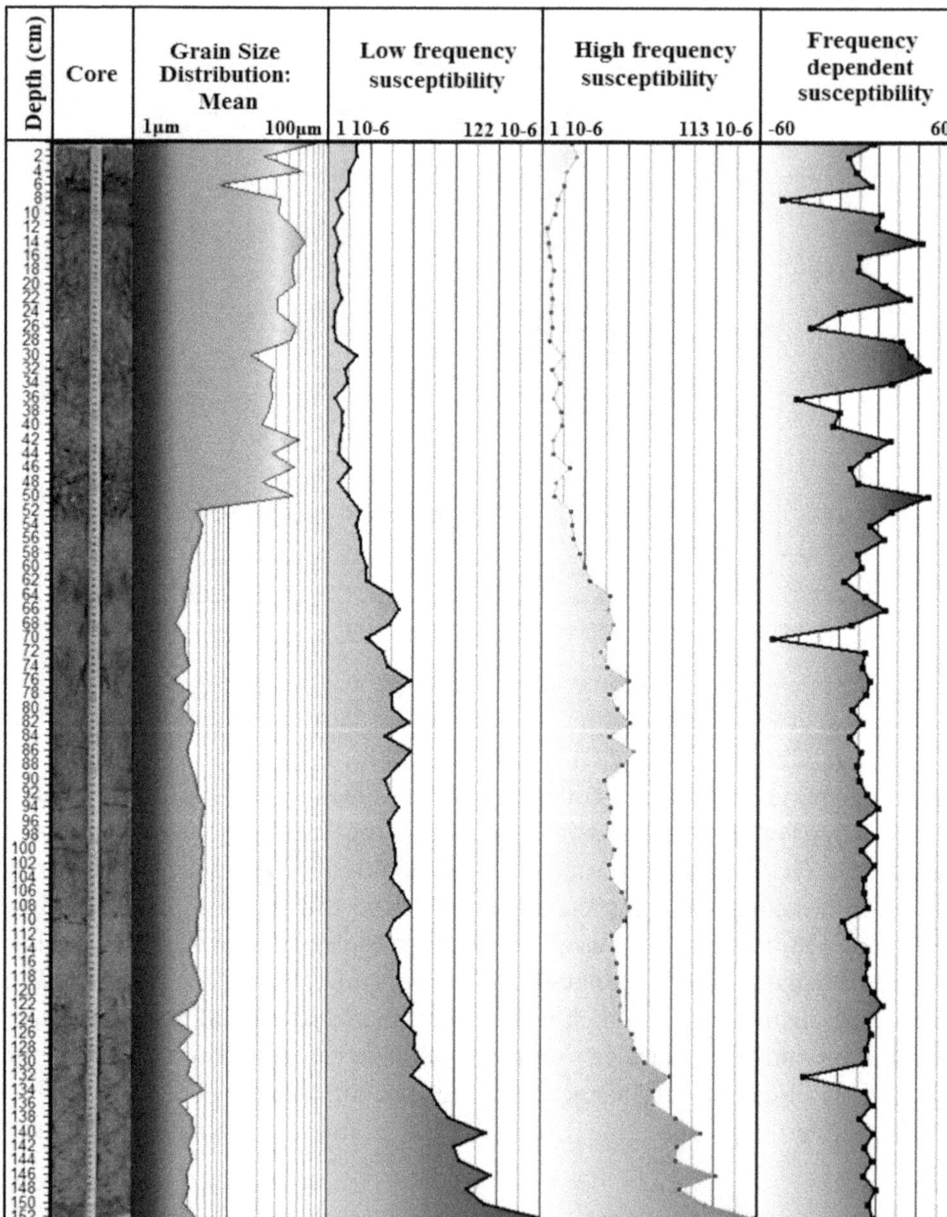

7. CONCLUSION

Compared with other direct methods, in forensics, geophysical techniques currently need to identify, characterize and distinguish hidden structures in order to obtain useful information and provide accurate and effective results for the research purposes in this field geophysical technology can minimize time and cost factors, and maximize the amount of data, information and knowledge available. It is time to classify environmental polluting activities as crimes. Further, the setting of standards to limit of criminal activities represents the urgency. Traditional crimes may damage a person or a small group while environmental crimes threaten the whole humanity. This chapter discussed advances in forensic geophysics. The focus was on the use of the magnetic susceptibility for the environmental forensic investigations. This tool proves relevant due to its fast and cheap manipulation. In Tunisia, environmental crimes are really neglected and wrongdoer are still damaging the ecosystems. As it is shown in this study pollution of Tunisian coasts is quite evident by using the magnetic susceptibility.

REFERENCES

Alajmi, R., Abdel-Gaber, R., & Haddadi, R. (2020). Molecular identification of forensically important beetles in Saudi Arabia based on mitochondrial 16 srRNA gene. *Entomological Research*, *50*(7), 343–350. doi:10.1111/1748-5967.12438

Amran, T. S. T., Amin, M. S. M., Ahmad, M. R., Sani, S., Masenwat, N. A., Abas, A. A., ... Adnan, M. A. K. (2020). Ground Penetrating Radar (GPR) Applications in Forensic. *Jurnal Sains Nuklear Malaysia*, *32*(1), 49–57.

Baldaino, J., Ommen, D. M., Saunders, C. P., Hietpas, J., & Buscaglia, J. (2021). Characterization and differentiation of aluminum powders used in improvised explosive devices–Part 1: Proof of concept of the utility of particle micromorphometry. *Journal of Forensic Sciences*, *66*(1), 83–95. doi:10.1111/1556-4029.14564 PMID:33006770

Berezowski, V., Mallett, X., Ellis, J., & Moffat, I. (2021). Using Ground Penetrating Radar and Resistivity Methods to Locate Unmarked Graves: A Review. *Remote Sensing*, *13*(15), 2880. doi:10.3390/rs13152880

Burnier, C., Massonnet, G., Coulson, S., DeTata, D., & Pitts, K. (2020). Characterization and classification of water-based compounds in condoms and personal hygiene products using GC-MS. *Forensic Science International*, *317*, 110513. doi:10.1016/j.forsciint.2020.110513 PMID:32987281

Cadwell, Z. (2020). *Pointing the Finger: Applying Extreme Environment Fingermark Development to Forensic Archaeology*. Academic Press.

Canata, R. E., Ferreira, F. J. F., Borges, W. R., & da Silva Salvador, F. A. (2020). Analysis of 2D and 3D GPR responses in the Federal University of Paraná Forensic Geophysics Controlled Site–a Case Study. *Brazilian Journal of Geophysics*, *38*(2). Advance online publication. doi:10.22564/rbgf.v38i2.2045

Colledge, D. (2020). *A review of intrusive search methods and excavation techniques for clandestine grave site recovery in forensic archaeology* (Doctoral dissertation). Murdoch University.

Damiata, B. N., Walker, J., Stansell, A., & Steinberg, J. M. (2020, November). On GPR surveying of historical cemeteries and ancient graveyards to aid forensic research. In *18th International Conference on Ground Penetrating Radar* (pp. 220-223). Society of Exploration Geophysicists. 10.1190/gpr2020-059.1

Davis, A., Whitehead, C., & Lengke, M. (2021). Subtle early-warning indicators of landfill subsurface thermal events. *Environmental Forensics*, 1–19. doi:10.1080/15275922.2021.1887973

De Angelis, D., Mapelli, G., Mazzullo, F. L., Lorenz, M. T., & Cattaneo, C. (2020). Possible applications of reflected UV photography in forensic odontology: Food for thought. *Legal Medicine*, *42*, 101641. doi:10.1016/j.legalmed.2019.101641 PMID:31751798

de Bruin, M., & Schmitz, P. M. U. (2021). The application of forensic geomorphology in rhinoceros poaching investigations in Africa. *The South African Geographical Journal*, *103*(1), 43–63. doi:10.1080/03736245.2020.1835700

de Caritat, P., Woods, B., Simpson, T., Nichols, C., Hoogenboom, L., Ilheo, A., ... Hoogewerff, J. (2021). Forensic soil provenancing in an urban/suburban setting: A sequential multivariate approach. *Journal of Forensic Sciences*. PMID:33955554

De Giorgi, L., Fragalá, G., Leucci, G., Mazzaglia, A., Malfitana, D., & Russo, S. (2020). Forensic geophysical survey to investigate illegal looting of a site of historical importance. *Archaeological Prospection*, arp.1796. doi:10.1002/arp.1796

Deveci, C., Atilgan, M., & Demirçin, S. (2021). The forensic aspect of thermal injuries during MRI: A case report with a review of literature. *Revista de Medicina Legala*, *29*, 49–52.

Dong, C. E., Webb, J. B., Bottrell, M. C., Saginor, I., Lee, B. D., & Stern, L. A. (2020). Strengths, limitations, and recommendations for instrumental color measurement in forensic soil characterization. *Journal of Forensic Sciences*, *65*(2), 438–449. doi:10.1111/1556-4029.14193 PMID:31524965

Ekštajn, H., Kružić, I., & Bašić, Ž. (2021). Forensic investigation of a mass grave at Ovčara, near Vukovar, of victims killed by the Serbian army in 1991. *ST-OPEN*, *2*, 1–26. doi:10.48188o.2.3

Elis, V. R., Almeida, E. R., Porsani, J. L., & Stangari, M. C. (2020, November). Ground-penetrating radar, resistivity, and induced polarization applied in forensic research in tropical soils. In *18th International Conference on Ground Penetrating Radar* (pp. 224-227). Society of Exploration Geophysicists.

Essefi, E. (2020). On the Geochemistry and Mineralogy of the Anthropocene. *International Journal of Water and Wastewater Treatment*, *6*(2), 1–14.

Essefi, E. (2021a). *Apport des Approches Géochimique et Sédimentologique au Fonctionnement des Systèmes Endoréiques et Exoréiques de l'Holocène à l'Anthropocène: Intérêts Environnemental et Géoéconomique*. Institut Supérieur des Sciences et Techniques des Eaux de Gabès.

Essefi, E. (2021b). Record of the Anthropocene-Great Acceleration along a core from the coast of Sfax, southeastern Tunisia. *Turkish Journal of Earth Science*, 1-16.

Essefi, E., Gharsalli, N., Kalabi, S., Ben Ameur, M., & Yaich, C. (2015d). Spectral analysis of a core from the Sebkha of Sidi Mansour, southernTunisia: The Holocene cyclostratigraphy. *J Geophys Remote Sens*, *4*(02), 141. doi:10.4172/2169-0049.1000141

Essefi, E., Gharsalli, N., Salhi, S., & Yaich, C. (2015b). Cyclicity of grain size parameters along a Core from the Gleysol of Sebkha Ennoual, southeastern Tunisia. *J Basic Appl Res Int, 15*(5–3), 124–134.

Essefi, E., Gharsalli, N., & Yaich, C. (2015a). Climatic significance aeolian sedimentary filling of sebkha Mhabeul, southeastern Tunisia: Contribution of the genetic approach. *J Basic Appl Res Int*, *15*(2–4), 140–152.

Essefi, E., Gharsalli, N., & Yaich, C. (2015c). Geophysical and geochemical study of the silico-evaporitic sedimentary filling of Boujmal wetland, eastern Tunisia: Inferring the climatic signal within groundwater noise. *J Basic Appl Res Int, 15*(2–1), 14–25.

Essefi, E., Jmaa, H. B., Touir, J., Tagorti, M. A., & Yaich, C. (2013). Variability to sedimentary dynamics and climatic conditions during the last two millennia at sebkha Souassi in eastern Tunisia. *Papers on Global Change IGBP*, *20*(1), 29–45. doi:10.2478/igbp-2013-0001

Essefi, E., Mefteh, S., Medhioub, M., & Yaich, C. (2014). Magnetic study of the heated and unheated sedimentary fillings of Sebkha Mhabeul, Southeast Tunisia: A geophysical method for paleoclimatic investigation and tephrochronological dating. *International Journal of Geophysics*, *2014*, 7–7. doi:10.1155/2014/908395

Fabian, K., Reimann, C., Kuzina, D., Kosareva, L., Fattakhova, L., & Nurgaliev, D. (2016, April). GEMAS: Unmixing magnetic properties of European agricultural soil. In EGU General Assembly Conference Abstracts (pp. EPSC2016-12495). Academic Press.

Ferrell, M. (2020). *Applications of Close-range Photogrammetry for Documenting Human Skeletal Remains in Obstructed Wooded Environments*. Academic Press.

Finkelstein, N. S., Rosengarten, H., & Levy, O. (2020). Photographic Image Comparison of Vegetation. *Journal of Forensic Sciences*, *65*(6), 2165–2169. doi:10.1111/1556-4029.14505 PMID:33460121

Fitzpatrick, R. W. (2009). Soil: forensic analysis. *Wiley encyclopedia of forensic science*, 1-14.

Fitzpatrick, R. W., & Baker, A. (2006). Magnetic susceptibility and low altitude aerial photography using a helium balloon: applications to forensic soil examinations. In Book of Abstracts (p. 19). Academic Press.

Franceschetti, L., Galante, N., Del Sordo, S., Casali, M. B., & Genovese, U. (2021). Forensic Considerations on the Two Major Civilian Terrorist Events occurred in Milan: A Retrospective Autopsy-Based Study. *Forensic Science International*, *326*, 110929. doi:10.1016/j.forsciint.2021.110929 PMID:34329954

Ghannem, N., Gargouri, D., Sarbeji, M. M., Yaich, C., & Azri, C. (2014). Metal contamination of surface sediments of the Sfax–Chebba coastal line, Tunisia. *Environmental Earth Sciences*, *72*(9), 3419–3427. doi:10.100712665-014-3248-z

Hachem, M., Sharma, B. K., El Naggar, A., Pilankar, I., & Anwar, N. (2020). Systematic Approaches For Soil Analysis In Forensic Investigation. In *Advances in Science and Engineering Technology International Conferences (ASET)* (pp. 1-5). IEEE.

Harvey, A. J., & Sekulla, A. (2021). Evidence of alcohol induced weapon focus in eyewitness memory. *Applied Cognitive Psychology*, *35*(5), 1263–1272. doi:10.1002/acp.3858

Jakubec, P., Qubaiová, J., Novák, M., & Růžička, J. (2021). Developmental biology of forensically important beetle, Necrophila (Calosilpha) brunnicollis (Coleoptera: Silphidae). *Journal of Medical Entomology*, *58*(1), 64–70. PMID:32829403

Kelly, T. B., Angel, M. N., O'Connor, D. E., Huff, C. C., Morris, L. E., & Wach, G. D. (2021). A novel approach to 3D modelling ground-penetrating radar (GPR) data–A case study of a cemetery and applications for criminal investigation. *Forensic Science International*, *325*, 110882. doi:10.1016/j.forsciint.2021.110882 PMID:34182205

Kootker, L. M., Plomp, E., Ammer, S. T., Hoogland, V., & Davies, G. R. (2020). Spatial patterns in 87Sr/86Sr ratios in modern human dental enamel and tap water from the Netherlands: Implications for forensic provenancing. *The Science of the Total Environment*, *729*, 138992. doi:10.1016/j.scitotenv.2020.138992 PMID:32361454

Koşaroğlu, S., Kamacı, Z., Erdoğan, S., Bektaş, Ö., & Büyüksaraç, A. (2021). Determination of historical graves by ground penetrating radar method: Sakarya Field Battle (August 23–September 13, 1921, Turkey). *Australian Journal of Forensic Sciences*, 1-19.

Leucci, G. (2020). *Advances in geophysical methods applied to forensic investigations: New developments in acquisition and data analysis methodologies*. Springer Nature. doi:10.1007/978-3-030-46242-0

Listos, P., Gryzinska, M., Panasiuk-Flak, K., Ciesielka, M., & Teresinski, G. (2021). Assessment of temperature changes in carcasses in the early post-mortem period using the spectrum of a thermal imaging camera. *Medycyna Weterynaryjna*, *77*(5), 253–257. doi:10.21521/mw.6535

Lockwood, E., & Masters, P. (2021). Application of viewshed analysis and probability mapping for search area determination based on the Moors murders on Saddleworth Moor, the United Kingdom. *Journal of Forensic Sciences*, *66*(5), 1770–1787. doi:10.1111/1556-4029.14799 PMID:34296762

Madzunya, D., Uushona, V., Mathuthu, M., & Heike, W. (2021). Rare earth elements in uranium ore deposits from Namibia: A nuclear forensics tool. *Journal of Environmental Radioactivity*, *237*, 106668. doi:10.1016/j.jenvrad.2021.106668 PMID:34116457

Mahaney, W. C. (2020). The Hannibal enigma of 218 BC: A forensic exercise of importance to historical archaeology. *Sci. Cult*, *6*, 7–24.

Malejko, J., Deoniziak, K., Tomczuk, M., Długokencka, J., & Godlewska-Żyłkiewicz, B. (2020). Puparial cases as toxicological indicators: Bioaccumulation of cadmium and thallium in the forensically important blowfly Lucilia sericata. *Frontiers in Chemistry*, *8*, 8. doi:10.3389/fchem.2020.586067 PMID:33330373

Mankoff, K. D., van As, D., Lines, A., Bording, T., Elliott, J., Kraghede, R., Cantalloube, H., Oriot, H., Dubois-Fernandez, P., Ruault du Plessis, O., Christiansen, A. V., Auken, E., Hansen, K., Colgan, W., & Karlsson, N. B. (2020). Search and recovery of aircraft parts in ice-sheet crevasse fields using airborne and in situ geophysical sensors. *Journal of Glaciology*, *66*(257), 496–508. doi:10.1017/jog.2020.26

Manrong, C., Lizhong, Y., Xiangfeng, N., & Bin, C. (2009). Application of environmental magnetism on crime detection in a highway traffic accident from Yangzhou to Guazhou, Jiangsu Province, China. *Forensic Science International*, *187*(1-3), 29–33. doi:10.1016/j.forsciint.2009.02.012 PMID:19303726

Mansegosa, D. A., Giannotti, P. S., Marchiori, J. I., Jofré, F. N., Aballay, F. H., & Aisa, C. F. (2021). The story of a homicide: The location, exhumation, and multidisciplinary analysis of a clandestine burial. *Forensic Science International: Reports*, *3*, 100165.

Mia, R., Panchal, V., & Guglot, Y. (2021). *The Role of 'Digital Forensic Photography'.* The Indian Criminal Justice System.

Molina, C. M., Pringle, J. K., Saumett, M., & Hernández, O. (2015). Preliminary results of sequential monitoring of simulated clandestine graves in Colombia, South America, using ground penetrating radar and botany. *Forensic Science International*, *248*, 61–70. doi:10.1016/j.forsciint.2014.12.011 PMID:25596556

Molina, C. M., Wisniewski, K. D., Drake, J., Baena, A., Guatame, A., & Pringle, J. K. (2020). Testing application of geographical information systems, forensic geomorphology and electrical resistivity tomography to investigate clandestine grave sites in Colombia, South America. *Journal of Forensic Sciences*, *65*(1), 266–273. doi:10.1111/1556-4029.14168 PMID:31424577

Morrison, R. D. (2020). Forensic techniques for establishing the origin and timing of a contaminant release. In *Expert Witnessing* (pp. 145–172). Routledge. doi:10.1201/9781003067764-10

Nde, S. C., Manjoro, M., & Mathuthu, M. (2019). Farm dam siltation and sediment source tracing in the Zeerust Swartruggens area in the north-west of South Africa. *International Journal of Hydrology Science and Technology*, *9*(3), 324–337. doi:10.1504/IJHST.2019.102322

Osbourne, M. G., & Christensen, L. S. (2020). Pathways to Child Sexual Offending: Applying Ward and Siegert's Pathways Model on Offenders Convicted of Penetrative Offenses on Children. *Sexuality & Culture*, *24*(6), 1756–1773. doi:10.100712119-020-09718-4

Pensieri, M. G., Garau, M., & Barone, P. M. (2020). Drones as an integral part of remote sensing technologies to help missing people. *Drones (Basel)*, *4*(2), 15. doi:10.3390/drones4020015

Pringle, J. K., Giubertoni, M., Cassidy, N. J., Wisniewski, K. D., Hansen, J. D., Linford, N. T., & Daniels, R. M. (2015). The use of magnetic susceptibility as a forensic search tool. *Forensic Science International*, *246*, 31–42. doi:10.1016/j.forsciint.2014.10.046 PMID:25460105

Pringle, J. K., Stimpson, I. G., Wisniewski, K. D., Heaton, V., Davenward, B., Mirosch, N., Spencer, F., & Jervis, J. R. (2021). Geophysical monitoring of simulated clandestine burials of murder victims to aid forensic investigators. *Geology Today*, *37*(2), 63–65. doi:10.1111/gto.12344

Qasim, M., Xiao, H., He, K., Noman, A., Liu, F., Chen, M. Y., Hussain, D., Jamal, Z. A., & Li, F. (2020). Impact of landfill garbage on insect ecology and human health. *Acta Tropica*, *211*, 105630. doi:10.1016/j.actatropica.2020.105630 PMID:32673623

Rocke, B., Ruffell, A., & Donnelly, L. (2021). Drone aerial imagery for the simulation of a neonate burial based on the geoforensic search strategy (GSS). *Journal of Forensic Sciences*, *66*(4), 1506–1519. doi:10.1111/1556-4029.14690 PMID:33576508

Różycki, S., Zapłata, R., Karczewski, J., Ossowski, A., & Tomczyk, J. (2020). Integrated Archaeological Research: Archival Resources, Surveys, Geophysical Prospection and Excavation Approach at an Execution and Burial Site: The German Nazi Labour Camp in Treblinka. *Geosciences*, *10*(9), 336. doi:10.3390/geosciences10090336

Sangwan, P., Nain, T., Singal, K., Hooda, N., & Sharma, N. (2020). Soil as a tool of revelation in forensic science: A review. *Analytical Methods*, *12*(43), 5150–5159. doi:10.1039/D0AY01634A PMID:33135702

Sant'Ovaia, H., Marques, G., Santos, A., Gomes, C., & Rocha, A. (2015). Magnetic susceptibility and isothermal remanent magnetization in human tissues: A study case. *Biometals*, *28*(6), 951–958. doi:10.100710534-015-9879-z PMID:26373856

Schneider, B., Knippel, E., & Miller, R. (2020). Utilizing ground-penetrating radar to assist in a forensic cold case investigation. *SEG Technical Program Expanded Abstracts*, *2020*, 3324–3330. doi:10.1190egam2020-3420999.1

Sharma, R. C., Kumar, S., Kumar, S., Mann, M., & Sharma, M. (2020). Photoacoustic remote sensing of suspicious objects for defence and forensic applications. *Spectrochimica Acta. Part A: Molecular and Biomolecular Spectroscopy, 224*, 117445. doi:10.1016/j.saa.2019.117445 PMID:31382229

Shina, H. G., Leeb, J., Yunc, Y. H., Yood, S. H., Jange, J., Ohf, S. H., ... Leea, J. (2021). *χ-separation: Magnetic susceptibility source separation toward iron and myelin mapping in the brain.* Academic Press.

Shvedchikova, T., Barone, P. M., & Moghaddam, N. (2021). Crimes in the Past: Archaeological and Anthropological Evidence. *Crimes in the Past*.

Swain, A. (2020). Big Data Challenges and Hype Digital Forensic: A Review in Health Care Management. *Big Data Analytics and Computing for Digital Forensic Investigations*, 43-63.

Vanderheyden, N., Verhoeven, E., Vermeulen, S., & Bekaert, B. (2020). Survival of forensic trace evidence on improvised explosive devices: Perspectives on individualisation. *Scientific Reports*, *10*(1), 1–11. doi:10.103841598-020-69385-1 PMID:32732923

Wallace, J. R., Scala, J. R., Gultepe, I., & LeBlanc, H. N. (2021). Bridging Disciplines: The Application of Forensic Science to Meteorological Evidence. *Pure and Applied Geophysics*, *178*(3), 693–704. doi:10.100700024-020-02646-2

Weiss, C., & Roy, B. (2020). President's Page: Near-surface geophysics for the future. *The Leading Edge*, *39*(5), 304–306. doi:10.1190/tle39050304.1

Xu, X., Du, C., Ma, F., Shen, Y., & Zhou, J. (2020). Forensic soil analysis using laser-induced breakdown spectroscopy (LIBS) and Fourier transform infrared total attenuated reflectance spectroscopy (FTIR-ATR): Principles and case studies. *Forensic Science International, 310*, 110222. doi:10.1016/j.forsciint.2020.110222 PMID:32193128

Zawadzki, J., Fabijańczyk, P., & Przeździecki, K. (2020). *Geostatistical Methods as a Tool Supporting Revitalization of Industrially Degraded and Post-Mining Areas*. Academic Press.

Chapter 3
Advances in Forensic Sedimentology

Elhoucine Essefi
University of Gabes, Tunisia

ABSTRACT

Forensic sedimentology is a relatively recently realized field. Sedimentological methods used to solve cases have evolved as the field has developed, beginning with simple identification of minerals and progressing to the examination of individual grains using highly advanced scanning electron microscopes. More simple methods, such as color analysis, are still used today, but in addition, forensic sedimentologists look at surface textures and grain size distribution. For instance, quartz grains were used in a forensic technique as sediment fingerprint. The particle size distribution is one of the important tests when analysing sediments and soils in geological studies. For forensic work, the particle size distribution of sometimes very small samples requires precise determination using a rapid and reliable method with a high resolution. FRITSCH laser granulometer offers rapid and accurate sizing of particles in the range 0.04–2000 µm for a variety of sample types, including soils, unconsolidated sediments, dusts, powders, and other particulate materials.

1. INTRODUCTION

Although the idea of using sediment for forensic purposes has been popular in fiction for more than a century (Lanier, 2017), forensic sedimentology had remained a novel field of study (Douglas et al., 2020; Madureira-Carvalho et al., 2020). Simple methods such as color analysis are still used today. In addition, forensic sedimentologists also examined the surface and grain structure via binocular (morphosocopy), optic

DOI: 10.4018/978-1-7998-8386-9.ch003

microscope and SEM (exoscopy). Several analytical techniques have been used such as high-tech scanning electron microscope which is used to examine individual crystal grains. Cases handled by sedimentologists involve environmental violations, murder, theft or kidnapping. The most famous cases are related to the prevention of war victims (Inkret et al, 2020). The application of forensic expertise or science in criminal investigation (Ruffell and McKinley, 2005) has long been at the forefront of research and development of scientific technology and methods. Geology plays a vital role in any forensic environmental survey (Hashmi et al, 2020). In spite of recent criticisms (Melo et al, 2020), sedimentology in particular can establish connections between people, objects, and places to determine the facts and possibilities of these connections. With the development of analytical techniques, forensic sedimentology has become a leader in scientific criminology (Frankl et al., 2021).

Historically speaking, the first known idea of forensic geology may have come from Sir Arthur Conan Doyle (1887). Then, geology was used to decipher real cases in 1904 (Ruffell and McKinley, 2005). In some cases, we can judge how long a person has been in a given site by the color and density of the sediment on his clothes. Other theoretical applications of geology in solving crimes, as well as other forensic techniques were recently discussed such as ballistics (Ruffell and McKinley, 2005; Sydow et al., 2020). The spread of crime and the spread of geography quickly became popular methods in well-known organizations, rather than just for personal use. In the early 20th century, governments and educational institutions recognized the potential of forensics and used their advantages. They established specialized forensic laboratories and trained scientists to use their expertise to fight crime (Ruffell and McKinley, 2005). The university has developed a forensic science plan, and national organizations such as the FBI have begun to use forensic science to regularly investigate cases. The importance of using sediment as evidence is still used today to establish connections between people, places and objects. Forensics sedimentology is basically applied on soil rather than rocks (Sangwan et al, 2020). Soil forensics research, studies and analyzes of traces of soil connected to a crime (Xu et al, 2020). In the context of forensic analysis, the term soil is referred to any of the material of geological origin that consists of removable elements, such as a sandy beach, mud or soil (Profumo et al., 2020). The ground can provide important information to tie a person to a place and clarify the dynamics of a crime, as it can be transferred from one place to another for filing on moving surfaces, such as shoes, tires or work tools. Many crimes are committed in the open where the 'environment, which is not a closed system, it is able to exchange items with those who participated in. Soil is a system made up, at the microscopic level, of several components that have been formed and developed during its long process of formation. This process is affected by many factors: the nature of the mother rock, the climate, the topography, the micro-organisms and vegetation and the current time. Under the above factors if they combine others due to man that affect the components of a soil with the help of various substances and materials (fragments of bricks, glass, chemicals, etc.), which are present in quantities proportional to the spatiotemporal character of the anthropogenic effect. Due to this complexity, there is a noticeable difference between soil samples, recording significant variations in the components and their proportions, even in small areas. The precise analysis of the terrain allows a characterization of the samples to the point that they can also distinguish whether they were collected a few meters away. For this evidence, the use of geologic and pedologic soil analysis is excellent as a support to forensic investigations.

In this chapter, the aim was to enumerate some recent advances of forensic sedimentology to overcome difficulties of using sediment signal for legal investigations.

2. SEDIMENTARY FORENSIC METHODS

Based on sample limited in size and number, sedimentologists are familiar with examining rock or sediment samples to infer past depositional environment in which they may be deposited as well as the source of sediments. For example, the aim may be to determining the source of a sediment sample or comparing two samples to see if they are from the same source. In most cases, the purpose of forensic sedimentologists is limited to exceptions where two or more specimens cannot be determined. These situations include physically matched samples, such as broken rocks or characteristic footprints in two halves, or mineralogical coincidences, such as several very rare particles commonly found in two samples (Pye and Croft, 2004). Forensic sedimentologists often look for exotic minerals to characterize sediment samples (Pirrie et al. 2009). However, the problem with this is that the number of samples is usually limited. The deposition and inherent tendency of these deposits are scarce (Bull and Morgan, 2006). For example, if a piece of shoes is found with dirt at the crime scene, then the piece of dirt found is everything that the sedimentologist will use. Sometimes this small amount is enough to hold a rare and unique mineral, but usually, the sample is too small to cover the full volume of the original mineral, or the source itself does not contain the mineral. They also identify or exclude specific sources or routes of transportation. For example, the presence of feldspar in the sample indicates that water transport is not taking place because feldspar is particularly prone to hydrolysis (Pitts et al., 2020). However, from a mineralogical point of view, rock deposits from different sources may have the same minerals and color scheme, so in addition to these qualities, other attributes must be considered. Bull and Morgan (2006) proposed that careful examination of the more abundant mineral quartz can help further distinguish and characterize sediments. These grains can be inspected with modern microscopes and classified according to their surface texture. Quartz is used for its excellent physical and chemical weather resistance and its richness worldwide (de Bruin and Schmitz, 2021). The surface texture of this stable mineral can tell the story of grain travel in the past few days, and it can also reveal the story of grain millions of years ago. Nonetheless, presenting forensic data to jury in courts should be done with ultimate caution. When submitting such sedimentological evidence to the jury, the original data should be simply and accurately explained and summarized without going into too much detail. Too many details can overwhelm the jury and obscure the points that need to be emphasized. This elimination method is usually a more reliable indicator, but unless it can be clearly determined that the two materials come from the same source, it can be found that there is a similarity between the type of grain on the suspect's packaging and the suspect's vehicle (Bull and Morgan, 2006). Probably the most common test performed by sedimentologists is particle size analysis.

2.1. Grain Size Distribution

The particle size distribution in a sediment sample can provide information about the source of the sediment and the transport mechanism to which the sediment is exposed (Mahdavi et al., 2020). Forensic sedimentologists use grain size analysis to compare or characterize sediments, rather than what traditional sedimentologists might do, such as predicting the paleoenvironment of sedimentation. The method of measuring particle size mainly depends on the type of sediment being analyzed. The particle size varies from sand to gravel, and settlement (sucking) is effective for silt and clay (Pye and Blott, 2004). Essefi (2013) suggested the use of a Fritsch ANALYSETTE MicroTec to measure the size of silt and clay, especially in the case of small sample sizes, which is usually the case in forensics. However, certain laser sorting tools are good at sorting sand into clay, making it the most versatile method (Pye and Blott, 2004). However, the results of each of these methods cannot be directly compared because they measure the size of each grain based on different properties. For example, mesh size or equivalent sphere diameter (Pye and Blott, 2004). However, the comparison between samples should always be combined with other methods, because the grain size strongly depends on its size sample and selective delivery (Pye and Blott, 2004; Morgan and Bull, 2007). Particle size is a fundamental property of any sediment, soil or dust deposit which can provide important clues to nature and provenance. For forensic work, the particle size distribution of sometimes very small samples requires precise determination using a rapid and reliable method with a high resolution. Fritsch laser granulometer offers rapid and accurate sizing of particles in the range 0.04–2000 μm for a variety of sample types, including soils, unconsolidated sediments, dusts, powders and other particulate materials. Reliable results are possible for sample weights of just 50 mg. Discrimination between samples is performed on the basis of the shape of the particle size curves and statistical measures of the size distributions. In routine forensic work laser granulometry data can rarely be used in isolation and should be considered in combination with results from other techniques to reach an overall conclusion. Recently, the National Engineering School of Sfax purchased a Fritsch ANALYSETTE MicroTec (Fig.1). This apparatus provides the highest technology of the grain size distribution measurement.

Figure 1. Fritsch ANALYSETTE 22 MicroTec
(Essefi, 2013)

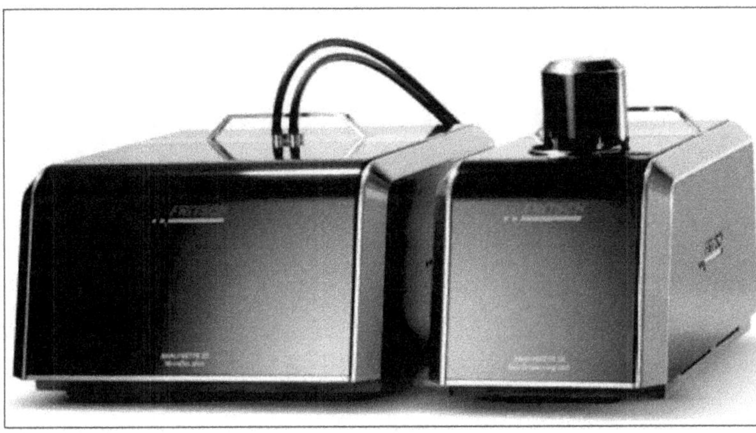

In terms of data computation, Fritsch apparatus is equipped with sophisticated software (MaScontrol) providing the calculation of different statistical parameters. In addition to numerous graphs and other options, MaScontrol also offers its user statistical values, which may contain useful information for evaluation of particle size distribution. The most important values are the "complete moments of the distribution". The samples H9 and H10 are silty belonging to the facies F4 and F3 respectively. The top of this period H9 is coarser (Mean: 6.58 µm); it is expressed by a coarse aeolian fraction (Fig.2.). The bottom of this period becomes finer (Mean: 6.58 µm); it is characterized by a primary mode (M: ca 2.90 µm) as indication of the fine hydraulic sedimentation.

Dudley (1976) recommends further evaluation of the color and pH of the sample, and Pye and Blott (2004) also recommend the use of chemical and mineralogical analysis. Master basic geological knowledge and carry out particle size analysis to obtain reliable conclusions.

Figure 2. Grain size distribution of the Dark Age along the core from sebkha Dkhila

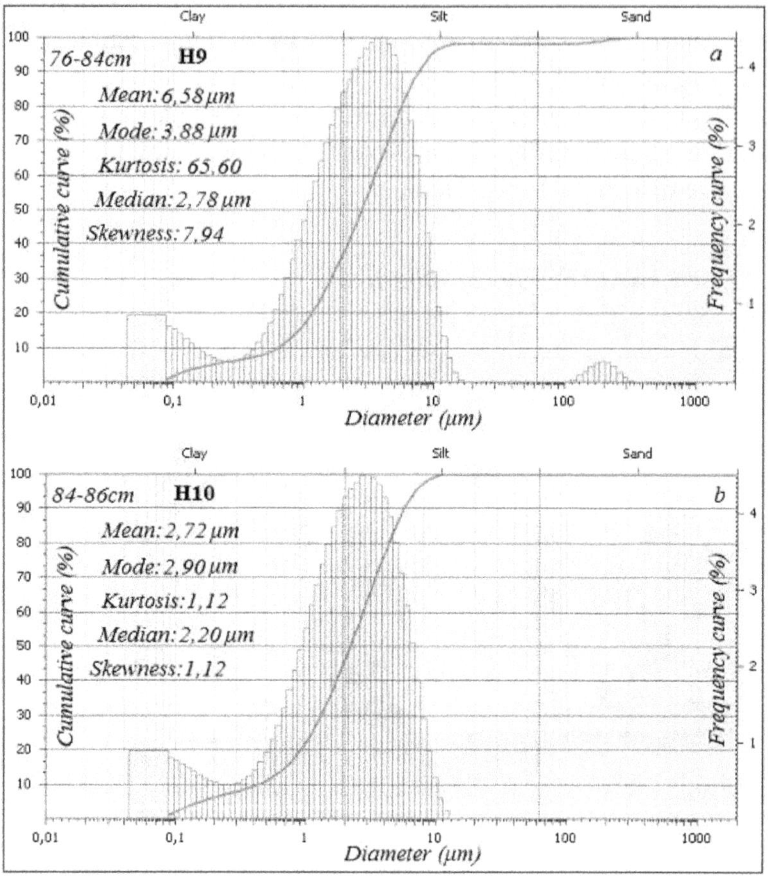

2.2. Grains Morphoscopy

The observation of four samples from a core from a Tunisian wetland shows a heterogeneous grain size distribution ranging from 10 µm to 400 µm (Fig.3). As examples, the sample of 2.5 cm is made up of 10 µm to 200 µm grains; the shape is sub-angles. On the other hand, the sample of 77 cm is made up of 40 µm to 400 µm round grains.

3. DIFFICULTIES OF FORENSIC SEDIMENTOLOGY

Although sedimentologists can use a variety of tools, forensic sedimentology is particularly challenging, because no analytical method can be worldwide popularized

Figure 3. Microscopic observations of some sediment samples

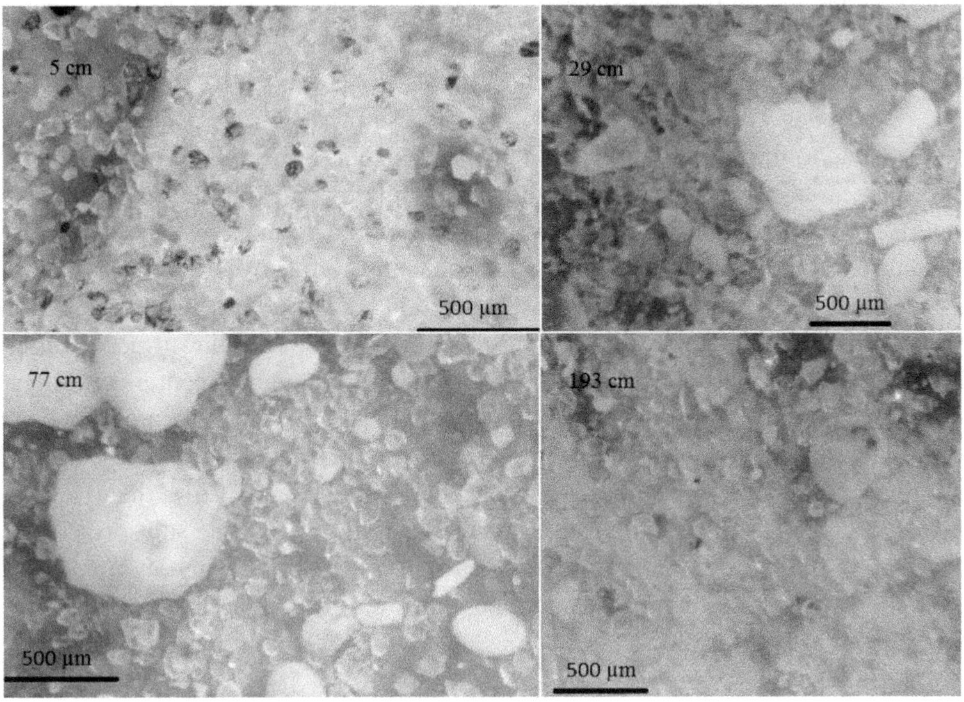

(Melo et al., 2020). This will affect the composition and analysis of the sediment. Comply with proper geological sampling requirements, resulting in poor sample quality or contamination. The collected samples must be representative of the source in order for the source to be accurately analyzed. In addition, samples can only be compared with similar samples (Morgan and Bull, 2007) For example, if a crime scene such as a grave has a hole, samples should be taken from the top soil and the sediment below because the bottom depth usually changes. Only compare soil samples with other surface soil samples, because comparisons between higher and lower soil samples may lead to false negative results and unreasonable conclusions (Morgan and Bull, 2007). One of the most common problems faced by forensic sedimentologists is the small sample size. Evidence usually contains only a small amount of sediment available for analysis, which severely limits the number and types of tests that can be performed, but Fritsch apparatus may overcome this handicap. In addition, at least half of the samples should be saved untested so that analyses redundancy may be confirmed (Agarwal et al., 2020). Therefore, forensic sedimentologists should carefully determine the order of sample analysis in advance to ensure that all necessary analyses are performed. As it is the case for all forensic analyses (García et al., 2020), forensic sedimentologists also need to consider the impact of humans on the samples. Applied sedimentology can be used in many ways to solve many different types of cases, because the versatility of geological knowledge makes it very useful in many situations. Sedimentologists are usually hired to deal with environmental incidents, such as illegal wetland drainage (Podhade et al., 2020) and engineering interruptions (Parise et al., 2020), such as highway collapse due to improper construction or maintenance. Unfortunately, in the most notorious cases, there is a tendency to harm people in some way, such as forensic experts are often asked to determine whether the suspect has visited the crime scene (Pye and Croft, 2004). Provide clues about where the offender was before visiting the crime scene, or compare it with the suspect's grain or their belongings to determine whether the suspect went to the crime scene. For example, certain surface textures of wind particles can indicate certain energy states and environmental conditions that caused them. Geologists use the same method to predict the structure of ancient rocks and sediments. Therefore, the concept is not necessarily a new concept, but a new application of the basic concept. In addition to analyzing the evidence provided, geologists often help find evidence. Irregularities of the soil surface and information on soil drainage can help locating buried objects such as corpses to avoid more aggressive, expensive and time-consuming methods such as excavation (Ruffell and McKinley, 2005).

4. CONCLUSION

Forensic sedimentology has slowly evolved over time from a core of associated ideas in a very novel way to at least one of the most important forms of crime-fighting investigation. It continues to utilize and revolutionize the fundamental concepts of sedimentology and drive innovation both within the research laboratory and within the research laboratory. Since each case is a completely new problem, inventive solutions must be developed to ensure that the correct conclusions are drawn within the case. However, these solutions are used by sedimentologists in every work area. However, the interpretation and presentation of these conclusions lies somewhere between ancient sedimentology and rhetoric. While ancient sedimentologists give their findings and theories to various scientists, rhetorical sedimentologists give their evidence to the average citizen before excessive judgment and should then leave no room for speculation. Any interpretation that makes rhetoric must be impartial and based primarily on fact, as it has a direct impact on the lives of those involved in the case.

REFERENCES

Agarwal, A., & Gupta, A. (2020). A maximum relevancy and minimum redundancy feature selection approach for median filtering forensics. *Multimedia Tools and Applications*, *79*(29-30), 21743–21770. doi:10.100711042-020-08994-4

Bull, P. A., & Morgan, R. M. (2006). Sediment fingerprints: A forensic technique using quartz sand grains. *Science & Justice*, *46*, 107–124. doi:10.1016/S1355-0306(06)71581

de Bruin, M., & Schmitz, P. M. U. (2021). The application of forensic geomorphology in rhinoceros poaching investigations in Africa. *The South African Geographical Journal*, *103*(1), 43–63.

Douglas, G. S., Graan, T. P., & Hardenstine, J. H. (2020). Forensic identification and quantification of oil sands-based bitumen released into a complex sediment environment. *Marine Pollution Bulletin*, *155*, 111141. doi:10.1016/j.marpolbul.2020.111141 PMID:32469767

Dudley, R. J. (1976). The particle size analysis of soils and its use in forensic science—The determination of particle size distributions within the silt and sand fractions. *Journal - Forensic Science Society*, *16*(3), 219–229. doi:10.1016/S0015-7368(76)71064-8 PMID:1022844

Essefi, E. (2013). *Wet aeolian sedimentology and sequence stratigraphy in Eastern Tunisia: implications for wet aeolian sedimentology and sequence stratigraphy on mars* [Ph.D. thesis]. National Engineering School of Sfax.

Frankl, A., Evrard, O., Verleyen, E., Cammeraat, E., & Stokes, A. (2021, March). How forensic science can lead the way in identifying culprit soil fingerprints in European mountains. In *MCAA Annual Conference* (p. 38). Academic Press.

García, M. G., Pérez-Cárceles, M. D., Osuna, E., & Legaz, I. (2020). Impact of the human microbiome in forensic sciences: A systematic review. *Applied and Environmental Microbiology*, *86*(22). Advance online publication. doi:10.1128/AEM.01451-20 PMID:32887714

Hashmi, T. A., Qureshi, R., Tipre, D., & Menon, S. (2020). Investigation of pesticide residues in water, sediments and fish samples from Tapi River, India as a case study and its forensic significance. *Environmental Forensics*, *21*(1), 1–10. doi:10.1080/15275922.2019.1693441

Inkret, J., Podovšovnik, E., Zupanc, T., Haring, G., & Pajnič, I. Z. (2021). Intra-bone nuclear DNA variability in Second World War metatarsal and metacarpal bones. *International Journal of Legal Medicine*, 1–12. PMID:33624158

Lanier, A. (2017). *Applications of forensic sedimentology*. Academic Press.

Madureira-Carvalho, Á., Ribeiro, H., Newman, G., & Brewer, J.,, M., Guedes, A., Abreu, I., ... Dawson, L. (2020). Geochemical analysis of sediment samples for forensic purposes: Characterisation of two river beaches from the Douro River, Portugal. *The Australian Journal of Forensic Sciences*, *52*(2), 222–234. doi:10.1080/00450618.2018.1533037

Mahdavi, A., & Siegel, J. A. (2020). Quantitative filter forensics: Size distribution and particulate matter concentrations in residential buildings. *Indoor Air*.

Marshall, J. R., Bull, P. A., & Morgan, R. M. (2012). Energy regimes for aeolian sand grain surface textures. *Sedimentary Geology*, *253-254*, 17–24. doi:10.1016/j.sedgeo.2012.01.001

Melo, V. F., Testoni, S. A., Dawson, L. A., & Salvador, F. A. D. S. (2020). Sand fraction is not suitable for forensic investigations in subtropical soils. *Revista Brasileira de Ciência do Solo*, 44.

Morgan, R. M., & Bull, P. A. (2007). The use of grain size distribution analysis of sediments and soils in forensic enquiry. *Science & Justice*, *47*, 125–135. doi:10.1016/j.scijus.2007.02.001

Morgan, R. M., Wiltshire, P., Parker, A., & Bull, P. A. (2006). The role of forensic geoscience in wildlife crime detection. *Forensic Science International*, *162*, 152–162. doi:10.1016/j.forsciint.2006.06.045

Parise, G., Mohla, D., Parise, L., & Lombardi, M. (2020, October). Forensic Implications in the Continuous Discontinuity of IoT Innovations. In *2020 IEEE Industry Applications Society Annual Meeting* (pp. 1-6). IEEE.

Pirrie, D., Power, M. R., Rollinson, G. K., Wiltshire, P. E., Newberry, J., & Campbell, H. E. (2009). Automated SEM-EDS (QEMSCAN®) mineral analysis in forensic soil investigations: testing instrumental reproducibility. In *Criminal and environmental soil forensics* (pp. 411–430). Springer.

Pitts, K. M., & Clarke, R. M. (2020). The forensic discrimination of quartz sands from the Swan Coastal Plain, Western Australia. *Forensic Science International: Reports*, *2*, 100130.

Podhade, D., Lal, S. B., Singh, S., Mehera, B., Khare, N., & James, A. (2020). Evaluating the Impact of Wetland Health on Wildlife Health by Soil and Water Quality Analysis. *International Journal of Current Microbiology and Applied Sciences*, *9*(9), 839–849.

Profumo, A., Gorroni, A., Guarnieri, S. A., Mellerio, G. G., Cucca, L., & Merli, D. (2020). GC-MS qualitative analysis of the volatile, semivolatile and volatilizable fractions of soil evidence for forensic application: A chemical fingerprinting. *Talanta*, *219*, 121304.

Pye, K., & Blott, S.J. (2004). Particle size analysis of sediments, soils and related particulate materials for forensic purposes using laser granulometry. *Forensic Science International, 144*, 19-27. doi:10.1016/j.forsciint.2004.02.028

Ruffell, A., & McKinley, J. (2005). Forensic geoscience: Applications of geology, geomorphology and geophysics to criminal investigations. *Earth-Science Reviews*, *69*, 235–247. doi:10.1016/j.earscirev.2004.08.002

Sacchi, E., & Nicosia, U. (2013). Forensic paleontology: A tool for "intelligence" and investigation. *Journal of Forensic Sciences*, *58*, 651–657. doi:10.1111/1556-4029.12084

Sangwan, P., Nain, T., Singal, K., Hooda, N., & Sharma, N. (2020). Soil as a tool of revelation in forensic science: A review. *Analytical Methods*.

Sydow, L., Johns, D. A., Zappitello, S. J., & Maurer, T. (2020). Environmental forensic investigation of mystery sediment plumes at Barton Springs. Academic Press.

Xu, X., Du, C., Ma, F., Shen, Y., & Zhou, J. (2020). Forensic soil analysis using laser-induced breakdown spectroscopy (LIBS) and Fourier transform infrared total attenuated reflectance spectroscopy (FTIR-ATR): Principles and case studies. *Forensic Science International*, *310*, 110222.

Chapter 4
Advances of Forensic Remote Sensing Applications in the Face of Transnational Organized Crime and Terrorism

Elhoucine Essefi
University of Gabes, Tunisia

ABSTRACT

This chapter aims to investigate advance and relevance of remote sensing in detecting the increasing transnational terrorist and crimes acts. This work should take into the widest definition of transnational crimes and terrorist activities and the link between. Geopolitics has created a favor climate for the setting of transnational crimes and terrorism at the Tunisian-Libyan borders. A possible future scenario is the fall of a military base with high technology arms in the hand of terrorist groups. Remote would be relevant by monitoring of terrorist mobility and their number evolution, arms quality and quantity within the base and the region, linked illegal activities funding terrorist groups (human trafficking from Africa to Europe, arms trade towards Mali, and smuggling).

1. INTRODUCTION

According to the academic definition of criminologists, a simple crime occurred, but it may be not recorded neither reported due to spatial and temporal constraints.

DOI: 10.4018/978-1-7998-8386-9.ch004

Further, transnational organized crimes and terrorism are of great complexity for two main reasons: victims fail to report crimes and law enforcement agents are unable to detect crimes, because transnational crimes take place in remote or hidden location or because of a lack of technology and staff to carry out a complete surveillance. Transnational organized crime and terrorism were considered two crimes with different aspects. Transnational organized crimes mainly focus on economic profit and acquire illegal markets. On the other hand, terrorism is motivated by an ideological background and an aim for a political revamp. Criminal terrorism which is defined as the systematic use of acts of terror was excluded from transnational organized crimes typology. Two separate bodies of criminology literature have emerged due to this distinction between organized crime and terrorism. Nonetheless, the hypothesis of transnational organized crimes and terrorism convergence has been recently defended (Kadir, 2020; Mirdad, 2020; Omelicheva and Markowitz, 2021; Petrich, 2021). Actually, some link between the two phenomena has been suggested. The idea raised in the 1980s with the discovery of narco-terrorism mixing drug trafficking and terrorist activities (Atsa'am et al., 2021; Rosen, 2021). A symbiosis ever seen before between terrorism and organized crime has recently taken place (Singh, 2021). Due to accelerated global interaction of the globalisation, transnational organized crime and international terrorism are flourishing. Other scholars (Fisabillah and Pujiyono, 2021) went further and suggested the theoretical possibility that in concrete cases, terrorism can change into organized crime and vice versa. A number of geopolitical and economic emerging conditions have led to metamorphosis of terrorism and organized crime (Perry, 2020). Also, clear structural similarities make collaboration advantageous for both of them. There not difference between transnational crimes wrongdoers and terrorists throwing bombs. In general, the state by its law enforcement is common enemy of both of them. In the underworld, both crimes secretly operate using the similar infrastructures for their activities. Further, they use the same networks of white-collar crime and corruption. Both use the same type of tactics: they engage in cross-border smuggling, money laundering, counterfeiting, kidnapping, extortion and various kinds of violence (Puttonen and Romiti, 2020). In their literatures, transnational organized crime wrongdoers and terrorists theorize to the use the mechanism of the political life to create the social and economic instable climate making their profitable activities feasible and to push their own agendas. These two types of complex crimes need an international collaboration to follow their spatial evolution without political limits and temporal metamorphosis without any obvious rule. Nonetheless, this collaboration is not quite evident because the confidence between countries does not ensure the organized work to face these issues. As a matter of fact, since the Cold War, terrorism was set and helped by official states. For these reasons, many countries refused to open their frontiers for any foreign military or security activity on their lands. Actually, the use of forensic remote sensing is a magic solution to face transnational crimes and terrorism (Dashnyam, 2020; Avtar et al., 2021). In inaccessible places, it basically expands the supervision of agents of the law. It may make up for a lack of manpower and reduces the exposure of these agents to dangerous circumstances on the ground. This chapter is meant to investigate the relevance of remote sensing in detecting transnational crimes and terrorist acts.

2. LIMITATIONS OF FORENSIC REMOTE SENSING

Three limitations concerning transnational organized crimes and terrorism investigations are discussed. Forensic remote sensing images are carried out by overflying helicopters (Han et al., 2020; Legleiter and Kinzel, 2020), aircraft or satellites can only detect crimes or crime's impacts that are visible from above and for sustained periods of time. Second, remote sensing may not record the economic, social, political and historical context of the crime scene in which the actions had taken place The continued and increasing use of remote sensing for these purposes brings us to the third limitation that we will mention here: the issue of validation (Lee and de Vries, 2021).

3. BIBLIOMETRIC ANALYSIS OF REMOTE SENSING OF TRANSNATIONAL CRIMES AND TERRORISM

The remote sensing of transnational crimes and terrorism is the use of airborne and satellite remote imagery for detect unreported or undetected cases. Even with moderate or low spatial resolution imagery, like Landsat (30-m resolution) (Redowan et al., 2020; Khalifa et al., 2021; Olayinka et al., 2021), transnational and terrorist crimes patterns may be easily detected since they occur over relatively large geographic areas. In spite of absence of anterior detection of these crimes, their evidence is generally visible for extended periods a Plethora of academic papers investigated remote sensing of transnational crimes including testing algorithms (Matczak, and Mazurek, 2021), detection techniques (Bi et al., 2020; Gruber, 2021) or spectral reflectances of illicit crops (Diaz et al., 2020). Before 2001, few studies document the use of remote sensing in the active reconnaissance of transnational crimes and terrorist acts (Fig.1). These topics were found through searches in the Google Scholar, Web of Science and Jstor search databases using a number of combined words and phrases including remote sensing, transnational crimes and terrorism. Transnational crimes studies dramatically increase since 2010 (Fig.1a); while terrorism studies tend to decrease from 2020 (Fig.1b). This is due to its merging within the concept transnational crimes (Makarenko, 2004; Shelley and Picarelli, 2005; Saul, 2017; Kadir et al., 2019).

Figure 1. Evolution of publications of (a) remote sensing and transnational crime (b) remote sensing and terrorism

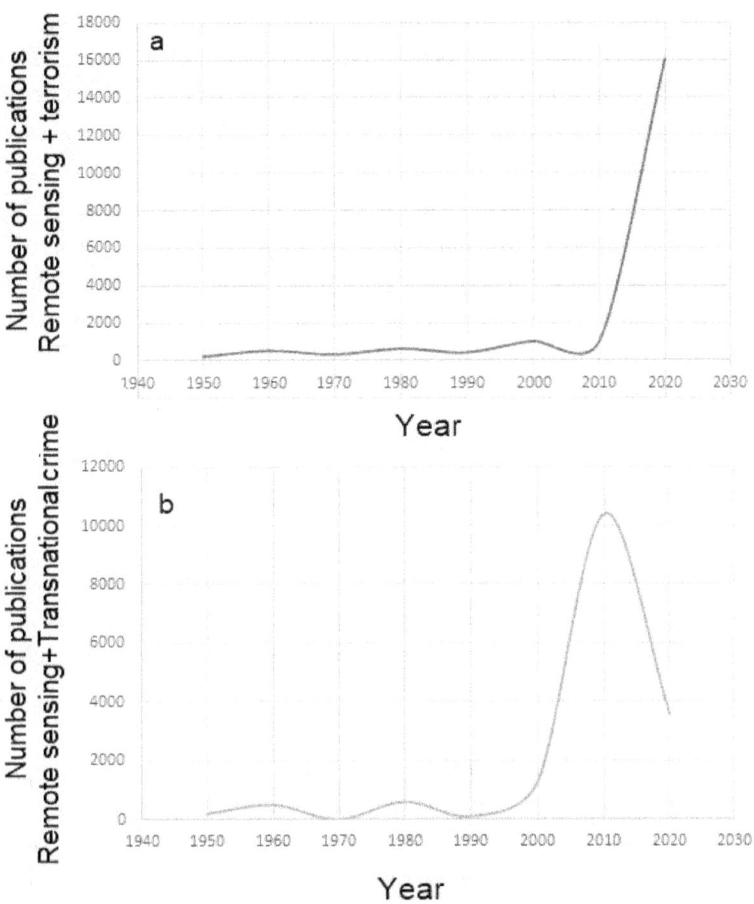

4. REMOTE SENSING OF HUMAN TRAFFICKING: THE FIRST STEP AGAINST TRANSNATIONAL CRIMES AND TERRORISM

As a matter of fact, the criminalization of human movements across national borders either for trade or migration may be considered as fraught with problematic economic, political, social, and militaristic issues and implications. In the recent years, some countries increased their borders measure. For instance, along the

U.S. Mexico border more regular patrols of the borderline by Homeland Security agents were initiated (Fleuriet and Castellano, 2020). This situation has led gangs of organized crime to increase their trade of illegal migration. These trades evolved toward massive human trafficking feeding the human resources of transnational crimes and terrorism (Welch, 2016; Okubo and Shelley, 2011; Palomino et al., 2020; Attah, 2020; Bergström, 2020). In face of these well-organized immigration waves, even the expensive infrastructure and personnel investments remain ineffective for complete surveillance of borders (Corbin, 2021; Yakubu and Idehen, 2021; Ali et al., 2021; Yendaw, 2021). After the worldwide flourishing of remote sensing techniques from the year 2000 onwards, the daily monitoring of borders has become an easy task to do (Mobariz and Kaplan, 2021; Mahmoudi et al., 2021; Caparó Bellido and Rundquist, 2021). Based aerial and satellite images (Desille, 2021), remote sensing can allow to target surveillance and enforcement efforts by revealing where human trafficking is taking place by displaying habitual paths through the desert, mountains and wetlands. The advanced technology of remote sensors affixed to light aircraft provide almost real-time detection of human movement across the landscape (Kaur, 2021).

5. GOOGLE EARTH CRIME DETECTION

Although a limited discussion of the use of Google Earth to detect crimes in the academic literature, it is widely discussed in the popular press. Added to refined remote sensing techniques, the free available remote sensing products, such as Google Earth, were also used to detect crimes (Xu et al., 2020). As a civil application, in many countries such as Greece, Italy, Argentina, India and the United States, Google Earth has been used to identify homes violating building codes, built swimming pools without permits and to compare declared home values with actual existing structures. This cheap and easy way to monitor the territory, in spite of its lack of accuracy, proves relevant to anticipate the transnational crimes and terrorist acts.

6. APPLICATION: THE SCENARIO OF AL-WATIYA MILITARY BASE UNDER THE DOMINANCE OF TERRORIST GROUPS

In our case, we can deal with the use of the remote sensing manage a possible scenario between Libya and Tunisia, which is 'al-Watiya military base under the dominance of terrorist groups'. As a matter of fact, both countries witnessed unprecedent political and security instabilities paving the way for terrorists acts and transnational crimes. In Tunisia, terrorist activities are restricted into isolated and limited with individual attempts (Cragin, 2021). While, in Libya, terrorist groups are more organized and control vast areas in the country at the expense of the local government (Beccaro, 2020). Added to external funding of some international intelligences, they receive their funds from transnational crimes including smuggling, human trafficking from Africa to Europe, arms and drug trades (Kubai, 2021). After the Turkish intervention in Libya, the game of power has changed on lands (Joffé, 2020) and new power struggles in the Mediterranean region started (Pierini, 2020). Some countries oriented their intelligences to collect press via terrorist groups. One of the repercussions of this intervention is setting of al-Watiya military base, which is already over the control of the central government, under the Turkish control and the setting of some terrorist groups in the vicinities of this base. A possible future scenario is the retraction of the Turkish troops. In this case, these groups are the more candidate to take the power on this base containing high technology armies. In doing so, they represent a real threat for Tunisia and Libya.

6.1. Al-Watiya Military Base and the Tunisian National Security

Al-Watiya Air Base also known as Okba Ibn Nafa Air Base is a military airport in western Libya. It is 27 Km east of the Tunisian border and 125 Km from Tripoli. A simple investigation of Google Earth images gives a clear idea about the base, which may contain more than 7000 fighters. The 10 Km^2 is protected basically by its geological surrounding against any terrestrial invasion.

Figure 2. Google Earth image of Al-Watiya military base: an overview of 10 Km^2 surface

Countries including the United Arab Emirates and Egypt back Haftar, while the Government of National Accord GNA is supported by Turkey. Sixteen members of the Turkish armed forces and more than 100 Syrian mercenaries were killed during fight in the Libyan capital, Tripoli. Al-Watiya is considered as a key and sole military airbase. After losing the main battle in western Libya in19 May 2020, Eastern Libyan forces pulled out Tripoli, in a major blow to their year-long campaign to seize the capital. Logistically Helped by Turkey, after weeks of attempts, the pro-GNA forces took the Al-Watiya military airbase west of the capital. According to the footage released by the GNA media, a Turkish-made drone "Bayraktar TB2" have targeted and destroyed a Pantsir-S1 system. Located in Yafren province, northwest of Libya, Al-Watiya is a large airbase, 140 km from Tripoli, with a capacity (estimated) of 7,000 to 10,000 troops. Al-Watiya plays an important role in controlling northwestern areas of Libya. Backed by Turkey, since 5th May 2020, the GNA forces had been attempting to seize this vital military airbase.

Figure 3. Destruction of Al-Watiya military base by drones
https://www.globalsecurity.org/index.html

6.1. Al-Watiya Military Under Terrorist Groups Control

Soon or later, the withdrawal of Turkish troops is coming either by the regional pression of the surrounding countries or by the internal resistance. Nonetheless, to fill the void within Al-Watiya military base, terrorist groups (especially the so-called

Syrian mercenaries) are the more candidate to take control of the base. Ever since, a remote sensing work should be carried out to anticipate any possible aggression on the surrounding countries, especially Tunisia. This work should take into the widest definition of transnational crimes and terrorist activities and the link between them including the daily remote sensing monitoring of terrorist mobility and their number evolution, arms quality and quantity (Bayraktar TB2, S-125 SAM System) within the base and the region, linked illegal activities funding terrorist groups (human trafficking from Africa to Europe, arms trade towards Mali, and smuggling). Further, the countries of the region (including Egypt, Algeria and Tunisia) should initiate anticipatory attacks against terrorist groups to limit their capacity to destabilize regional security. Further, logistic remote sensing preparation to rapidly, safely, and massively react to any terrorist act. In doing so, Al-Watiya military base, which is considered in our case out of the control of the central government in Tripoli, is a legitimate target for the surrounding countries, especially Tunisia. As a matter of fact, the base is 27 Km from the Tunisian borders. So, even with traditional arms, terrorists could reach parts of southern Tunisia. And, an open question remains about possible inherited high technology arms such as Bayraktar TB2 and S125 SAM System.

6.3. Al-Watiya Military Base as Military Target: Remote Sensing Monitoring

Considered as out of control of the central government and threatening surrounding countries, Al-Watiya military base should be treated as legitimate target under deep remote sensing investigations. Further, a data base within a Geographic Information System (GIS) should be ready for any possible intervention. The pivotal role Remote Sensing (RS) and Geographic Information System (GIS) in military operations against terrorist has been recently discussed (Majumdar, 2021). The use of RS-GIS and its applications in defense purposes has revolutionized the way in which these forces function and operate against official armies (Ekim et al., 2021) and non-traditional terrorist troops (Psomiadis et al., 2021; Radil et al., 2021). Military forces use GIS in a variety of applications including cartography (Bachagha et al., 2020), intelligence (Panigrahi and Tripathy, 2021), battlefield management (Zakiev et al., 2020), terrain analysis (Pundir and Garg, 2021), military installation management (Cullis, 1995) and monitoring of possible terrorist activity (Majumdar, 2021).

7. CONCLUSION

Since maps inception, with the ease of satellite images access and analysis, the use of crime mapping remote sensing technologies has become a relevant tool in criminology. Especially in transnational crimes and terrorist activities, these technologies have decreased the dark figure of crime, shedding light on wrongdoers that otherwise would not be recognized. The accuracy assessments may make remotely sensed crime data more serviceable in courtroom settings and law enforcement planning strategies. In this chapter, high resolution geospatial technology for transnational and terrorism crimes are encouraged. This chapter is meant to investigate the relevance of remote sensing in detecting transnational crimes and terrorist acts. This work should take into the widest definition of transnational crimes and terrorist activities and the link between them including the daily remote sensing monitoring of terrorist mobility and their number evolution, arms quality and quantity (Bayraktar TB2, S-125 SAM System) within the base and the region, linked illegal activities funding terrorist groups (human trafficking from Africa to Europe, arms trade towards Mali, and smuggling). Further, the countries of the region (including Egypt, Algeria and Tunisia) should initiate anticipatory attacks against terrorist groups to limit their capacity to destabilize regional security. Further, logistic remote sensing preparation to rapidly, safely, and massively react to any terrorist act. The remote sensing become active detection of transnational crimes and terrorism. It is worth noting that first order validation is not always feasible. On the other hand, second order validation techniques that may facilitate validation in difficult situations.

REFERENCES

Ali, Y., Khan, A., Khan, G., Khan, A. U., & Ahmad, Z. (2021). Evaluation of effectiveness of fence on a country's border: A case study of Pak-Afghan border fence. *Journal of International Migration and Integration*, *22*(2), 769–790. doi:10.100712134-020-00774-1

Atsa'am, D. D., Wario, R., & Alola, U. V. (2021). Why Do Terrorists Attack Other Terrorists? Knowledge Discovery from the Global Terrorism Database. *The International Journal of Intelligence, Security, and Public Affairs*, 1-14.

Attah, C. E. (2020). Opinion Piece–Terrorism, Human Trafficking and Conflict-Related Sexual Violence: Examining the Nexus. Journal of Human Trafficking. *Enslavement and Conflict-Related Sexual Violence*, *1*(2), 117–133. doi:10.7590/266644720X16061196655016

Avtar, R., Kouser, A., Kumar, A., Singh, D., Misra, P., Gupta, A., Yunus, A. P., Kumar, P., Johnson, B. A., Dasgupta, R., Sahu, N., & Besse Rimba, A. (2021). Remote Sensing for International Peace and Security: Its Role and Implications. *Remote Sensing*, *13*(3), 439. doi:10.3390/rs13030439

Bachagha, N., Wang, X., Luo, L., Li, L., Khatteli, H., & Lasaponara, R. (2020). Remote sensing and GIS techniques for reconstructing the military fort system on the Roman boundary (Tunisian section) and identifying archaeological sites. *Remote Sensing of Environment*, *236*, 111418. doi:10.1016/j.rse.2019.111418

Beccaro, A. (2020). ISIS in Libya and beyond, 2014–2016. *Journal of North African Studies*, 1–20. doi:10.1080/13629387.2020.1747445

Bergström, F. (2020). *Trafficking as Terrorism: A Theory-testing Case Study on the Islamic State's Human Trafficking of the Yazidi*. Academic Press.

Bi, F., Zhang, J., Pang, F., Bian, M., & Wang, Y. (2020). Suburban Building Detection from Optical Remote Sensing Images Based on a Deformation Adaptability Model. *Photonirvachak (Dehra Dun)*, *48*(6), 831–839. doi:10.100712524-020-01117-4

Bovenkerk, F., & Chakra, B. A. (2005). Terrorism and organized crime. In Forum on Crime and Society (Vol. 4, No. 1 and 2, pp. 3-16). Academic Press.

Caparó Bellido, A., & Rundquist, B. C. (2021). Semi-Automatic Fractional Snow Cover Monitoring from Near-Surface Remote Sensing in Grassland. *Remote Sensing*, *13*(11), 2045. doi:10.3390/rs13112045

Corbin, S. L. (2021). *Ineffective US immigration policies and deterrence mechanisms at the US-Mexico border: The story of Mexico and the northern triangle*. Academic Press.

Cragin, R. K. (2021). Preventing the next wave of foreign terrorist fighters: Lessons learned from the experiences of Algeria and Tunisia. *Studies in Conflict and Terrorism*, *44*(7), 543–564. doi:10.1080/1057610X.2019.1568005

Cullis, B. (1995). *An Exploratory Analysis of Responses to Geographic Information System Adoption on Tri-Service Military Installations*. Army Engineer Waterways Experiment Station Vicksburg MS Tri-Service Cadd/Gis Technology Center.

Dashnyam, Z. (2020). Joint Statement on Strengthening the Strategic Partnership between Mongolia and India. *Mongolian Journal of International Affairs*, *21*, 145–150.

Desille, A. (2021). On the Use of Visual Methods to Understand Local Immigration Politics. In *Visual Methodology in Migration Studies* (pp. 67–84). Springer. doi:10.1007/978-3-030-67608-7_4

Diaz, E., Yokosuk, M. O., Myers, T. L., & Johnson, T. J. (2020, September). Preliminary parameter studies to form pellets of hard materials for optical constants derived via single-angle reflectance spectroscopy. In *Counterterrorism, Crime Fighting, Forensics, and Surveillance Technologies IV* (Vol. 11542, p. 115420H). International Society for Optics and Photonics. doi:10.1117/12.2574250

Ekim, B., Sertel, E., & Kabadayı, M. E. (2021). Automatic Road Extraction from Historical Maps Using Deep Learning Techniques: A Regional Case Study of Turkey in a German World War II Map. *ISPRS International Journal of Geo-Information, 10*(8), 492. doi:10.3390/ijgi10080492

Fisabillah, A., & Pujiyono, U. R. (2021). *Policy to Eradicate Crime Funding of Terrorism as Transnational Organized Crime*. Academic Press.

Fleuriet, K. J., & Castellano, M. (2020). Media, place-making, and concept-metaphors: The US-Mexico border during the rise of Donald Trump. *Media Culture & Society, 42*(6), 880–897. doi:10.1177/0163443719890539

Gruber, K. (2021). *Remote Sensing Techniques for Middle Eastern Archaeology-K.* Gruber.

Han, Y., Ma, S., Zhang, F., & Li, C. (2020, July). Object Detection of Remote Sensing Airport Image Based on Improved Faster R-CNN. *Journal of Physics: Conference Series, 1601*(3).

Joffé, G. (2020). Libya: The new geopolitical arena. *Journal of North African Studies, 25*(5), 681–688. doi:10.1080/13629387.2020.1790110

Kadir, N. K. (2020). Cyberterrorism's Dilemma: Renewal of Conventional Terrorism. *International Journal of Global Community, 3*(2), 97-110.

Kadir, N. K., Judhariksawan, J., & Maskun, M. (2019). Terrorism and cyberspace: A phenomenon of cyber-terrorism as transnational crimes. *FIAT JUSTISIA: Jurnal Ilmu Hukum, 13*(4), 333–344. doi:10.25041/fiatjustisia.v13no4.1735

Kaur, H. (2021). Machine learning and border security. *PalArch's Journal of Archaeology of Egypt/Egyptology, 18*(1), 3467-3475.

Khalifa, A., Bashir, B., Çakir, Z., Kaya, Ş., Alsalman, A., & Henaish, A. (2021). Paradigm of Geological Mapping of the Adıyaman Fault Zone of Eastern Turkey Using Landsat 8 Remotely Sensed Data Coupled with PCA, ICA, and MNFA Techniques. *ISPRS International Journal of Geo-Information, 10*(6), 368. doi:10.3390/ijgi10060368

Kubai, A. (2021). The "Africa Rising" Paradox, Human Trafficking, and Perilous Migration Across the Sahara and the Mediterranean to Europe. In The Governance, Security and Development Nexus (pp. 355-370). Palgrave Macmillan.

Lee, C., & de Vries, W. T. (2021). Testing and Validating the Suitability of Geospatially Informed Proxies on Land Tenure in North Korea for Korean (Re-) Unification. *Remote Sensing, 13*(7), 1301. doi:10.3390/rs13071301

Legleiter, C. J., & Kinzel, P. J. (2020). Inferring surface flow velocities in sediment-Laden Alaskan rivers from optical image sequences acquired from a helicopter. *Remote Sensing, 12*(8), 1282. doi:10.3390/rs12081282

Mahmoudi, L., Doumari, S. A., Safarianzengir, V., Maleki, R., Kianinejad, S., & Kianian, M. (2021). Monitoring and Prediction of Dust and Investigating its Environmental Impacts in the Western Half of Iran and the Eastern Borders of Turkey and Iraq, Using Remote Sensing and GIS. *Photonirvachak (Dehra Dun), 49*(4), 713–724. doi:10.100712524-020-01224-2

Majumdar, S. (2021). The Role of Remote Sensing and GIS in Military Strategy to Prevent Terror Attacks. *Intelligent Data Analytics for Terror Threat Prediction: Architectures, Methodologies, Techniques and Applications*, 79-94.

Makarenko, T. (2004). The crime-terror continuum: Tracing the interplay between transnational organised crime and terrorism. *Global Crime, 6*(1), 129–145. doi:10.1080/1744057042000297025

Matczak, G., & Mazurek, P. (2021). Comparative Monte Carlo Analysis of Background Estimation Algorithms for Unmanned Aerial Vehicle Detection. *Remote Sensing, 13*(5), 870. doi:10.3390/rs13050870

Mirdad, M. A. (2020). Taliban insurgency and transnational organized crime nexus. *Masyarakat, Kebudayaan dan Politik, 33*(3), 266-277.

Mobariz, M. A., & Kaplan, G. (2021). Moving Borders: Mapping and Monitoring Amu Darya River Dynamics Using Remote Sensing Data and Techniques. *Geodetski list, 75*(1), 29-44.

Okubo, S., & Shelley, L. (Eds.). (2011). Human security, transnational crime and human trafficking: Asian and Western perspectives (Vol. 6). Routledge.

Omelicheva, M. Y., & Markowitz, L. P. (2021). Rethinking Intersections of Crime and Terrorism: Insights from Political Economies of Violence. *International Studies Review*, viab025. doi:10.1093/isr/viab025

Palomino, W., Barreto, R., & Cerón, A. (2020). Drug Trafficking and Terrorism as Transnational Crimes in Latin America. In Maritime Security: Counter-Terrorism Lessons from Maritime Piracy and Narcotics Interdiction (pp. 29-46). IOS Press.

Panigrahi, N., & Tripathy, S. (2021). Design Criteria of a UAV for ISTAR and Remote Sensing Applications. *Photonirvachak (Dehra Dun), 49*(3), 665–669. doi:10.100712524-020-01249-7

Perry, B. (2020). "Button-Down Terror": The Metamorphosis of the Hate Movement. In Experiencing Social Research (pp. 281-302). Routledge.

Petrich, K. (2021). The Crime–Terror Nexus. In Oxford Research Encyclopedia of International Studies. OUP.

Pierini, M. (2020). New power struggles in the Mediterranean. *Carnegie Europe*, (July), 30.

Psomiadis, E., Tomanis, L., Kavvadias, A., Soulis, K. X., Charizopoulos, N., & Michas, S. (2021). Potential dam breach analysis and flood wave risk assessment using HEC-RAS and remote sensing data: A multicriteria approach. *Water (Basel), 13*(3), 364. doi:10.3390/w13030364

Pundir, S. K., & Garg, R. D. (2021). Development of rule based approach for assessment of off-road trafficability using remote sensing and ancillary data. *Quaternary International, 575*, 308–316. doi:10.1016/j.quaint.2020.07.017

Puttonen, R., & Romiti, F. (2020). *The Linkages between Organized Crime and Terrorism*. Academic Press.

Radil, S. M., Irmischer, I., & Walther, O. J. (2021). Contextualizing the Relationship Between Borderlands and Political Violence: A Dynamic Space-Time Analysis in North and West Africa. *Journal of Borderlands Studies*, 1–19. doi:10.1080/08865655.2021.1968926

Redowan, M., Phinn, S. R., Roelfsema, C. M., & Aziz, A. A. (2020). CLASlite unmixing of Landsat images to estimate REDD+ activity data for deforestation in a Bangladesh forest. *Journal of Applied Remote Sensing, 14*(2), 024505. doi:10.1117/1.JRS.14.024505

Rosen, J. D. (2021). The Drug War in Colombia. In The US War on Drugs at Home and Abroad (pp. 17-43). Palgrave Macmillan.

Saul, B. (2017). The legal relationship between terrorism and transnational crime. *International Criminal Law Review, 17*(3), 417–452. doi:10.1163/15718123-01703001

Shelley, L. I., & Picarelli, J. T. (2005). Methods and motives: Exploring links between transnational organized crime and international terrorism. *Trends in Organized Crime*, *9*(2), 52–67. doi:10.100712117-005-1024-x

Singh, A. (2021). *An Interlink between White Collar Crimes and Terrorism*. Neolexvision Blogs. https://www. aequivic. in/post/an-interlink-between-white-collar-crimes-and-terrorism

Weisburd, D., Savona, E. U., Hasisi, B., & Calderoni, F. (Eds.). (2020). *Understanding recruitment to organized crime and terrorism*. Springer. doi:10.1007/978-3-030-36639-1

Welch, S. A. (2016). Human trafficking and terrorism: Utilizing national security resources to prevent human trafficking in the Islamic state. *Duke J. Gender L. & Pol'y*, *24*, 165.

Xu, J., Xiao, W., He, T., Deng, X., & Chen, W. (2021). Extraction of built-up area using multi-sensor data—A case study based on Google earth engine in Zhejiang Province, China. *International Journal of Remote Sensing*, *42*(2), 389–404. doi:10.1080/01431161.2020.1809027

Yakubu, A., & Idehen, R. (2021). *Border Patrol And Surveillance In The Control Of Transnational Terrorism In Nigeria And Niger*. Academic Press.

Yendaw, E. (2021). Cross-Border Migration of Itinerant Immigrant Retailers in Ghana. *Journal of International Migration and Integration*, 1–21. PMID:34025218

Zakiev, E. S., Pankov, S. V., & Kalabay, K. B. (2020). Application of geoinformation systems in the armed forces and other military formations in the Republic of Kazakhstan. *Vojnotehnički glasnik, 68*(2), 356-382.

Chapter 5
Cross-Layer Learning:
A Deep Learning-Based Forensic Framework for IoT Systems

Tushar Mane
Symbiosis Institute of Technology, Symbiosis International University (Deemed), India

Ambika Pawar
Symbiosis Institute of Technology, Symbiosis International University (Deemed), India

ABSTRACT

Deep learning-based investigation mechanisms are available for conventional forensics, but not for IoT forensics. Dividing the system into different layers according to their functionalities, collecting data from each layer, finding the correlating factor, and using it for pattern detection is the fundamental concept behind the proposed intelligent system. The authors utilize this notion for embedding intelligence in forensics and speed up the investigation process by providing hints to the examiner. They propose a novel cross-layer learning architecture (CCLA) for IoT forensics. To the best of their knowledge, this is the first attempt to incorporate deep learning into the forensics of the IoT ecosystem.

INTRODUCTION

Along with the exponential growth of the IoT, there has been a surge in security and privacy breaches as well. A survey conducted by Hewlett Packard depicts 70% of the investigated IoT applications contain security and privacy breaches (Rawlinson,

DOI: 10.4018/978-1-7998-8386-9.ch005

2014). The well-known Mirai botnet attack, which caused severe Denial- of- Service (DoS) for thousands of IoT devices in America is another notable case (*Mirai (malware) - Wikipedia*, n.d.). As the IoT systems are being deployed rapidly, it is quite clear that there is a need for research in the IoT forensics. Although existing forensic techniques and tools are still found to be useful in some phases of the IoT forensics, there is still a desperate need to upgrade existing tools, methodologies, and legislation to tackle the characteristics of IoT. IoT security altogether a different type of security approach, as it deals with the light-weight computing devices which are connected to the cloud. IoT security architecture and communication standards are discussed thoroughly in (Silva et al., 2019). Challenges and research opportunities for IoT security are highlighted in (Ryan & Watson, 2017) . Traditional security methodologies are not sufficient to secure IoT security (Fernandes et al., 2017). IoT security understanding is unquestionably valuable in the field of IoT forensics, nevertheless, tracing the root cause, collecting evidence, and correlating it in the end to end IoT system entails aid of the machine intelligence.

Deep learning is proving to be more robust and accurate in comparison with the rest of the machine learning techniques due to the support of larger datasets (Q. Zhang et al., 2018). Deep linking in IoT environments is achievable with the help of deep learning (Fadlullah et al., 2017). Deep Linking Protocol enables IoT objects and applications to interact involuntarily with each other. Smart home example (Li et al., 2018) gives an idea of how things in a smart home can collaborate to form a system. As of now, deep learning has been applied in Network Forensics only. We believe Deep learning assisted investigation of end-to-end IoT applications will certainly help speed up the investigation process with fewer errors and leftovers. Notable contributions of our study are as follows:

- We provide a compact description of prerequisite areas of deep learning assisted IoT forensics, such as- IoT, IoT security, digital forensics, deep learning, and its application in the aforementioned area.
- Further, we do a comparative study of existing IoT Forensic models and highlight the uniqueness of the proposed model.
- A Comprehensive study on contributions which inspired us to device Deep Learning assisted IoT Forensics.
- We coin Cross-Layer Learning Architecture (CLLA) for IoT Forensics. It studies the correlation between the layers to analyze the attacks. By establishing links it hints probability and priority value of each layer to the investigator.
- Thorough research directions, opportunities, and challenges are presented to encourage researchers in a related area.

The composition of the paper is as follows. Section Two is preparatory and sets up the context of the paper. The third Section reviews existing IoT Forensics models and do a comparative analysis with the proposed model, based on attributes of IoT forensics. Section Four sets up the model with a corresponding related work. Section Five proposes the holistic approach to investigate the IoT systems, while Section Six discusses research opportunities in the commencing field of deep learning assisted IoT forensics. We conclude the article in Section Seven.

BACKGROUND

Internet of Things

Internet is composed of the entities, which can sense or act, communicate their state to a processing device, which in turn synchronizes with the remote server is called the Internet of Things. It has been there since the inception of the internet, and automation. But, it started gaining more attention with the increase in energy-efficient computation and enhancements in wired and wireless networking. Embedded devices are stable at the micro-level and heading towards Nano (Miraz et al., 2018). Every object around us can now be supplemented by computation and communication. Cloud computing can be considered as a resilient broker between all Things for corresponding use cases. To name a few, use cases can be applications in smart homes, smart cities, retail markets, logistics, redefined industrial automation, vehicular systems, healthcare, agriculture (Hassija et al., 2019). Birds eye view of IoT is shown in fig. 1.

IoT Security

Resource-constrained environments restrict the execution of conventional security mechanisms due to a variance in computation power, communication mechanisms, and power consumption. Also, there is no community accepted implementation guidelines for the IoT Security and Privacy, which would help the developers to design and develop the security systems. Secure architecture should be devised and amended on a timely basis due to heterogeneous and dynamic environments. Side-Channel Attacks, Hardware Trojan, and Physical Attack are concerning factors (Mohamad Noor & Hassan, 2019). Other security factors include Botnets and malicious nodes. Considering the characteristics of the IoT, the use of lightweight cryptography mechanisms is recommended by researchers (Dhanda et al., 2020). Eavesdropping is commonly used for privacy breaches. Secure bootstrapping is a process via which communication between two IoT objects is configured and

Figure 1. Internet of Things ecosystem

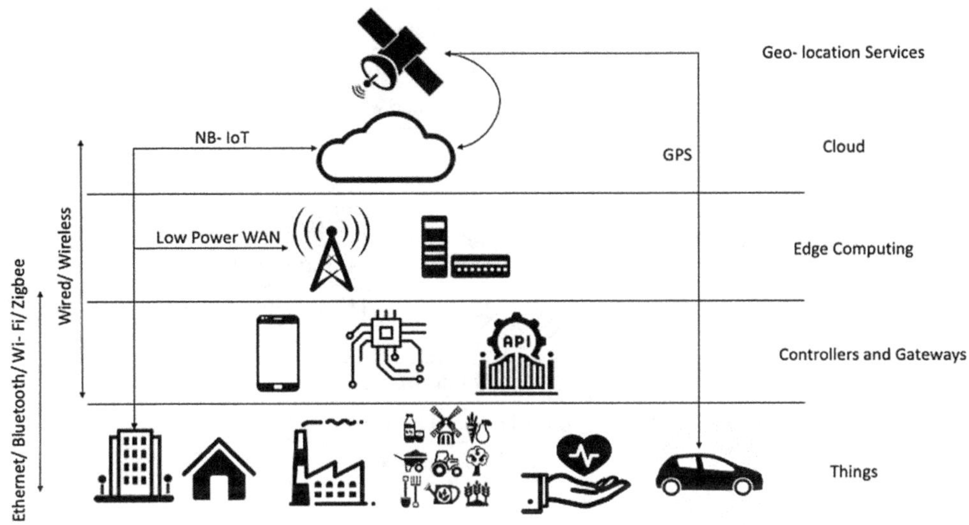

initiated, so the importance of bootstrapping should be considered while designing IoT applications (Abdul-Ghani & Konstantas, 2019). Fig. 2 shows categorization of IoT vulnerabilities.

IoT Forensics

To identify an attack, collect evidence, determine the purpose, reform the sequence of events, by analyzing respective the physical, logical entities and proving an intention in the Court of Law is called as the forensics. Forensics is always case centered and time-bound process. When applied to the digital world, it involves extraction, preservation, and analysis of digital evidence (Stamm et al., 2013). IoT Forensics is nothing but a subclass of digital forensics that involves the IoT Objects and their state as evidence and in the particular timeline (Atlam et al., 2020). Objects can be physical or logical. It is an emerging research area due to the extensive challenges involved in it (Yaqoob et al., 2019). A massive number of Things, supported by cloud generate big data and are interconnected by heterogeneous networks. Also, the evidence is present across the layers. So, the traditional forensic approach and techniques have to be altered to suit the IoT environment. Phases of digital forensics are depicted in fig. 3.

Figure 2. Possible vulnerabilities on IoT systems

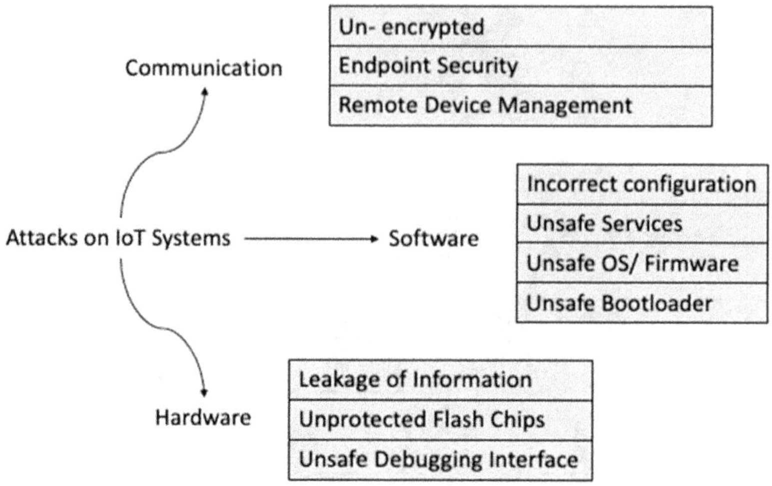

Figure 3. Phases of digital forensics

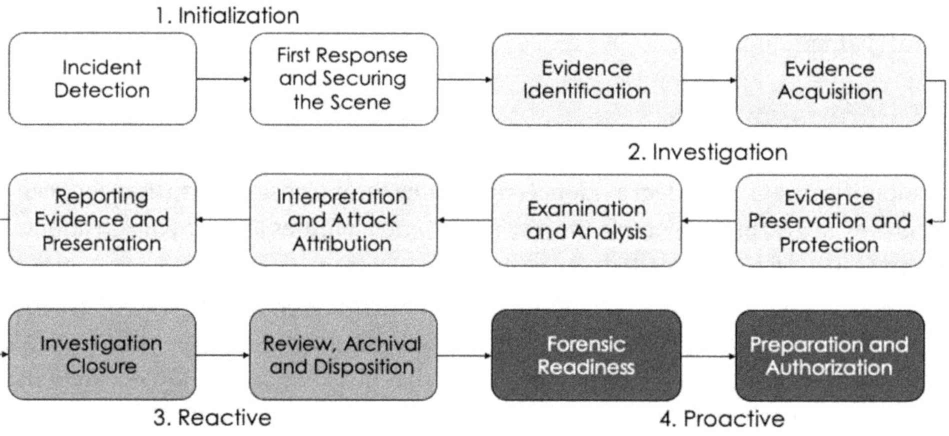

Deep Learning

Deep learning, a subcategory of artificial intelligence, which has started gaining tremendous attainment in recent years, and now being incorporated in various areas such as image processing, natural language processing, robotics, genetics, bioinformatics, cybersecurity to name a few (Dargan et al., 2020). The branch of machine learning called neural networks have shown more performance and accuracy, the developments in the neural networks have to lead to deep learning. Unlike machine learning, it keeps forming complex features on its own, based on simple inputs with multiple layers and adjustment in weights. Fig. 4 differentiates between machine learning and deep learning.

Figure 4. AI ecosystem, machine learning, and deep learning

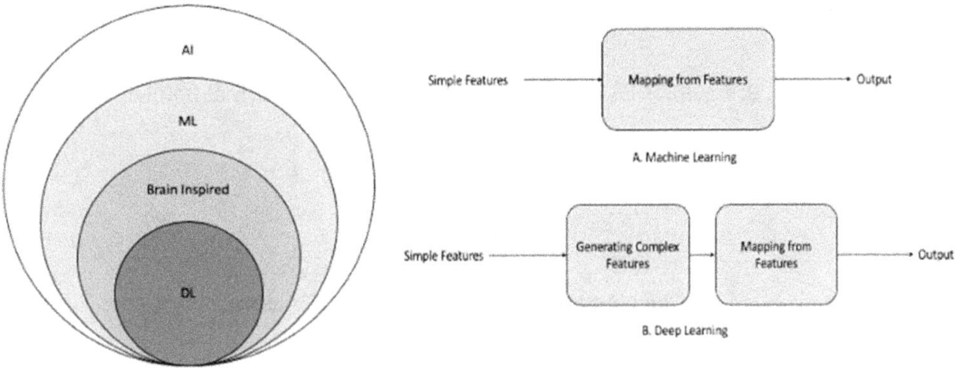

Cybersecurity is getting the assistance of deep learning in intrusion detection, intrusion prevention systems, network flow analysis, deep packet inspection, and many others. Zero-day attacks are variations in the known attacks. The ability to learn from the complex features makes deep learning the perfect fit for detecting unknown malicious behaviors. It has also made its way in application security and information security (Koroniotis et al., 2019).

Application of Deep Learning in IoT Security

IoT Objects, networking services, cloud services, and mobile applications are vulnerable from the IoT perspective. Domain research includes Wi-Fi fingerprinting, side-channel attacks, developing an anti-jamming scheme, incorporating cognitive radio capability, intrusion detection, anomaly detection, malware detection, distributed denial of service prevention, botnet detection, and many others using Deep learning methods (Al-Garadi et al., 2020). Availability of the datasets, learning from raw data, increasing algorithm performance are some of the major research concerns (Hussain et al., 2020). Learning is self- governing, intersected and interactive environments is being considered as a foremost challenge while designing the secure architectures. Edge nodes are drastically benefitting the IoT ecosystem by reducing latency. The availability of the services at the edge has improved user experience hence increasing applications. Hence the security of edge nodes has become a pivotal attribute of

IoT security (K. Zhang et al., 2020). Fig. 5 shows how ML or DL can prove useful in IoT security.

Figure 5. Applicability of ML/ DL in IoT security

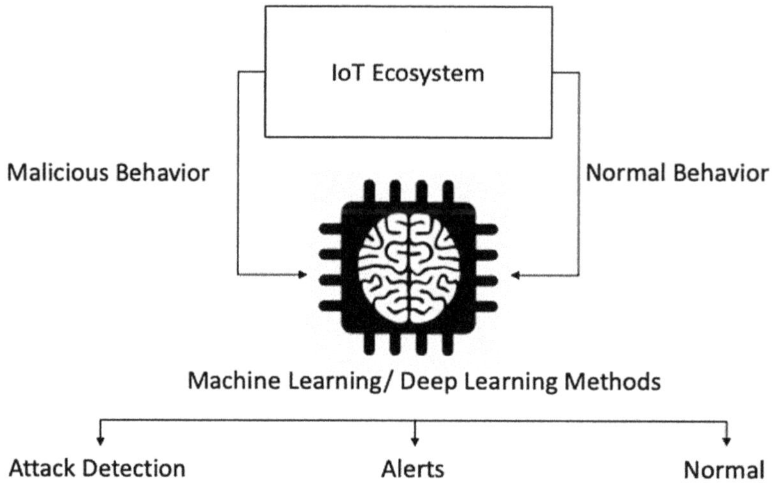

IoT Forensics and Deep Learning

Amplifying the forensic value of the potential evidence is called forensic readiness (Yu et al., 2019). intrusion detection systems, security information, and event management systems can be useful assets doing the same.

Figure 6. Applicability of ML/ DL in IoT forensics

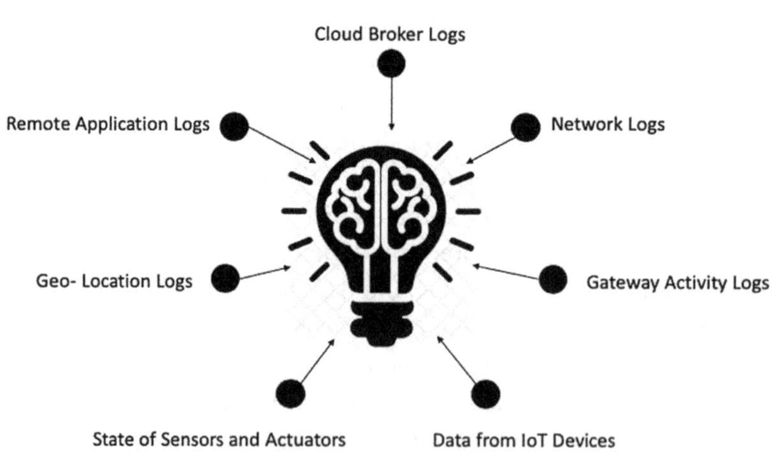

Deep learning is being used to improve forensic readiness with its application in intrusion detection systems. It is also getting attention in network traffic classification, botnet detection, and deep packet inspection. The accuracy of deep learning in the detection and classification of malware is notable (Koroniotis et al., 2020). But the other areas of forensics such as evidence management and analysis are yet to be explored. Investing IoT applications from an end to end perspective is the most vital phase in an investigation which we explore in our work. It includes cross-layer forensics, which states, how the state of change in one layer can affect the other, with the help of deep learning architecture. Fig. 6 illustrates how deep learning can be enabled in IoT forensics using variety of data sources.

EXISTING IOT FORENSICS MODELS

A generic investigation model, DFIF- IoT, is proposed by (Kebande & Ray, 2016), which is composed of pro-active, forensic, and reactive layers, with the pro-active layer focusing on the forensic readiness, the forensic layer performing an evidence extraction, and the reactive layer listening to an abnormal event. The reactive layer triggers the investigation in the case of a suspicious event. As per (Valjarevic & Venter, 2015), gaining the authorization, detailed documentation, conserving chain of custody, physical investigation, and interaction with the physical investigation are the concurrent processes that increase admissibility of evidence. DFIF- IoT integrates these processes to increase the effectiveness of the IoT forensics.

The ISO/ IEC 27043 based holistic model devised by (Sadineni et al., 2019), and the DFIF-IoT are similar in structure, except the holistic model includes and explains all the digital investigation modules at the micro-level in the context of the IoT, as indicated in fig. 7. Further, authors have applied the proposed model to a smartwatch and analyzed the scenario theoretically.

PRoFIT (Nieto et al., 2017) is the privacy-aware IoT forensics methodology that incorporates the privacy principles defined in ISO/IEC 29100:2011 (*ISO - ISO/IEC 29100:2011 - Information technology — Security techniques — Privacy framework*, n.d.) into the investigation process. Data sharing is based solely on the consent of the device owner. It follows a collaborative approach to generate the targeted crime scene based on available information. Context-based collection and information sharing phases are added in the investigation process. Policies are studied in isolated and non- isolated environments concerning personal and non- personal devices. The software lifecycle of PRoFIT middleware is discussed to explain its compliance with privacy policies. Finally, the theoretical set up of a coffee shop is used to demonstrate the application of the model.

Figure 7. DFIF- IoT and ISO/ IEC 27043

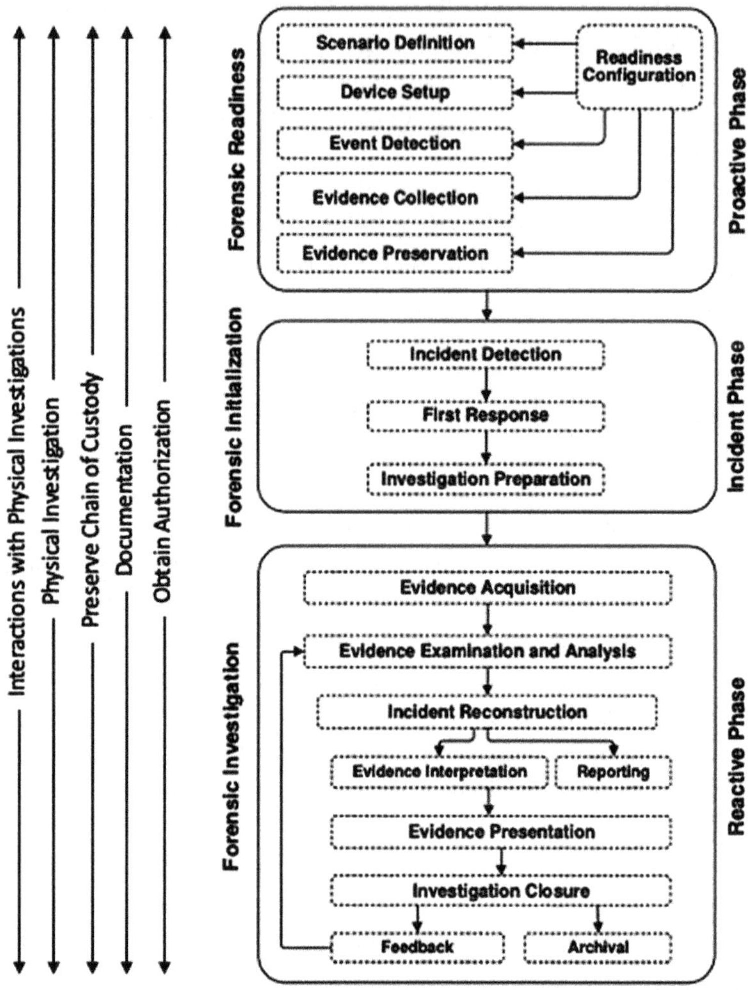

Forensically irrelevant data is filtered out using machine learning by (Verma et al., 2018). Further, the degree of relevance is calculated for the available data sources. Both automated and manual queries are required to complete the entire picture of an investigation. Though researchers attempt to preserve privacy and maintain the integrity of evidence, the contribution could not automate the whole evidence analysis process as presented in fig. 8.

(Meffert et al., 2017) demonstrated the importance of state acquisition from the IoT devices in the FSAIoT model. Controllers take care of state acquisition (read-only), timestamp logging, secure storage along with integrity. The controller can

Figure 8. Privacy and relevance based mechanisms in IoT forensics

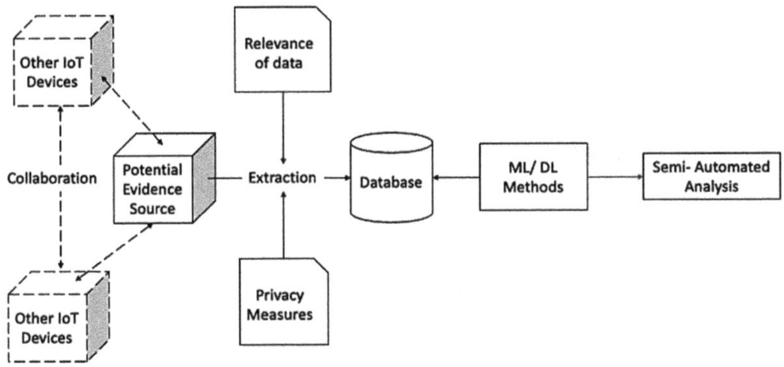

communicate with the device, cloud, and other controllers in an environment. An experimentation scenario consisting of motion sensors and IP cameras are used to monitor a server room. Results have validated the proposed model. (Perumal et al., 2015) used a top-down approach to formulate an investigation model for the IoT. Researchers have decomposed the connectivity into three zones, internal network, middle network, and outside the network. Contribution suggests considering the IoT protocols, cloud/ fog computing resources, router, and the gateway services for the robust evidence collection.

Smart environment forensics is modeled by IoTDots (Babun et al., 2018). Applications are analyzed for the detection and storage of relevant information using the modifier and analyzer modules. The database module is responsible for the storage of relevant data. the proposed framework processes and constructs evidence using the machine learning algorithms. However, the model renounces the privacy for relevance. Investigation methodologies for various applications in the IoT such as wearables, smart homes, and smart city are examined by (Zia et al., 2017). Generalized investigation techniques are then formed for each of the use cases to form application-specific models. Later, thing forensics, network forensics, and cloud forensic methodologies are discussed briefly as a layered approach to the investigation process. (Hossain et al., 2017) proposed an investigation framework for the Internet of Vehicles consisting of the provenance model, fog gateway, and a forensic service. Pedestrians' smartphone, Road Side Units (RSU), and vehicles are the actors.

Provenance model is composed of actors, interactions, events and stories, fig. 9. Actors and interactions between them form events. The events form stories, and ultimately stories result in the interaction provenance model. The fog gateway is responsible for interactions while a forensic service module maintains the tamper-proof stories using the blockchain technology. The prototype of the model, with a case study

Figure 9. Provenance model for Internet of Vehicles

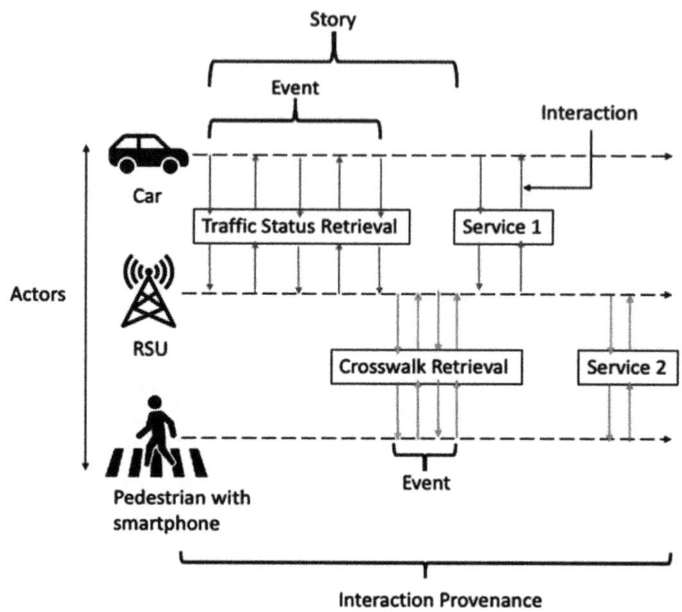

and implementation, shows how the model can provide real-time forensic value with low energy consumption. (Feng et al., 2017) considered smart city environments and the autonomous vehicles and developed the model focusing on the confidentiality and the integrity of the acquired forensic data using asymmetric cryptography and hashing. Data Image is formed using data sent by vehicles, data sent by the smart city data hub, and data collected manually by the investigator. Encrypted and Hashed data image is then created by using hashing and AES algorithms.

(Karie et al., 2019) delivered first official glance at Cyber Forensics through the glass of Deep Learning. The proposed deep learning framework tries to find potential data sources that would upturn the cyber forensic examination. Potential data sources would be input to Deep Learning Forensic Engine, where the actual analysis will happen. Phases of investigation are arranged to effectively adapt the proposed inspection model, making it more compelling.

(Sathwara et al., 2018) presented IoT Forensics Ecosystem, which depicts IoT Layers and possible artifacts of each layer. Forensic aspects are discussed in three parts. Cache and memory analysis is proposed at the IoT Platform segment. Log analysis is recommended as a part Network Effects component of the ecosystem. According to the casted model, the Market Expectations division should focus on Service Layer Agreement Inspection and Fingerprint Collection. Further, challenges before investigators and analyzers in different forensic phases are briefed. A zone-

Figure 10. Fog based IoT forensics architecture

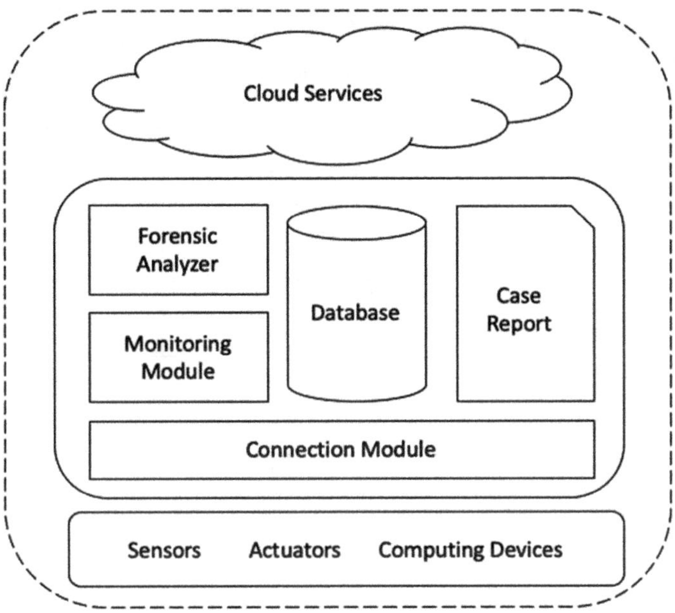

wise, time-sequenced model, Improved IoT Forensics, is proposed by (Harbawi & Varol, 2017), which divides IoT systems under investigation into three zones, namely personal network zone, gateway zone, and cloud zone. The sequence of investigation should start from a personal network zone, and end up in the cloud zone. More restrictions get imposed on the investigation as it proceeds to the upper zones. Things in zones get filtered based on if it is connected to the thing(s) of interest. Evidence is backed up before further analysis. The final investigation report is prepared in detail and precautionary measures are taken to avoid similar cases in the future. Fog based IoT Forensic Framework (Al-Masri et al., 2018), fig. 10, is a generalized approach which helps IoT forensics with the help of computation power on fog nodes. Communication module takes care of the connectivity of IoT devices with fog nodes or cloud instances, based on configuration. Monitor Manager continuously monitors traffic flowing to and from IoT devices and logs it in the database.

In case of any suspicious activity, it raises a flag that triggers a forensic analyzer module. The forensic analyzer then immediately collects volatile data from IoT devices and stores in a database. Other data from IoT devices is cloned and stored in the cloud. It then issues a restart signal to the targeted IoT device. The case report module prepares a document that is analyzed by a forensic expert to mark the event as an attack or threat.

Table 1. Comparative analysis of existing IoT forensic models

Feature/ Model	Generic Approach	Forensic Readiness	Privacy-Preserving	Integrity and Security	Redefined Phase(s)	Layered Approach	Machine Intelligence	Case Study
DFIF- IoT	·	·			·			
FSAIoT		·				·		·
Holistic	·	·			·			·
IDFIM	·	·			·	·		
PRoFIT			·					·
DF 2.0	·				·	·		·
IoTDots			·		·		·	
Application-Specific		·			·		·	
DLCF	·				·		·	
IoT Forensic Ecosystem	·					·		
Improved IoT Forensics	·	·	·		·	·		
Trust- IoV	·	·						·
AAV		·		·				
FoBI	·	·			·	·		·
IoT- FR	·	·		·		·		·
THIS WORK	·	·	·	·	·	·	·	·

Organizational processes and IoT security processes are associated with the forensic readiness process (Kebande et al., 2020) to make forensics more robust in IoT environments. Forensics strategies, policies, and scenarios are pre-defined with organizational processes. IoT security processes like key management, certificates, authentication, authorization, and access control are necessary for preventing or as investigating. Readiness processes include a description of IoT architecture and areas which has got forensic value. Incident monitoring, detection, and analysis are fundamental phases in any forensic process. Finally, various forensic tools can be used to extract information with the assistance of IoT security processes.

Figure 11. Log storage and retrieval

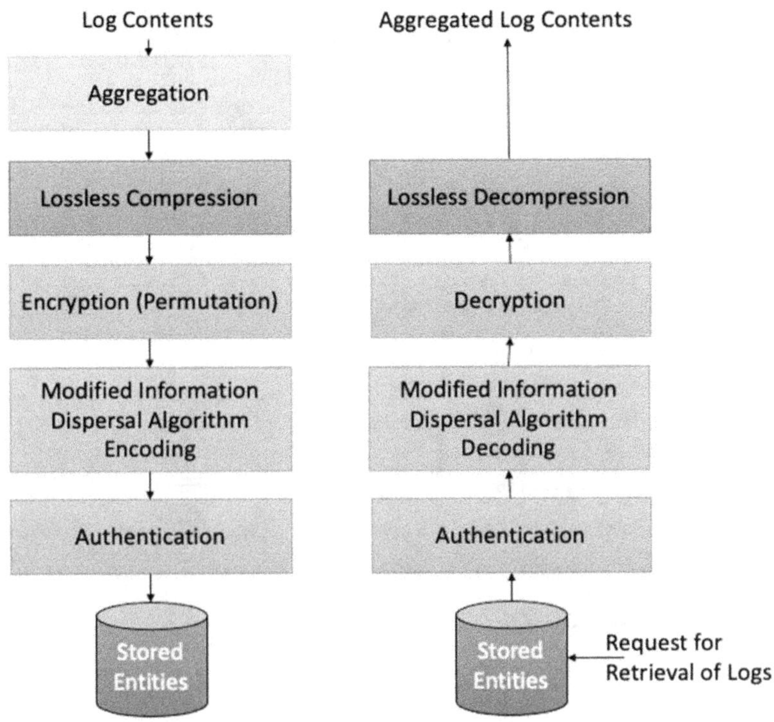

RELATED WORK AND MODEL SETUP

Distributed Logging Mechanism for IoT Forensics

IoT forensics helps in reducing the attacks and improving defense mechanisms, provided you have sufficient logs to trace the root cause of the malicious activity. Sophisticated attackers use an anti-forensic approach and erase log files disabling the root cause detection. (Noura et al., 2020) proposed a scheme to secure log files, increase availability, and ensuring the integrity by aggregation, compression, and encryption. Fog nodes are utilized for distributed storage. A trial has been conducted in two modes, where all fog nodes are interconnected and another being not. Former trial distributes 'n' fragments to 'n' aggregation devices, i.e. IoT devices, while later one transmitting only to fog node, which in turn scatters those to 'n' associated fog nodes.

Fig. 11 depicts the mechanism behind storing and retrieving logs from fog nodes. Further researchers have proposed, dynamic key-based lightweight cryptographic algorithm, MIDA, demanding low storage, and computation requirements.

Figure 12. IoT layered architecture and sources of interest, PR- privacy required

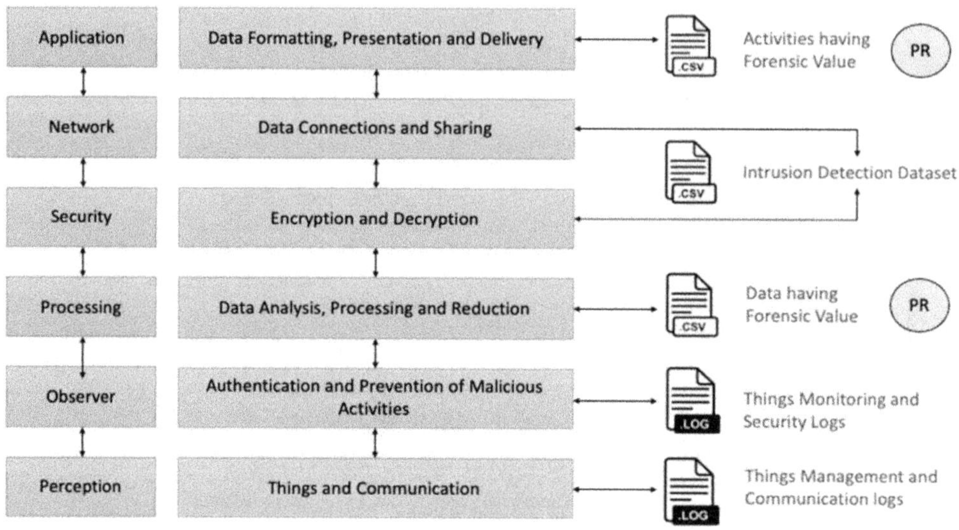

Layered IoT Architecture

Muhammad Burhan, et. al proposed secure layered IoT architecture (Burhan et al., 2018), which conforms to the principle of 'System with Separation of Duties'. We have depicted it in a format that identifies data sources of interest enabling the learning process of our proposed framework. Digital objects and their communication happening at perception layer is monitored by observer layer for threats and attacks before sending data to processing layer for analysis, processing, and reduction purpose. Only authenticated digital objects are allowed to send or receive data. Communication over the network is secured by security layer by appropriate cryptographic mechanisms, as network layers takes care of all data connections for sharing. Formatted data is presented to various applications at the application layer, where all user activities initiate and end. We extend the architecture by adding a logging system as explained in the previous subsection which stores logs separately at each layer with security and integrity measures in place, as posed in fig. 12.

Though Data and User Activities have got forensic value, as privacy requirements are imperative, an exclusive privacy-preserving mechanism is desirable.

Cross-Layer Learning for IoT Forensics

Bit-level information residing at the physical level is often unnoticed during the digital forensic process. Rondeau et. al have proposed the Cross-Layer Industrial

Figure 13. Cross-layer forensic investigation model

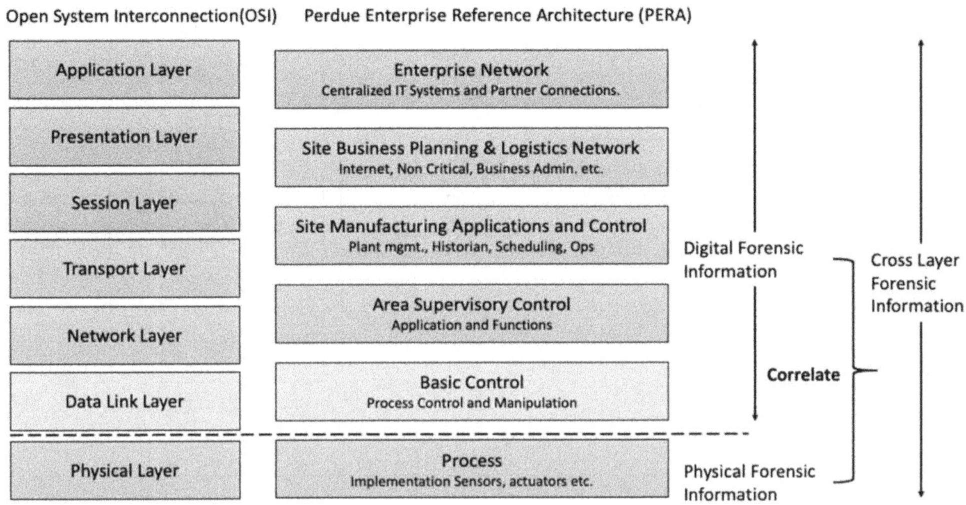

Internet of Things framework (Rondeau et al., 2019) which attempts to bind cyber forensics with physical devices forensics, especially industrial instruments such as controllers (PLCs, RTUs). Writing erroneous information in inaccurate registers can fail the entire system, so the possibility of such attacks should also be reviewed while investigating critical infrastructures. Layers in the OSI model are mapped to layers in Industrial IoT systems (PERA model), fig. 13. Authors have correlated digital forensic information with forensic data obtained at the physical layer, resulting in a proposed cross-layer forensics model. Possible forensic data sources at various layers are shared by research to aid investigators in examining Supervisory Control and Data Acquisition systems. The proposed framework intends to correlate and learn with available information in every layer concerning the configurable time frame.

Long Short Term Memory in Context of IoT Systems

(W. Zhang et al., 2018) have employed Long Short Term Memory (LSTM) for analyzing Industrial IoT equipment and attempted to solve key glitches such as, reduction of redundant data without affecting the accuracy of forecasting abilities, classification of outliers in sensor data, extending memory of states of the equipment, and predicting it's running status based on data archival.

Outliers are abolished by pre-processing while normalizing help in scaling the extreme data values. Overlong time series data resulted from 33 sensors of power equipment is merged to reduce its multidimensionality based on strong correlation. The reduction of dimensions helps to minimize the time complexity of the algorithm.

Figure 14. LSTM for time series data

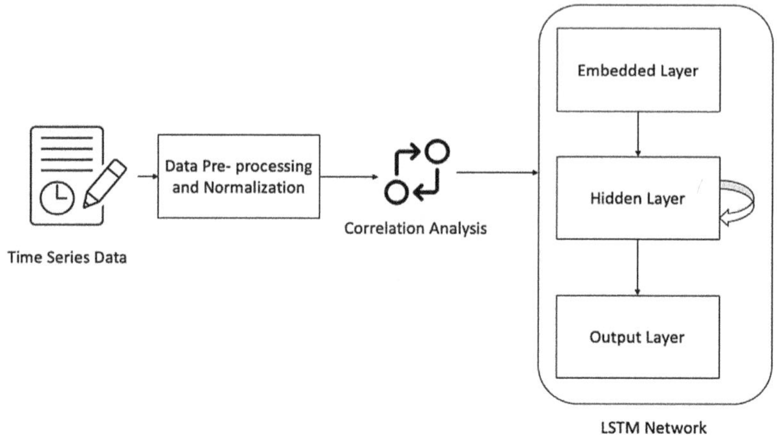

Single longwinded time series is broken down into several small-time series of fixed length using an embedded layer. These are then passed to a hidden layer for model training. Output layers predict the sensor values and hence the condition of the equipment. Overall architecture is depicted in fig. 14.

Table 2. Model setup

Work	Perception	Extension
Distributed Logging	Security and Integrity of Logs	Representation and integration with third-party entities
Layered IoT Architecture	Representation of End to End IoT Systems	Privacy-preserving data extraction
Cross-Layer Forensics	Correlation between Layers	Automating the correlation
IIoT Object Analysis using Deep Learning	Application of Deep Learning for Analysing Time Series Data	Architecture suiting availability of data

CROSS-LAYER LEARNING ARCHITECTURE

Differentiating Factor

Event monitoring and security logs alone could not always help systems uncovering the anomalies. Zero-day attacks or sophisticated attacks can only be discovered by correlation. The consistency of two states at a particular time ensures the system is safe. Our work focuses on the resemblance of two layers at a specific point in time.

Figure 15. Storage and representation in cross-layer architecture

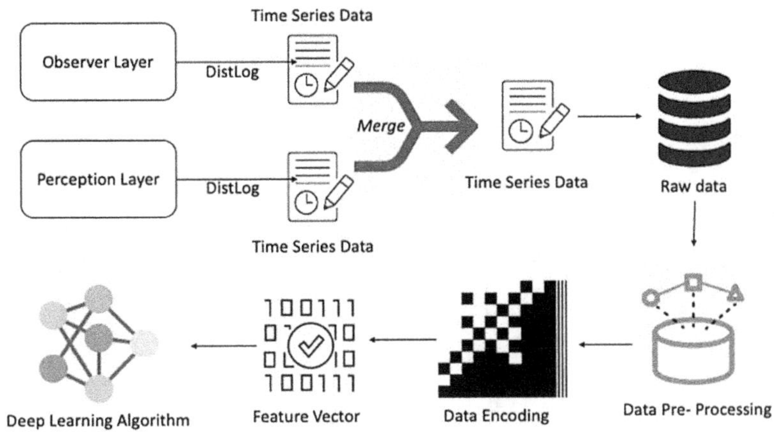

For example, if the thing is registered at a certain time, then logs should reflect the registration event at that particular time. Malicious nodes can communicate with communication systems without getting registered. Data from selected layers can be merged and represented in a definite format so that the program can reveal ambiguities in an automated way.

Secure Storage and Representation

We log the data using a distributed logging mechanism as stated in the model setup section. Two different layers will have their own logged data at specific intervals in time. Then these two-time series are merged based on timestamps. It is represented in the precise form with optimized storage which can be easily readable and processed. The raw data stored in a comma-separated value format is represented in a form of a feature vector to the algorithm after various data pre-processing and encoding techniques, fig 15.

Proposed Methodology

Recurrent Neural Network (RNN), fig. 16, classifies or predicts events based on a sequential set of events. Unlike traditional Deep Learning architectures, they need a context. RNNs use looped architecture which persists the context by having multiple instances of the same neural network, passing data to and forth to successors and predecessors respectively.

But as dependencies between two neural networks start increasing, gradients start fading, and hence making RNNs weaker to function. A variant of RNN called

Figure 16. Unrolling recurrent neural networks

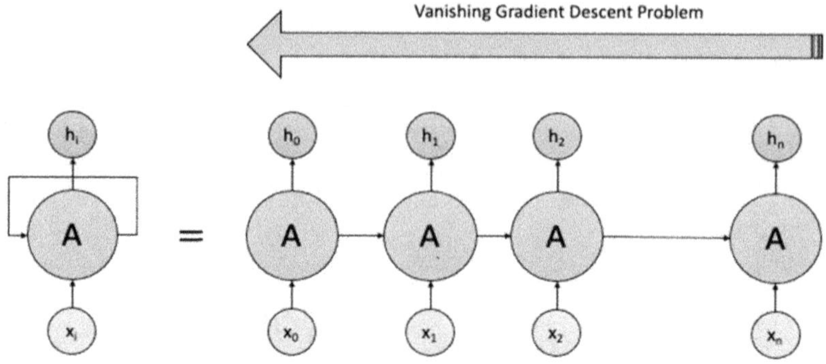

Figure 17. Single LSTM cell

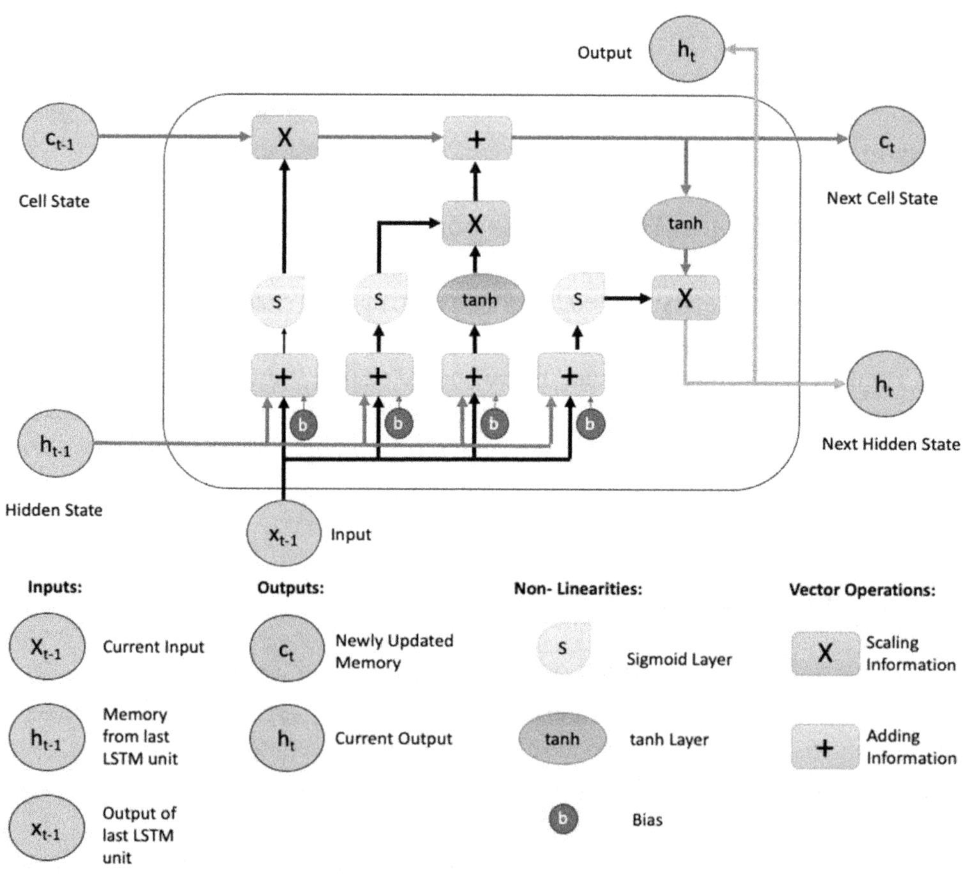

Figure 18. Bidirectional LSTM

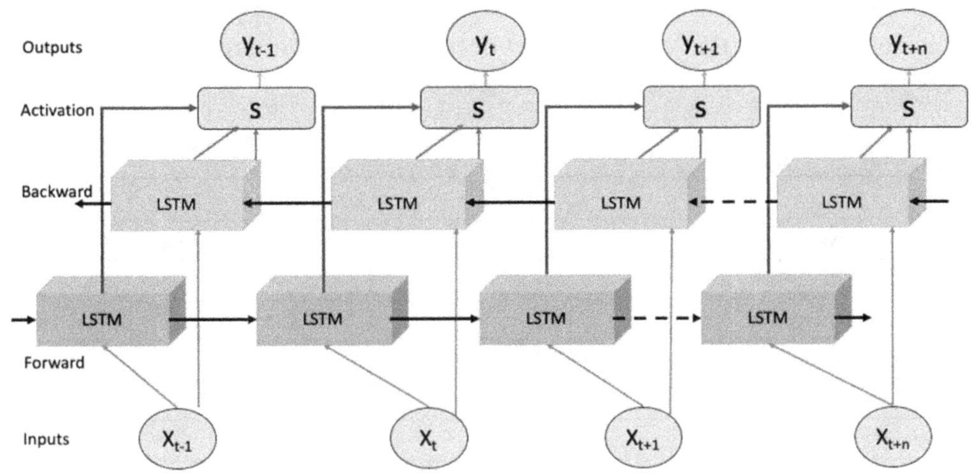

LSTM, fig. 17, is used to solve the problem of prolonged dependencies by retaining the states of neural networks for a longer time.

In Unidirectional- LSTM, successors are dependent on predecessors to classify or predict, on the other hand, in the case of Bidirectional- LSTM, fig 18, states are persisted in both the directions forward as well as backward for better efficacy. It is a well-proven model for keeping track of long term dependencies (Kiperwasser & Goldberg, 2016).

In our proposed model we consider perception layer data as forwarding pass and observer layer data as a backward pass. Observer layer authenticates things and monitors communication between things to ensure security.

Case Study: Malicious Node Injection

For illustration, we demonstrate how the system will help investigators in discovering malicious node injection attacks. Assumed activities are enlisted in table 3.

Observer layer authenticates things and communication between them to find malicious behavior. It can be thought of as a thing management layer too. Using forward pass at perception layer, it is discovered that the thing cannot get registered unless it authenticates itself in the first place. Registration request for Object 003 is sent without authenticating it. The forward pass was expecting AR003 before RR003.

When benign thing with device ID 003 tries to authenticate itself later, using backward pass observer layer comes to know that Object 003 has already been registered and hence it throws authentication error. Forward pass discovers 003 is not authenticated while backward pass verifying it has approved the registration request. This inconsistencies between the states raises alert for malicious activity, fig 19. Investigator analyses the activities within the timestamps and finds out the malicious node Injection attack. Expert accesses data from processing layer and checks out for possible damage that has been occurred.

Table 3. Sequential correlated data, Suffix 'X'- Device ID

Name of Event	Request	Response
Authentication Request	ARX	-
Authentication Response	-	ASX/ AEX
Registration Request	RRX	-
Registration Response	-	RSX/ REX
Objection Modification Request	OMRX	-
Object Modification Response	-	OMSX/ OMEX
Scale Up Request	SURX	-
Scale Up Response	-	SUSX/ SUEX
Scale Down Request	SDRX	-
Scale Down Response	-	SDSX/ SDEX
Deletion Request	DRX	
Deletion Response	-	DSX/ DEX
Communication Request	CR-X-Y.Y	-
Communication Request Response	-	CR-S-X-Y.Y/ CR-E-X-Y.Y

Figure 19. Case study
PRC- Perception, OB- Observer, PR- Processing, B- Backward Pass, F- Forward Pass

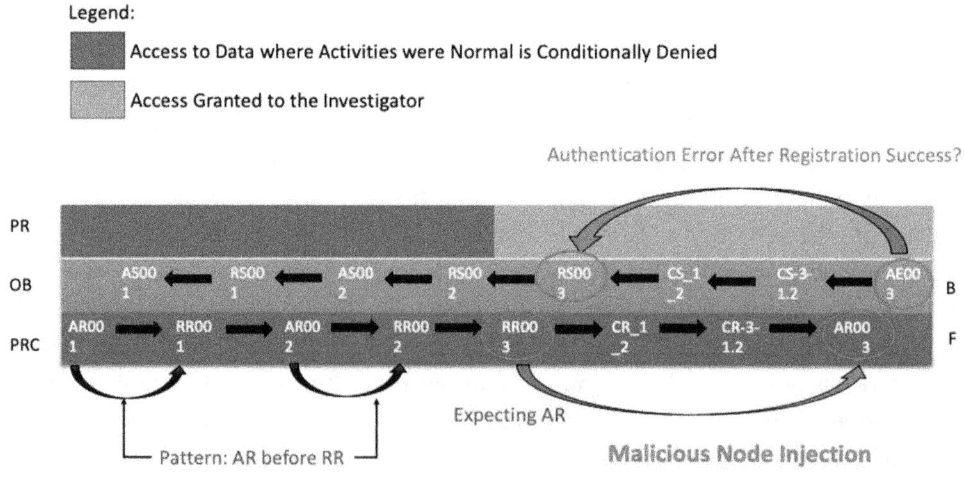

Figure 20. Re-defined forensic phases

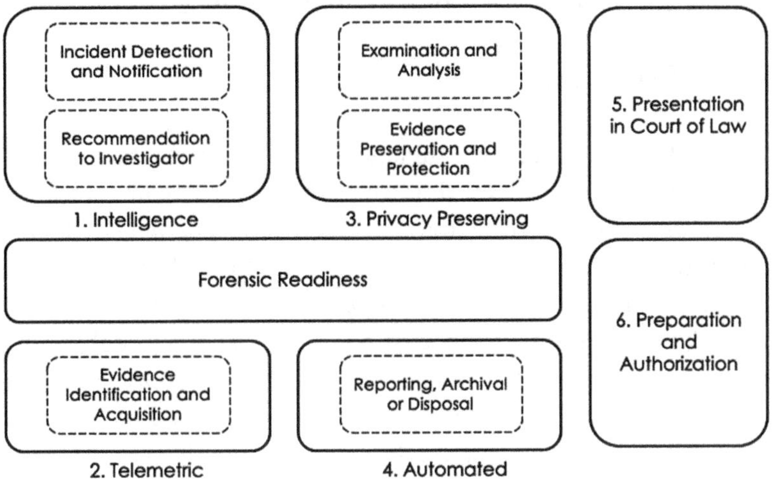

Observer layer is updated with new countermeasures. Telemetric Data Acquisition (TDA) and Forensic Readiness (FR) are preferred in maximum possible scenarios.

Preserving Privacy

With data available from both the layers, we can cross verify the activities with the help of forward and backward pass and investigate suspicious activities. Privacy can be preserved by providing access only to the data from the processing layer which is stored within the timestamps where distrustful activities are detected.

Refinement in Forensic Investigation

We introduce intelligence, telemetric data acquisition, privacy preservation, and automation as key factors. The major refinement is to have forensic readiness in all the above phases, fig 20. As soon as an incident is detected and recommended to the investigator, investigators can take quick actions and decide the business continuity management plan. Cross case learning is another important part of forensic readiness in any of the mentioned phases.

DEEP LEARNING ASSISTED IOT FORENSICS: CHALLENGES AND RESEARCH DIRECTIONS

Availability of Datasets

An adequate number of datasets are available for traditional digital forensics, but the only handful of datasets are available for IoT forensics, that too under the category of Intrusion Detection Systems. The community lacks a variety of datasets for different layers in IoT systems. Privacy concern is a big hurdle for making datasets available, which can be easily handled by data manipulation and augmentation techniques. Timely up-gradation and maintenance of datasets is another problem that arises due to the lack of platform and publisher support. A commonly approved secured and configurable platform is a vital requirement. However lossless compression techniques, standard guidelines for creating and sharing datasets (which makes it authentic) are some of the challenges in question.

The Need for Specialized Tools

Currently, conventional tools are used for the extraction of data from IoT devices. It can be helpful up to a certain level. A hybrid approach, involving multiple tools can make IoT forensics possible but time-consuming. A dedicated set of newly developed protocols for cutting edge embedded products which are connected to edge and cloud makes IoT systems complex. Extracting data from the device might not be sufficient. Targeting extraction keeping all data connectivity in mind is necessary and hence appropriate tool development. The need for a concurrent approach is inevitable. Plugins in existing tools can make extraction and analysis much feasible.

Deep Learning and Existing IoT Solutions

The availability of datasets is a major shortcoming as already discussed. But further, if we consider applying Deep Learning enabled IoT Forensics in already deployed IoT systems, the preparation of a compelling dataset is a titanic task. If functional test cases are available along with corresponding data, we can build a solid dataset that can mimic the exact behavior of the deployed system, otherwise, conceptual system simulation or emulation to prepare the dataset would need extra efforts and time. Further, data augmentation techniques, i.e. preparing a large enough dataset based on limited data is a questionable task. As research in the blend of IoT Forensics and Deep Learning is in a very infant stage, lack of quality literature to support the proposed theory is another crucial challenge. We strongly believe the use of re-enforcement learning in IoT forensics would make the system drastically intelligent with the trade-off of training time.

CONCLUSION

The article enlightens the importance of IoT Forensics as the dependability of human beings on IoT increases exponentially. IoT systems are dynamic and complex and composed of numerous interconnected fragments analogous to the Open System Interconnection model. The proposed Cross-Layer Learning model builds on top of Distributed Logging, Layered IoT Architecture, Cross-Layer Forensics, and Deep Learning Architecture. Applying Deep Learning on a single layer does not give a clear picture as IoT systems are interconnected. The idea behind the model is to fill this gap by merging data of two or more layers and automate the correlation. We chose Bi- LSTM to find long-term dependencies between the data and cross verify obtained data from an investigation point of view. It helps investigators by recommending suspicious portions at the same time preserving privacy. A case study is presented to demonstrate the notion. Development of a Cross-Layer Learning Architecture is in progress and we further aim to develop a tool for extracting data from all the stated layers making application of Deep Learning more feasible and hence examining the specified model in greater detail.

REFERENCES

Abdul-Ghani, H. A., & Konstantas, D. (2019). A Comprehensive Study of Security and Privacy Guidelines, Threats, and Countermeasures: An IoT Perspective. *Journal of Sensor and Actuator Networks*, *8*(2), 22. doi:10.3390/jsan8020022

Al-Garadi, M. A., Mohamed, A., Al-Ali, A. K., Du, X., Ali, I., & Guizani, M. (2020). A Survey of Machine and Deep Learning Methods for Internet of Things (IoT) Security. *IEEE Communications Surveys and Tutorials*, *22*(3), 1646–1685. doi:10.1109/COMST.2020.2988293

Al-Masri, E., Bai, Y., & Li, J. (2018). A Fog-Based Digital Forensics Investigation Framework for IoT Systems. *2018 IEEE International Conference on Smart Cloud (SmartCloud)*, 196–201. 10.1109/SmartCloud.2018.00040

Atlam, H. F., El-Din Hemdan, E., Alenezi, A., Alassafi, M. O., & Wills, G. B. (2020). Internet of Things Forensics: A Review. *Internet of Things*, *11*, 100220. doi:10.1016/j.iot.2020.100220

Babun, L., Sikder, A. K., Acar, A., & Uluagac, A. S. (2018). *IoTDots: A Digital Forensics Framework for Smart Environments*. https://arxiv.org/abs/1809.00745

Burhan, M., Rehman, R., Khan, B., & Kim, B.-S. (2018). IoT Elements, Layered Architectures and Security Issues: A Comprehensive Survey. *Sensors (Basel)*, *18*(9), 2796. doi:10.339018092796 PMID:30149582

Dargan, S., Kumar, M., Ayyagari, M. R., & Kumar, G. (2020). A Survey of Deep Learning and Its Applications: A New Paradigm to Machine Learning. *Archives of Computational Methods in Engineering*, *27*(4), 1071–1092. doi:10.100711831-019-09344-w

Dhanda, S. S., Singh, B., & Jindal, P. (2020). Lightweight Cryptography: A Solution to Secure IoT. *Wireless Personal Communications*, *112*(3), 1947–1980. doi:10.100711277-020-07134-3

Fadlullah, Z. M., Tang, F., Mao, B., Kato, N., Akashi, O., Inoue, T., & Mizutani, K. (2017). State-of-the-Art Deep Learning: Evolving Machine Intelligence Toward Tomorrow's Intelligent Network Traffic Control Systems. *IEEE Communications Surveys and Tutorials*, *19*(4), 2432–2455. doi:10.1109/COMST.2017.2707140

Feng, X., Dawam, E. S., & Amin, S. (2017). A New Digital Forensics Model of Smart City Automated Vehicles. *2017 IEEE International Conference on Internet of Things (IThings) and IEEE Green Computing and Communications (GreenCom) and IEEE Cyber, Physical and Social Computing (CPSCom) and IEEE Smart Data (SmartData)*, 274–279. 10.1109/iThings-GreenCom-CPSCom-SmartData.2017.47

Fernandes, E., Rahmati, A., Eykholt, K., & Prakash, A. (2017). Internet of Things Security Research: A Rehash of Old Ideas or New Intellectual Challenges? *IEEE Security and Privacy*, *15*(4), 79–84. doi:10.1109/MSP.2017.3151346

Harbawi, M., & Varol, A. (2017). An improved digital evidence acquisition model for the Internet of Things forensic I: A theoretical framework. *2017 5th International Symposium on Digital Forensic and Security (ISDFS)*, 1–6. 10.1109/ISDFS.2017.7916508

Hassija, V., Chamola, V., Saxena, V., Jain, D., Goyal, P., & Sikdar, B. (2019). A Survey on IoT Security: Application Areas, Security Threats, and Solution Architectures. *IEEE Access: Practical Innovations, Open Solutions*, *7*, 82721–82743. doi:10.1109/ACCESS.2019.2924045

Hossain, M., Hasan, R., & Zawoad, S. (2017). Trust-IoV: A trustworthy forensic investigation framework for the internet of vehicles (IoV). *Proceedings - 2017 IEEE 2nd International Congress on Internet of Things, ICIOT 2017*, 25–32. 10.1109/IEEE.ICIOT.2017.13

Hussain, F., Hussain, R., Hassan, S. A., & Hossain, E. (2020). Machine Learning in IoT Security: Current Solutions and Future Challenges. *IEEE Communications Surveys and Tutorials*, *22*(3), 1686–1721. doi:10.1109/COMST.2020.2986444

ISO - ISO/IEC 29100:2011 - Information technology — Security techniques — Privacy framework. (n.d.). Retrieved October 2, 2020, from https://www.iso.org/standard/45123.html

Karie, N. M., Kebande, V. R., & Venter, H. S. (2019). Diverging deep learning cognitive computing techniques into cyber forensics. *Forensic Science International: Synergy*, *1*, 61–67. doi:10.1016/j.fsisyn.2019.03.006 PMID:32411955

Kebande, V. R., Mudau, P. P., Ikuesan, R. A., Venter, H. S., & Choo, K.-K. R. (2020). Holistic digital forensic readiness framework for IoT-enabled organizations. *Forensic Science International: Reports*, *2*, 100117. doi:10.1016/j.fsir.2020.100117

Kebande, V. R., & Ray, I. (2016). A Generic Digital Forensic Investigation Framework for Internet of Things (IoT). *2016 IEEE 4th International Conference on Future Internet of Things and Cloud (FiCloud)*, 356–362. 10.1109/FiCloud.2016.57

Kiperwasser, E., & Goldberg, Y. (2016). Simple and Accurate Dependency Parsing Using Bidirectional LSTM Feature Representations. *Transactions of the Association for Computational Linguistics*, *4*, 313–327. doi:10.1162/tacl_a_00101

Koroniotis, N., Moustafa, N., & Sitnikova, E. (2019). Forensics and Deep Learning Mechanisms for Botnets in Internet of Things: A Survey of Challenges and Solutions. *IEEE Access: Practical Innovations, Open Solutions*, *7*, 61764–61785. doi:10.1109/ACCESS.2019.2916717

Koroniotis, N., Moustafa, N., & Sitnikova, E. (2020). A new network forensic framework based on deep learning for Internet of Things networks: A particle deep framework. *Future Generation Computer Systems*, *110*, 91–106. doi:10.1016/j.future.2020.03.042

Li, H., Ota, K., & Dong, M. (2018). Learning IoT in Edge: Deep Learning for the Internet of Things with Edge Computing. *IEEE Network*, *32*(1), 96–101. doi:10.1109/MNET.2018.1700202

Meffert, C., Clark, D., Baggili, I., & Breitinger, F. (2017). Forensic state acquisition from internet of things (FSAIoT): A general framework and practical approach for IoT forensics through IoT device state acquisition. *ACM International Conference Proceeding Series, Part F1305*. 10.1145/3098954.3104053

Mirai (malware) - Wikipedia. (n.d.). Retrieved October 17, 2020, from https://en.wikipedia.org/wiki/Mirai_(malware)

Miraz, M., Ali, M., Excell, P., & Picking, R. (2018). Internet of Nano-Things, Things and Everything: Future Growth Trends. *Future Internet*, *10*(8), 68. doi:10.3390/fi10080068

Mohamad Noor, M., & Hassan, W. H. (2019). Current research on Internet of Things (IoT) security: A survey. *Computer Networks*, *148*, 283–294. doi:10.1016/j.comnet.2018.11.025

Nieto, A., Rios, R., & Lopez, J. (2017). A Methodology for Privacy-Aware IoT-Forensics. *2017 IEEE Trustcom/BigDataSE/ICESS*, 626–633. doi:10.1109/Trustcom/BigDataSE/ICESS.2017.293

Noura, H. N., Salman, O., Chehab, A., & Couturier, R. (2020). DistLog: A distributed logging scheme for IoT forensics. *Ad Hoc Networks*, *98*, 102061. doi:10.1016/j.adhoc.2019.102061

Perumal, S., Norwawi, N. M., & Raman, V. (2015). Internet of Things(IoT) digital forensic investigation model: Top-down forensic approach methodology. *2015 Fifth International Conference on Digital Information Processing and Communications (ICDIPC)*, 19–23. 10.1109/ICDIPC.2015.7323000

Rawlinson, K. (2014). *HP Study Reveals 70 Percent of Internet of Things Devices Vulnerable to Attack*. HP Inc. https://www8.hp.com/us/en/hp-news/press-release.html?id=1744676#.X4prekIzbBJ

Rondeau, C. M., Temple, M. A., & Lopez, J. (2019). Industrial IoT cross-layer forensic investigation. *Wiley Interdisciplinary Reviews: Forensic Science*, *1*(1), e1322. doi:10.1002/wfs2.1322

Ryan, P., & Watson, R. (2017). Research Challenges for the Internet of Things: What Role Can OR Play? *Systems*, *5*(1), 24. doi:10.3390ystems5010024

Sadineni, L., Pilli, E., & Battula, R. B. (2019). *A Holistic Forensic Model for the Internet of Things*., doi:10.1007/978-3-030-28752-8_1

Sathwara, S., Dutta, N., & Pricop, E. (2018). IoT Forensic A digital investigation framework for IoT systems. *2018 10th International Conference on Electronics, Computers and Artificial Intelligence (ECAI)*, 1–4. 10.1109/ECAI.2018.8679017

Silva, J., Rodrigues, J., Al-Muhtadi, J., Rabêlo, R., & Furtado, V. (2019). Management Platforms and Protocols for Internet of Things: A Survey. *Sensors (Basel)*, *19*(3), 676. doi:10.339019030676 PMID:30736424

Stamm, M. C., Wu, M., & Liu, K. J. R. (2013). Information Forensics: An Overview of the First Decade. *IEEE Access: Practical Innovations, Open Solutions*, *1*, 167–200. doi:10.1109/ACCESS.2013.2260814

Valjarevic, A., & Venter, H. S. (2015). A Comprehensive and Harmonized Digital Forensic Investigation Process Model. *Journal of Forensic Sciences*, *60*(6), 1467–1483. doi:10.1111/1556-4029.12823 PMID:26258644

Verma, R., Govindaraj, J., & Gupta, G. (2018). DF 2.0: Designing an automated, privacy preserving, and efficient digital forensic framework. *Annual ADFSL Conference on Digital Forensics, Security and Law*. https://commons.erau.edu/adfsl/2018/presentations/12

Yaqoob, I., Hashem, I. A. T., Ahmed, A., Kazmi, S. M. A., & Hong, C. S. (2019). Internet of things forensics: Recent advances, taxonomy, requirements, and open challenges. *Future Generation Computer Systems*, *92*, 265–275. doi:10.1016/j.future.2018.09.058

Yu, Y., Barthaud, D., Price, B. A., Bandara, A. K., Zisman, A., & Nuseibeh, B. (2019). LiveBox: A Self-Adaptive Forensic-Ready Service for Drones. *IEEE Access: Practical Innovations, Open Solutions*, *7*, 148401–148412. doi:10.1109/ACCESS.2019.2942033

Zhang, K., Cao, J., Liu, H., Maharjan, S., & Zhang, Y. (2020). Deep Reinforcement Learning for Social-Aware Edge Computing and Caching in Urban Informatics. *IEEE Transactions on Industrial Informatics*, *16*(8), 5467–5477. doi:10.1109/TII.2019.2953189

Zhang, Q., Yang, L. T., Chen, Z., & Li, P. (2018). A survey on deep learning for big data. *Information Fusion*, *42*, 146–157. doi:10.1016/j.inffus.2017.10.006

Zhang, W., Guo, W., Liu, X., Liu, Y., Zhou, J., Li, B., Lu, Q., & Yang, S. (2018). LSTM-Based Analysis of Industrial IoT Equipment. *IEEE Access: Practical Innovations, Open Solutions*, *6*, 23551–23560. doi:10.1109/ACCESS.2018.2825538

Zia, T., Liu, P., & Han, W. (2017). Application-Specific Digital Forensics Investigative Model in Internet of Things (IoT). *Proceedings of the 12th International Conference on Availability, Reliability and Security*, 1–7. 10.1145/3098954.3104052

Chapter 6
Fire Investigation and Ignitable Liquid Residue Analysis

Sachil Kumar
https://orcid.org/0000-0002-2681-3937
Naif Arab University for Security Sciences, Saudi Arabia

Anu Singla
Bundelkhand University, India

Ruddhida R. Vidwans
Jain University, India

ABSTRACT

A fire investigation is a difficult and challenging task. An investigator's basic task at a fire scene is two-fold: first, to ascertain the origin of the fire and, second, to closely investigate the site of origin and try to determine what triggered a fire to start at or near that spot. Usually, an investigation would begin by attempting to obtain a general view of the site and the fire damage; this may be achieved at ground level or from an elevated location. Following this, one may examine the materials available, the fuel load, and the condition of the debris at different locations. Surprisingly, the science of fire investigation is not stagnant, and each year, more information to assist investigators in determining the location and cause of a fire by diligent observation of the scene and laboratory study of fire debris is released. This chapter is split into two sections. The first section discusses the general procedures to be used during a fire investigation, and the second section discusses laboratory analysis of ignitable liquid residue analysis.

DOI: 10.4018/978-1-7998-8386-9.ch006

PART 1

Fire Investigation

Fire investigation is the study of fire-related incidents using fire dynamics. When a tragic fire happens, experts must decide where and how the fire began, as well as whether it was accidental or deliberate. If the fire was started on intentionally, the investigator's results could lead to felony charges. If there were deaths as a result of the fire, an individual may be charged with murder.

If they suspect that a fire was lit deliberately, they want to see some signs of an ignitable liquid residue (ILR). Spotting the source of the fire allows the investigator to gather residue that could contain substances suggestive of, say, fuel or hydrocarbons that are trademarks of other ignitable liquids which can be used to cause fires.

INTRODUCTION

Prior to covering the subject of how a fire investigation is led, it is first important to comprehend how a fire examination is directed. This is the initial step of the logical methodology, depicted in National Fire Protection Association (NFPA) 921 as "Perceive the Need" (Reno et al., 2000). The principal assignment of most fire-fighting agents is to figure out the origin of the fire and how it started, yet the cause frequently has numerous more profound implications, contingent upon the unique situation. In case of notwithstanding the reason for a fire, one should comprehend the reason for fire demise. There may be a basic reason assurance, for example, an obvious and very common cooking fire, while investigating the cooking fire the following inquiry emerges "For what reason did the casualty neglect to get away from the fire area?", "Was there some kind of problem with the person in question?", "Is it safe to say that he or she was debilitated?", "Was there a major issue with the structure of the fire area?", "Whether the ways out were obstructed?", "Did the smoke caution function properly, and if not, why not?", "Did the fire spread excessively fast, and provided that this is true, why?" Fire specialists are frequently asked to decide for what good reason a fire spread in the manner that it did (Hall, 2006).

It is critical to arrive at the fire incident as soon as possible because it includes a variety of volatile substances such as fuel traces, smoke, eyewitnesses for questioning, and even the victim, and most notably, to preserve and protect the crime scenes from intrusive spectators.

FIRE SCENE INVESTIGATION

Seize the Scene of Crime

As an investigator reaches the crime scene, he/she must seize the crime scene and evacuate the fatalities and rescue the injured victims. The NFPA921 has described a 6-step approach to analyze the origin and cause of the fire. If an investigator tries to answer these, then the arsonist can solve the case successfully (Redsicker and O'Connor, 1996).

Each of these approaches will now be discussed and explored in detail one by one.

Recognizing the Problem

The first and most critical step in the investigation would be to ascertain where the fire originated. This can be accomplished by studying the patterns of the fire at various locations and interacting with eyewitnesses. Also, an investigator should be able to establish the resources that might be needed to investigate the crime scene. Sometimes if an investigator reaches on the scene of the fire before the fire is extinguished, there are chances that he can recognize the pattern of the fire which can act as a very important point in solving the case.

Defining Problem

In this case, a thorough investigation into the origin and cause of the fire must be undertaken using the facts at his disposal and by processing the crime scene. The analyst should prepare a strategic plant and report with preliminary findings and description of the loss in the fire.

Collect Data

The direct observations by visiting inside the spot, photography, evidence markings and its collections elaborates the facts and information about the fire crime scene. All the evidences and data that is collected from the scene is a subject to verification for its authenticity. The evidences and data collected from the scene are debris from the crime scene, photographs of the smoke and the fire patterns on the wall, fuel containers if available and suspected, construction details, documentation of the property, etc. All the collected evidences and data must follow the proper chain of custody in order to prevent any mishandling and loss of the essential evidences.

Analyze the Data

On the basis of the observations which are based on the evidences from the fire scene analyze the data to develop the probable hypothesis. This process is known as inductive reasoning. Using this approach in the investigation can help in evaluating the damages in fire patterns, heat and flame vectoring, arc mapping, and fire engineering and modeling analysis. Under this step the assessment of histories of similar kinds of fires, training and understanding of fire chemistry and dynamics, fire testing experience of an investigator, and other experimental data is very crucial in forming the base for the hypothesis.

Developing a Working Hypothesis

In simple words, it is the supposition statement which is put forward after carefully observing and analyzing all the evidences and facts of the fire scene. In this step, an analyst can generate a working hypothesis, which can help in solving the case by checking the facts and eliminating unnecessary clauses. This hypothesis can give an idea about the room dimensions, location of competent ignition source, impact of room conditions like if it was a closed room or open room.

Testing of the Working Hypothesis

This is the last step in investigation of the fire scene. In this step, one needs to apply a theory of deductive reasoning which helps in forming the conclusion based on the previously known an collected facts from the scene of crime and the laboratories. This step leads to eliminating all the false hypothesis and will lead to confirm your hypothesis by eliminating the all-other reasonable origins and causes for the fire. Keep on working in the direction to obtain new information on the case from the witnesses and modify the hypothesis to reach on the final and accurate conclusion. Repeat steps of analyzing the data, developing a working hypothesis, and its testing to arrive on balanced working hypothesis. After testing all the theories formed in the hypothesis one should not eliminate any fact or the considerations which cannot be eliminated conclusively as it can prove as an important viable element which can be used as an evidence. As we all are aware that one cannot investigate the fire incidence as other crime scenes as one cannot set another same type of building on fire to check various aspects of it, hence the hypothesis should be made by proper investigation and obtaining reliable data to develop and test the hypothesis. Because these formed hypotheses will be subjected to multiple trials in the courts while legal proceedings. While giving the testimony the examiner should be able to explain all the possible hypothesis to the court. This type of evaluation is sometimes referred to as abductive reasoning. The result of this product is to select the most accurate hypothesis that fits all the available evidences and data. This procedure of fire investigation is followed in most of the cases, as it is considered as the newest and modern method of fire investigation.

Selection of Final Hypothesis

The judgement should be based on the available facts, evidences, and research the hypothesis can be authoritatively presented as a conclusion of the opinion.

Examination of the Fire Scene

The exact and accurate examination of the fire scene cannot be demonstrated as setting another fire is not feasible. But with the help of the visible identification of degree of fire damage will help in objective identification (Gorbett et al., 2010).

In an investigation of a fire scene, it is very important to go systematically while processing the fire scene. An investigator should look for the exterior damage, fire patterns, and other evidences, and then start investigating inside area for searching for the room where fire originated, and then try to locate the area and point of origin of fire, and finally reach out to the cause of fire (Redsicker and O'Connor, 1996).

The processing of exterior and interior crime scene will now be elaborated systematically. The fire scene must be thoroughly overhauled as mishandling and delay in the evidences and observations leads to failure in crime scene reconstruction (David M. Banwarth et al., 2015). This process of overhauling is primarily done for identifying the hidden areas of sparks which may be covered in the debris.

Exterior Fire Scene Analysis

1. The investigation of fire scene can begin with interactions between fire fighters, police officials, and first responders on the scene. The investigator should search into any forced entries of fire fighters into locked buildings or rooms. This section will go through any more relevant issues that an investigator can look into to learn more about the case.
2. The actions on the fire scene should be observed and investigated for the questions like were there any suspicious vehicle parked nearby or travelling from nearby in last few days? Was there any fighting among the people who lived in that area? And so on.
3. Now, search the exterior crime scene and see whether the whole structure is involved in the fire or not, whether the available (if available) fuel has a consistency with the structure of fire, whether flames visible, what area of the

building is involved, if fire is extended out of the window or not, if yes then which side of the window was involved, try to know the colour of the flame and smoke with the help of eye witness, try to know the condition of the room or building if the doors were closed or opened, if it is latched, try to find out if the entries and exits were blocked, check if the fire alarm was rang or no and what time it was extinguished, try to look out for any unusual odours.

4. Check for any blockages in the way of hydrants, standpipes, and sprinklers.
5. Examiners should look around the structures to observe the type of damages that has occurred near the vents, windows, search for the "V" pattern for the fire, check for the lowest and highest flame patterns on the exterior side of the building. Try to find out the materials used in the structures, if they are made up of wood, how much wood was used, etc. Take the videos and photographs of the survey of the external area of the building.
6. When you externally analyse the site of fire, you also have to consider and understand the occupancy of the premise involved in the fire.

Internal Fire Analysis

1. To begin with the internal investigation, one should identify the highest point of fire's extension. At this point on, the initiation of internal fire examination should be commenced.
2. The internal surveillance should be carried out with the proper gears, since there may be toxic fumes or odors present inside the building. Also, an investigator should look for the solidity of the floor that you are walking on as it might be heavily damaged (wooden floors, stair case, etc) due to fire and might not be able to take the weight on it and is prone to collapse.
3. Where there is extensive damage, heavy machinery would be used to clear the debris. While searching the crime scene, you must also pay attention to the pattern in which the glass wares, steels, and other metals are bent or damaged. This will give you an indication of the temperature that the fire might have reached.
4. Failure of steel structures can be analysed by observing the expansions in Steel I-beams caused by fire heat. Check for the bulged portion inside a building as well as drilled holes inside the building.
5. Check for the forced entry into the buildings and try to identify or locate the suspected points where the entries are not made by fire fighters but by any suspected intruder/s.
6. Write down all of these points and observe all of the evidences and photographs, and mentally reconstruct the scene of fire and try to find out certain questions such as- how the fire initiated, whether it loos natural or manmade, how did it

extended, whether any personal belongings were removed or not, try to find out the flow of flames inside the room, if fuel is available then try to find out the answer if the available fuel is consistent with the fire progression.

7. Try to find out whether the progression of fire is natural or artificial. Look for any unusual fire burn patterns which may be indicative of use of liquid accelerant.
8. To find out the origin of the fire, follow the burn trails as they will take you to the place where the fire started. This will also help in reaching to the surface that has maximum damage.
9. And it is well studied that fire lasts for longer time near abundant sources of the oxygen (Bowman et al. 2009). Searching this location can provide valuable information regarding the point of origin and cause of the fire.
10. According to NFPA 921, Chapter 4, Fire Patterns the "fire language" that includes physical signs and substances that can reveal the cause of fire. To conduct the effective analysis and reach the near to accurate result needs to understand the fire language which includes lines or areas of demarcation (between any two types of flame patterns, between the affected area and unaffected or less affected area), spalling (flaking of concrete or discoloration of the bricks), oxidation (materials that cannot burn can produce a clear demarcation of fire patterns that may be useful for fire and arson investigation, colour change of an objects is also an indicator of the time of the heat that object is exposed to), heat indicators (objects like molten or fused glass or metals can be temperature determinants, as the objects which are near to the fire shows more damage), thermal indicators of metal, calcination (the changes occurs during the fire on the concrete/wall surfaces), various features of glass (windows pane, smoke stains on glass doors and windows), depth of char (indicates length of time that the wood is being exposed to the heat or fire), heat shadowing (the area or an object is covered by another object that can help in determining the point of origin), and protected areas.

Collection and Preservation of Fire Evidences

Various types of evidence are collected by fire investigators during the process of any fire scene investigation.

Evidence collected from fire scenes would be diverse and specific to the incident location. Evidence suspected of having ILRs is the most obvious type of chemical evidence collected at a fire scene (David Gossman). Other evidence that may be collected for ILR analysis involves comparison substrate samples, liquid samples for comparison, solid materials or suspect's clothes and shoes (Lentini, 2013). Suspected ignition sources, like explosive devices or malfunctioning appliances, are often collected in an attempt to test the field investigator's hypothesis about the origin and cause of the fire (Lentini, 2013). Samples can be obtained from soils, sumps, depressions, and areas of standing water starting at the point of origin and extending to a radius defined by the forensics expert. Waste fire water may contain chemical residues that could provide insight (David Gossman).

Primary Evidence

Primary evidence should be collected in an area or areas suspected of having been exposed to some type of liquid accelerant and should be analyzed to identify the accelerant. In situations where only a small amount of accelerant is used, the investigator should look for an irregular pattern of localized disruption in the area of origin. An accelerant odor, as well as intermixed light, medium, and heavy horizontal burn patterns, are several reliable markers (NIJ (National Institute of Justice) 2009).

Common Sampling Errors

One of the most common mistakes made when collecting evidence to be analyzed is collecting inadequate samples. Other mistakes include using ineffective sample preservation methods and obtaining no reference sample. Sampling from the wrong area(s) or materials will result in a negative analysis.

Evidence Collection Areas

Getting to know what to collect or what not to collect is a critical ability in evidence collection. Accelerant liquids burn better than most surfaces they are poured onto. Better, stronger samples are likely to be found in secured areas and within absorbent materials. The lowest and also most insulated areas within a burn pattern are the best places to collect. Porous plastic, cloth, synthetic fibers, cardboard, or paper in close contact within the pattern will absorb any possible liquid which might have been used to expedite the fire. Liquids will accumulate or pool within cracks, tears, lines, or deck drains, making them ideal storage areas for residue. Liquids can also pools along loadbearing support columns and at the base area of bulkheads. Areas of deeply charred wood, gray ash areas, or edges of holes burnt through wooden decks are less desirable due to the complete burn of material. The center of any burn pattern is not a desirable area due to intense burn of fuel. In general, any region that is subjected to the most heat or hose streams is a less suitable collection location (NIJ (National Institute of Justice) 2009).

Comparison Samples

Materials or objects that are assumed to be almost identical to comparable accelerant debris samples but do not contain accelerant traces are considered comparison samples. The aim of such samples is to classify and reduce or remove sources of interference in such samples' analysis.

Examples of Comparison Samples

New, unused evidence containers. The majority of evidence containers show no chromatographic interference. However, certain containers, such as some types of plastic bags and specially coated containers, can emit background vapors that may interfere with chromatographic analysis of data. As comparison tests, new kinds of containers and new lots of such container types, such as plastic bags, should be sent to the laboratory on a regular basis for chromatographic examination. Similarly, collection materials may sometimes be a cause of interference. Some absorbent fabrics used to manage gasoline spills, for example, are unsuitable for evidence gathering since they contain residues of contaminants.

Comparable material samples. It is preferable to gather samples of materials that are similar to those present in the accelerant debris matrix except that they contain no accelerant. These materials, like carpet or wood trim, often may give off pyrolysis products, which may mask accelerant patterns during analysis. For of questioned fire debris sample, a comparison sample should be taken. Since they include residues of chemicals, absorbent fabrics used to manage gasoline spills are unsuitable for evidence collection.

Comparable accelerant samples. Investigators frequently need to find comparisons between samples of identified flammable and combustible liquids gathered at the site, those discovered in barrels or retrieved from other sites near the fire scene, and flammable and combustible liquid residues found in fire debris. Flammable and combustible liquid residues in debris samples can originate on the scene or be transported there from another source. Insecticides, for example, often use a petroleum-based carrier. Comparing debris samples with liquid samples from these sites, as well as comparative samples of products similar to those found in fire debris, can be a useful way of discriminating between what was at the scene before the fire and what was carried to the scene to initiate or spread the fire.

Fire investigators also seek comparisons when comments, facts, or conditions suggest that an accelerant might have originated from a particular location, such as a suspect's car or a local gas station. Forensic science experts may also distinguish between samples of identical materials as coming from clearly separate origins. If no inconsistencies are detected, the best argument that can be made is that the samples could have come from the same source. This is primarily attributed to the petroleum industry's marketing methods and the near-impossibility of paying for any potential outlet. Another possibility is that the samples are from the same or separate groups of petroleum goods, which can be identified using gas chromatographic pattern recognition. For such comparisons, liquid-to-liquid samples are ideal. Using liquid and residue samples, determining a particular source is far more complex. A material's positive identification may rely heavily on comparison samples. Where it is unclear if a reference sample is needed, one should be collected and submitted, or the laboratory must be contacted.

Trace Evidence

Trace evidence characterization and comparison may provide convincing evidence of a suspect's involvement with an arson scene, or of one scene or system with another. Glass pieces, torn matches, cigarette butts, fabric wicks, tape segments, or other unit components may be linked to objects in a suspect's possession through physical matching or torn or shattered edges, examination of the composition and layout of the materials present, or identification of incidental trace evidence like hairs, fibers, or stains.

Toolmarks present on doors, locks, valves, windows, or sprinkler or alarm system components at an arson scene may be identified as having been produced by a specific tool recovered from a suspect.

Shoe impressions left at a fire scene may provide forensic leads on the number of people involved or the type(s) of footwear represented, and they can be compared to shoes retrieved from a perpetrator. Latent fingerprints can still remain on the scene's objects or surfaces.

Evidence Packaging

Evidence wrapping is critical for preserving and securing evidence between the moment it is collected and the time it can be analyzed in a lab. It can take weeks or months for evidence to be examined by a forensic science lab in some cases. As a result, fire investigators must exercise extreme caution when packing and storing evidence in order to preserve the integrity of the sample and avoid the release, degradation, or contamination of these residue as flammable/combustible liquids, insecticides, cleaning liquids, and so on (NFPA 921; Home, Evidence Collection).

Metal Containers

As per laboratories, the safest containers for liquid accelerants and hazardous products are fresh, contaminant-free, vapor-tight metal cans with tight-fitting lids (Williams & Sigman, 2007). The benefits of wide openings in relation to total container width, vapor-tightness, rigid shape, light weight, longevity, lack of appreciable noise disturbance during analysis, availability, and number of available sizes are cited by laboratories.

The container should be packed loosely and loaded to no more than two-thirds of its capacity, with the remaining one-third left as airspace between the top of the sample and the lid. When the sample allows it, such as furniture, linoleum, paper, cardboard, or linen, it should be wrapped before being placed in an evidence jar. This is referred to as a "chimney roll". A photograph and a fixed line drawing must be used to document the evidence. After inserting the sample in the jar, tap the lid tightly onto the container top to seal it. Avoid putting so much weight on the lid; you don't want to distort the seams on the container's edges. A strong general rule is to always put the mark on the container's side rather than the lid. During laboratory tests, lids can be removed and lost or mixed up.

Special Evidence Bags

Many fire investigators have noticed that Kapak bags are an excellent replacement for metal containers. These self-sealing bags have a polyester outer layer and a polyethylene inner layer. Despite their benefits, these bags are prone to tearing when used with debris samples containing sharp items such as glass or nails.

Glass Containers

Glass containers, such as glass jars, pharmacy bottles, and so on, may be used to hold a small volume of a liquid residual or a liquid sample for comparison. Mason@ jars are preferred by some fire investigators for collecting and preserving evidence (Hine, 2004; International Association of Arson Investigators, 1999). These containers, with adequate care and wrapping, can be excellent for collecting and preserving accelerant residue samples. The screw-top, vapor-tight lids fit well, and the big mouth makes inserting and removing samples easy. Furthermore, the sample is noticeable, allowing inspection without opening the bottle. The most noticeable drawback to glass containers is the possibility of breakage during shipping and/or storage. Furthermore, certain petroleum products and volatiles found in fire debris can weaken the rubber lid seal of the container, causing vapors to escape. Inserting a sheet of aluminum foil between the jar opening and the cover is one proposed solution to this issue.

Other Common Containers

Other possible evidence containers comprise envelopes, paper bags, plastics bags, or containers designed for specific forms of physical evidence (International Association of Arson Investigators, 1999). The form of container chosen is determined by the size and structure of the physical evidence, laboratory policies and practices, and the type of exam or examination being performed.

Labelling Evidence

It is necessary to label the container and its contents after it has been sealed. There are many types of evidence labeling available, each with an identifying information field. Attach these stickers to the container's side rather than the top or bottom.

Document the exact description of the evidence, the location of the evidence with dimensions, the evidence number, who retrieved the evidence, the date and time of retrieval, and the investigator's name, level, and organization. Other details to report include the location of the fire scene, the fire incident number, and any witnesses that were present when the evidence was retrieved. If the sample is being used for comparison purposes, this should be indicated on the evidence label.

Evidence Storage

Evidence storage is critical in the chain of custody. Keeping evidence for a law enforcement department is normally the only way to get "secured" storage and is more easily accepted by courts. Just keep evidence in a fire investigator's office as a last resort; it is often difficult to show that the evidence was safe. Storage should be held in a closed cabinet and in a locked office area at the very least. With the delays in processing evidence becoming longer in many parts of the country, adequate storage is becoming increasingly necessary. Some agencies may find themselves responsible for evidence samples due to delays in shipment, laboratory backlogs, or samples returned from one laboratory awaiting shipment to another laboratory. In such cases, fire investigators should be prepared to store, document, and preserve the evidence in the same way as would occur while awaiting analysis at the laboratory. The American Society for Testing and Materials Standard (ASTM E1492-92), Standard Practice for Receiving, Documenting, Storing and Retrieving Evidence in a Forensic Science Lab, outlines the basic procedures used by the forensic science community to maintain chain of custody and to preserve evidence samples before and after forensic examination.

Evidence Transportation

The admissibility of evidence in courts of law is largely dependent on the efforts of the fire investigator to ensure adequate protection, i.e., chain of custody from the time of identification and processing, through storage, during shipping to the laboratory, and up to the date of court appearance.

The best way is to transport evidence to a laboratory for examination on one's own. Where practicable, the evidence should be sealed, initialed, and transported to the laboratory by the same individual who recovered it. This eliminates the possibility of the evidence chain being broken, as well as handling by other agents or staff. Evidence may need to be shipped (mailed) to a laboratory for analysis or processing. When this becomes important, any care must be taken to maintain the chain of custody.

Choose a sturdy shipping container (box) wide enough to accommodate any of the evidence containers. Never transport evidence from several investigations in the same container. All evidence containers must be securely sealed to prevent vapor release and must be properly classified and labeled. Each evidence container should be adequately secured from physical harm and safely sealed in the shipping case. Seal the shipping box and label it "evidence." Put a transmittal letter in an envelope and tape it to the package, labeled "invoice." Cover the sealed delivery box in exterior wrapping paper and secure it with sturdy shipping tape. Ship the sealed shipping box via any commercial courier service.

Reconstruction of the Scene

During the way toward drawing, reporting, and getting altogether acquainted with the fire scene, the specialist gets an opportunity to mention objective facts about the fire and to form an arrangement for reconstruction. This quite often includes garbage evacuation and may require a few hours or long stretches of demanding active work to eliminate the trash that covers the curios that will permit an assurance of root and cause. Despite the fact that there is hefty hardware that can achieve this flotsam and jetsam evacuation, beginning with substantial gear will quite often annihilate more proof than it reveals. Such a large number of examinations have been undermined by wildcats and excavators utilized by investigators searching for "pour designs" or spilling. The causative specialist habitually arrives in a heap of garbage in the parking area, with no conceivable method to stop mine where it was situated before it was run over by the hefty hardware. The issue with substantial hardware, proficient however it very well might be, is that as opposed to getting out an entryway or a corridor, it brings down the dividers on one or the other side.

At some point after the consumed roofing material is conveyed outside, intriguing ancient rarities will start to be uncovered. It is important to stop occasionally and report this likely proof as it is found. It is often important to move things far removed with the goal that trash clearing can proceed, at that point move those things back to their unique area to comprehend their job, assuming any, in the advancement of the fire. In a professional flotsam and jetsam expulsion activity, the trash is moved once.

When the flotsam and jetsam have been moved far removed, things ought to be set once more into their unique areas, if conceivable. Tables and seats habitually leave "impressions", secured territories on floors that consider the furniture to be by and large repositioned. Use alert with detached things like tables, be that as it may, since it is feasible to get them repositioned precisely in reverse.

When the remaking is however finished as it seems to be going to get, any segments of the scene that have been changed ought to be re-photographed, giving specific consideration to consume designs that show development or force of the fire, fuel bundles, and start sources.

A typical misstep in the documentation of the scene is the inability to make in general efforts of the territory where close-ups will be taken. Photos of consume designs or of little items that may be start sources are only not as valuable if there is no photo that puts the subject of the nearby into its unique situation.

Probably the main antiquities are those that permit the assortment of successive information. The fire scene "records" every development of the fire starting with one spot then onto the next, yet frequently these examples don't assist us with realizing when a specific occasion happened, or in which request at least two occasions happened. Examples that give successive information are the most educational. Once in a while, it is the direction of consume designs on an article that permits surmising to be drawn with respect to the item's position when the example was made. Fallen items, for example, bits of the roof may display smoke or warmth harm on the top or base surface. It is critical to look at this "flotsam and jetsam" as it is being taken out. Now and then, the errand of moving trash is enormous to the point that emphasis on subtleties, for example, this can be lost.

The proof is currently uncovered and has been documented. The fire examiner is given what is probably going to be the entirety of the actual proof there is to see. It is trusted that the work that has been done will take into consideration the improvement of a dependable speculation concerning where the fire began, and what may have caused it.

PART 2

What Is Ignitable Liquid Residue?

ILR represents the portion of an ignitable liquid that did not burn during a fire. Based on chemical analyses, a scientist may decide whether ILR is present, while evidence from the scene can help to determine how the ILR was used or whether it may be present at the scene.

The highly ignitable fuel or liquid that is used in this world is none other than-gasoline, which is not at all a new to us. In arson investigations, several different varieties of ignitable liquids may be encountered; however, gasoline is the most common ignitable liquid.

Understanding the Ignitable Liquid Chemistry

Fire, as everyone knows is a reaction between combustible substance i.e., ignitable liquid, gas or solid, and an oxidizer that results in the evolution of heat and light. To illustrate, recall that when we are near a fire, we experience a lot of heat, and when we move away from the fire, the amount of heat we feel reduces. Will you figure out why? Yes, you are correct in assuming that the fire is hot. But then the question arises: what causes the fire to be so hot? As a hydrocarbon with a high potential energy reacts with oxygen, kinetic and light energy are released, resulting in carbon dioxide and water with a lower potential energy. As a result, when a hydrocarbon burns, it emits carbon dioxide and water. These hydrocarbons varies from petroleum, diesel, butane, pentane, derivatives of hydrocarbons, etc. (Schmidt-Rohr, 2015).

Some of the models like fire triangle, fire pentagon, fire tetrahedron, fire life cycle, etc. exists. However, the fire triangle and fire tetrahedron are two models that are widely accepted worldwide (Redsicker and O'Connor, 1996).

FIRE TRIANGLE AND FIRE TETRAHEDRON

Fire Triangle: Fuel, Oxidizer, Activation Energy

Air contains oxygen. This oxygen serves as an oxidizer, and when it gets into contact with the slight combustion, it causes a massive burn. Although, this process is not as simple as it is explained theoretically. The combustion involves three steps i.e. initiation, propagation, and termination.

This is basically a chain reaction steps which initiates with the formation of the free radical of the oxygen. After this, in second stage this free radical reacts with

Figure 1.

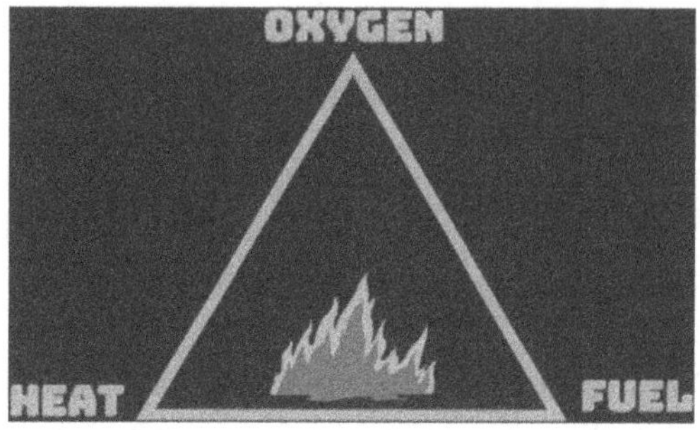

water to form two more free radicals of the OH. These OH radicals then reacts with CO which is formed due to partial combustion and forms water and hydrogen free radical, which then forms water by reacting itself to the previously formed OH radicals (Bell, 2013). Addition of other oxidizing agents, such as KMnO4, KCLO3, adds oxygen in the fire.

In the case of fire, the fuel should act as a material which can get oxidized by the oxygen or other oxidizing agents. The commonly encountered fuels in the crime scenes are cotton, wool, wood, plastic paint rubbers, the most commonly used refined fuels like petrol, diesel, kerosene, ethanol or spirit etc. These fuels have high amount of carbon and hydrogen, which is the very significant for fire to continue for the longer time (Siegel et al., 2000).

Fire Quadrangle

It is also known as the fire tetrahedron. This fire tetrahedron has four fire aspects, and if any of them is absent, the propagation of the fire is hindered. Although the

Figure 2.

$CO + O_2 \longrightarrow CO_2 + \cdot O$	Initiation
$\cdot O + H_2O \longrightarrow \cdot OH + \cdot OH$	Propagation
$CO + \cdot OH \longrightarrow CO_2 + \cdot H$	
$\cdot H + \cdot OH \longrightarrow H_2O$	Termination

principle of the fire progression or initiation remains the same, the main difference is the steady chain reactions occurring between the fuel or the combustible substance and the oxidizer (Lilley, 1996), and thermal energy that is produced in this reaction is applied individually. Whereas in fire triangle, it is applied to overall triangle.

Both the models for the fire works well for studying the nature of the fire and development of its extinguishers. This section will focus on studying more on the fuel property of the fire onset (Stauffer, 2007).

Classes of Fires

Fires are usually classified into five groups as per NFPA 10, US standards for fire classification (Fires classification, 2019):

- Class A fires belongs to the fire caused by combustible materials like paper, wool, wood, plastics, etc.
- Class B fires belongs to the fire caused by combustible or inflammable liquids like solvents, lacquer, alcohols, tar, oils, solvent based paints, inflammable gases, etc.
- Class C fires are caused by high energy electronic appliances. In later stages, it might appear as Class A and Class B fires.
- Class D fires are caused due to inflammable metals like zirconium, sodium, lithium, potassium, magnesium.
- Class K fires are caused due to cooking oils or fats used in kitchen for example wax.

According to the European norms, the fire is classified into 5 classes, similar to NFPA 10, with the exception that flammable liquids and flammable gases are classified into two different classes, Class B contains only flammable liquids, and Class C contains only flammable gases. The fire due to electrical appliances is merged into class F (Fires classification, 2019).

Combustibles or Fuels

Fuels are reducing agents that can exist in any forms such as gaseous, liquids, and solids.

Gaseous Fuel

Gaseous fuels are easier to handle than liquid fluids. They are also easy to ignite. The fire starts very easy as the gaseous molecules of fuels and the oxidizer i.e., oxygen present in atmosphere gets mix faster. The mixing of these two gases decides the compactness and intensity of the burning (Lackner et al., 2013). Flame stability is an ability of the fuel to burn easily, continuously without any hinderance till the flow of the ignitable gas is cut off (Malmgren and Riley, 2012). Natural gas is the most common gaseous fuel, and methane is the most toxic and highest in natural gas composition (Faramawy et al., 2016). Natural gas is colorless and produces a non-luminescent blaze when burned at a low velocity. Because of its high combustion temperature and calorific content, natural gas does not emit any harmful gases or soot particles when burned. In the case of natural gas, the air-to-fuel ratio can be stoichiometric, which causes the fuel to ignite (Sherman and Columbus, 1933).

Natural gas or pentane butane gas usually contributes to the accidental and intentional fires or arsons. The advantage behind using the gaseous fuels is that the traceability of the gas evidence after the incidence is very low as all the gas particles are burnt or dispersed in environment. The only evidence that can be obtained is the containers in which it is carried. However, if the examination or the sample collection takes time, analyzing the container containing the flammable gas becomes difficult.

Solid Fuels

Solid fuels are burning materials such as wood, charcoal, cow dung, and so on. They are more difficult for transportation than gaseous fuels or liquid fuels. The size and character of the fuel decides how easily it can catch up the fire in air or its combustibility (Gupta et al., 2015). In simple terms, the higher the mass of the fuel more the potential loss will be seen because of conduction. Flammable solids are the solids which can be ignited due to the friction (Remember that explosives are not considered as flammable solids). The solid fuels can be pyrolyzable or non-pyrolyzable. The vapors released by its chemical decomposition supports fire to continue (Redsicker and O'Connor, 1996). Pyrolysis is the process in which the solid fuel is broken down into small molecules. It is an endothermic process (A. et al., 2011). Pyrolysis takes place in the absence of oxygen to release combustible or non-combustible gases. In this process, the biomolecules like lignin, cellulose, hemicellulose, etc. remains partially burnt which leads to the formation of bio- chars, bio-oils, syngas, etc.

Liquid Fuels

Before to go any further, let's get familiar with a few words that will come up when learn about the various facets of ignitable liquids. This section would go through different properties of ignitable liquids terminology such as melting point, boiling point, S.G (Specific gravity), flash point, vapor mass, fire point, and so on.

1. Melting point: defined as a temperature when the solid turns into liquid and fusion occurs (Howard and Wilson, 1948).
2. Boiling point: It is the temperature at which the substance (liquid or solid) starts boiling and hence forms or turns into a gaseous phase.

Why Melting And Boiling Points Are Important to Study With Regards to IL

Lower the boiling temperature of the flammable liquid, more the chances of the liquid being flammable as it (lower boiling point) indicates higher volatility, for example, branching in the carbon chains, addition of cyclic compounds, and more double and triple bonds decreases the melting and boiling point of the liquid and hence make it more efficiently ignitable (McCain, 2017).

Aromaticity, addition of more carbon in the compounds, and addition of the alcohol in the compound increases the melting and boiling points of the combustible substances or combustible liquids (Bhimabhai, 2018).

3. Specific Gravity: Specific gravity is the weight of that particular liquid with respect to the equal amount of water at the standard temperature and pressure conditions.

Why Is Specific Gravity Important While Studying IL

Specific gravity is an important aspect while studying the chemistry of ignitable liquid because all the ignitable liquids have lesser specific gravity than that of water (SG of water is 1) (Stauffer et al., 2008).

4. Vapor Density: It is the property of the vapor of the substance that tells whether the vapors released by them (in our case the substances are ignitable substances) is heavier or lighter than air. It is calculated as the ratio of molecular mass of the ignitable substance with that of molecular weight of air which is approximately 29amu.

Why Vapor Density Important to Study With Regards to IL

Consider methane gas; its vapor density is 0.55, which is lesser than 1, and hence it will rise up in the air, and when vapor density is nearby to 1, it will get mixed in the air without any clear demarcation; for example, vapor density of CO, Ethylene, and nitrogen is 0.97, and hence it gets dissolved in the air. Whereas the substances whose density is above 1, such as kerosene, decane, heptane, these gases will settle down and can travel longer distances till any hindrances (Stauffer et al., 2008).

5. **Flammability Limits:** It tells how quickly the ignitable liquid releases the vapors in air and attains a point where it can catch a fire or ignite or explode. All the ignitable substances have flammability limits within two ranges i.e., Lower Flammability limit and upper flammability limit. Lower flammability limit is the one below which the concentration of the vapors does not ignite. Upper flammability limit is the one above which the ignition is not possible even though the ignitable vapors are present. The reason behind this is the unavailability of oxygen, as the fuel vapors increases, the availability of oxygen decreases and hence makes ignition difficult as the presence of oxygen is mandatory for its occurrence (Ludwig, 1999).
6. **Vapor pressure**- All the liquids have an ability to create a pressure on the surface of liquid until it reaches the state of equilibrium. As this equilibrium is attained at that particular time then pressure of the vapor is called vapor pressure of the liquid. In other words, it is the point at which the rate of condensation is equal to the rate of evaporation.

Why Vapor Pressure Is Important to Study With Regards to IL

Vapor pressure is directly proportional to the temperature, because as the temperature increases, so does the vapor pressure. When the vapor pressure of the liquid reaches 760mmHg, its reaches its boiling point, and the corresponding temperature is called boiling temperature. At is point, the fluid is said to attain the lower flammability limit (DeHann and Icove, 2012).

7. **Flash Point**: It is a minimum temperature at which the vapor of an ignitable substance can be ignited. At this temperature the substance attains its lowest flammability limit. But at this point the flame will not sustain as the energy dissipated by the flame is not sufficient to vaporize other molecules of the liquid. Lower the flash point, the substance is highly ignitable.

Why Flash Point Is Important to Study With Regards to IL

According to NFPA-11 the ignitable liquids are classified into 3 different classes on the basis of flashpoint temperatures.

Class I- These are the flammable liquids which has flash points below 38°C or 100°F (for e.g., Diethyl Ether, Petroleum ether, p-xylene etc.)

Class II- These are the combustible liquids which has flash points between 38°C and 60°C (100°F–140°F) (for e.g., Diesel, petrol, cleaning solvents, etc.)

Class III- These are the combustible liquids which has flash points above 60°C (140°F) (combustible liquids) (for e.g., Oil based paints, mineral oil, linseed oil, etc.)

8. **Fire point**- It is the lowest temperature beyond which enough vapors are produced to sustain the fire without any external application of the fire source. It is always few degree above the flash point (Lilley, 1997).

Why Fire Point Is Important to Study With Regards to IL

Flash points and flame points are dependent on size of the ignition source, time it (ignition source) is exposed to source, rate of IL heating, degree of movement of air above the fuel.

9. **Ignition and autoignition temperature**- Ignition temperature is the temperature at which the ignitable or flammable liquids produce flammable vapors, only after reaching this temperature the fuel can sustain the fire. Whereas, the ignition temperature depends on various aspects like presence of oxygen, temperature surrounding to the fuel, partial pressure, particulate size, etc. (Basu, 2013).

Autoignition temperature is the temperature that is required to produce flames with any source of ignition. This type of ignition sometimes referred to as spontaneous ignition (Almirall and Furton, 2004).

10. **Minimum Ignition energy**- It is the lowest energy of the ignition source or a spark which is required at which the flame will generate on the fuel (Holleyhead, 1996).

Why Studying Minimum Ignition Energy Is Important to Study With Regards to IL

This energy is very low for the majority of ignitable fluids, and it is dependent on the vapors and air mixture (DeHann and Icove, 2012).

METHODS OF ANALYSIS OF IL'S FROM FIRE DEBRIS

Gas Chromatography-FID Methods

In the 1960's, the knowledge of GC/MS was applied for analysis of fire debris (Lucas, 1960). This helps in identification of various hydrocarbons that are used in the fire. The separation will occur irrespective of mobility of mobile phase (Bodle and Hardy, 2007).

GC/MS is a well-known technique used by forensic chemists in the investigation of arson and fire cases. By replacing every other detector with MS, the chances of interference are reduced. The detection of the IL's is dependent on the volatility of the substances, temperature, and absorptive characteristic of the substances or the fire accelerants. The evidence of IL from the scene of crime should be collected and sent for analysis quickly, as with the passage of the time the concentration of the IL might undergo environmental weathering and might vanishes soon. Once the samples are evaporated completely it is impossible to find out the accelerant or the IL used for setting up the fire (Lim et al., 2018).

Solid Phase Microextraction (PAWLISZYN et al., 1997): Polymeric organic phase is widely used as an extracting phase. Polymers such as dimethyl siloxane and polyacrylate are crosslinked and attached with the rods. These polymers protect optical fibers which are made up of fused silica (Fused silica is chemically inert). These fibres are very delicate to handle as they are fragile in nature. Hence, these fibres are incorporated into the syringe like handling device. While sampling, by pushing the plunger, the fibre comes out of the needle into the vial which will allow the analyte extraction on the fibre. This fibre is then taken inside by pulling the plunger up. This allows the transfer of an analyte into the GC without using any solvents. There are two types of sampling methods in SPME, the first one is a direct sampling, in this type the fibre is immersed in the sample. In this the liquid sample is agitated so that the analytes will get attached to the fibre. The direct method is used for the analysis for certain explosive mixtures for example an immediate debris collection from soil post blast into water-acetonitrile solution, as in one of the cases the PETN peak was observed (Furton et al., 2000). The second method is a head space method, in which the fibre is suspended above the sample and volatile analytes are allowed to extract onto the fibre. While analyzing the ILRs, this method is used widely.

These attached analytes are then desorbed thermally into the injector port of the GC for analysis (Pawliszyn, 1997). In a study conducted for determination of efficiency of SPME-GC analysis for IL analysis in fire debris, the gasoline and its components were determined on the basis of the elution time of each gasoline component. Initially the standard gasoline was run under GC to identify the target compounds like toluene, ethylbenzene, propyl benzene, 1,2,3-trimethylbenzene and 1,3,5- trimethylbenzene, which will confirm the presence of gasoline in the subsequent analysis. On the various controlled experiments, it was concluded that more IL on any ignitable substrates will give rise to good peak for identification with respect to the standard samples, the thickness of the substrate on which the fuel burning will also give readable results. But, if the samples from fire crime scene are collected after 5-6 hours the chances of getting the peaks for identification becomes challenging (Dhabbah et al., 2014).

SPME proved to be one of the high result yielding, sensitive technique than previously used ACS which was being used widely for the analysis of the substances in fire debris analysis (Furton et al., 1995).

GC-GC Analysis: As GC itself serves as a suitable instrumental technique to identify the ignitable liquids from the fire debris, it's another derivative i.e., GC-GC makes it even more sophisticated instrument that analyses a single sample twice in a system. This allows to detect a minor compositional difference in the ignitable liquids (Frysinger et al., 2002). This Type of chromatography is also known as 2-D gas chromatography. This type of chromatography also allows to track the changes in ignitable liquids which occurs during weathering.

GC-DMS: In this technique, the analytes of the ILs gets separated on the basis of its ionic mobilities under high and low electric fields. Apart from the charge, the ionic mobility also depends upon size and shape of the ions (Lu et al., 2009). Using DMS as a detector for GC reduces the cost of analysis as it is of lower cost, provides high selectivity, and faster analysis (Lu and Harrington, 2007).

GC-PID: This detector works on the principle that many volatile organic components can absorb photos which has equal energy of theirs or higher than ionization potential of their compounds. The organic compounds need photos at around 5–20 eV of far UV rays. In this, the monochromatic light of specific intensity or energy is emitted from the discharge lamp that contains an inert gas or a mixture of inert gas at lower pressure. The carrier gas passes through thermostatic chamber where ionization takes place and the electrodes applied to each sides takes the ions to develop a charge and hence the signal is amplified to detect the actual gaseous ions (Poole, 2005).

Detection of IL by UV: As discussed earlier, the most commonly method of determining IL debris from scene of crime is GC or GC-MS. However, since these are destructible techniques, other debris material such as pyrolysis products can cause interference when obtaining the chromatogram peaks. Aside from these, low boiling substances can evaporate before the samples are even collected (Evans-Nguyen, 2019). However, derivative spectroscopy can convert the analysis graph from zero order spectrum to higher order spectrum which leads to obtain distinction between two or more components that have similar lambda max (Lawrence, 1983).

This technique can be used for preliminary screening to rule out the availability of certain components (Bumbrah et al., 2018).

CONCLUSION

As per the various literatures, analysis of fire scene investigation is one of the most challenging, though not impossible, tasks. This challenge is made more difficult by the reconstruction of a fire scene and the determination of the point of origin of a fire. However, some methods of interacting with onlookers, careful external and internal examination of the fire scene, tracing the fire patterns like V patterns (if the vertical surface is near the point of origin of fire), U patterns (indicates the pool of liquid which could be formed of accidental or intentional spillage of the fuel), and so on may help to arrive at near-accurate conclusions or reasons for the fire. Searching for the protected areas, conditions of the objects especially metallic and glass objects. Careful observation of the window panes can also give hint about the fire propagation and enhancements. Apart from careful observation of the fire scene, an examiner should also notice and try to differentiate the entries made on the scene before (by any culprit, suspects or the residents of that building) and after (firefighters and fire investigators) the termination of the fire. An examiner should first check for any container or the source that could be responsible for the fire and collect all the evidences from the scene and should transport it to the fire laboratory with the proper packaging and chain of custody. In the laboratory, the analysis should be performed unbiasedly and with the proper and appropriate use of the instruments, as discusses in the main text. The interpretation of the results must be performed by an experienced individual who has dealt with these analyses before. These laboratory analyses may prove very useful in the construction of fire scene to determine the point of origin and development of the fire.

Hence, it is very important to conduct the systematic fire scene analysis because the opinion formed after the analysis has to stand the court trials and also important for the insurance claims by the victims or affected people.

REFERENCES

Almirall, J. R., & Furton, K. G. (2004). *Analysis and Interpretation of Fire Scene Evidence*. CRC Press.

Banwarth. (2015). *Fire Service Features of Buildings and Fire Protection Systems*. Academic Press.

Basu, P. (2013). Biomass Characteristics. In *Biomass Gasification, Pyrolysis and Torrefaction* (pp. 72–73). Elsevier Inc., doi:10.1016/B978-0-12-396488-5.00003-4

Bhimabhai, K. S. (2018). Comparative Study of Aliphatic and Aromatic Compounds. *The Pharma Innovation Journal, 7*(9), 175–77. www.chemguide.co.uk

Bodle, E. S., & Hardy, J. K. (2007). Multivariate Pattern Recognition of Petroleum-Based Accelerants by Solid-Phase Microextraction Gas Chromatography with Flame Ionization Detection. *Analytica Chimica Acta, 589*(2), 247–254. doi:10.1016/j.aca.2007.03.006 PMID:17418188

Bowman, D. M. J. S., Balch, J. K., Artaxo, P., Bond, W. J., Carlson, J. M., Cochrane, M. A., D'Antonio, C. M., DeFries, R. S., Doyle, J. C., Harrison, S. P., Johnston, F. H., Keeley, J. E., Krawchuk, M. A., Kull, C. A., Marston, J. B., Moritz, M. A., Prentice, I. C., Roos, C. I., Scott, A. C., ... Pyne, S. J. (2009). Fire in the Earth System. *Science, 324*(5926), 481–484. doi:10.1126cience.1163886 PMID:19390038

Bumbrah, G. S., Sarin, R. K., & Sharma, R. M. (2018). Derivative Ultraviolet Spectrophotometric Studies on Ignitable Liquids. *Journal of Forensic Chemistry and Toxicology, 4*(1).

David GossmanL. D. (n.d.). https://www.hgexperts.com/expert-witness-articles/fires-and-explosions-evidence-gathering-and-preservation-47077

DeHann, J. D., & Icove, D. J. (2012). *Kirk's Fire Investigation* (7th ed.). Julie Levin Alexander.

Dhabbah, A. M., Al-Jaber, S. S., Al-Ghamdi, A. H., & Aqel, A. (2014). Determination of Gasoline Residues on Carpets by SPME-GC-MS Technique. *Arabian Journal for Science and Engineering, 39*(9), 6749–6756. doi:10.100713369-014-1233-1

Evans-Nguyen, K. (2019). Forensic Analysis of Fire Debris and Explosives. In *An Introduction to Instrumentation Used in Fire Debris and Explosive Analysis*. Academic Press.

Evidence Collection. (n.d.). https://www.iaaievidenceguide.com/

Faramawy, S., Zaki, T., & Sakr, A. A. E. (2016). Natural Gas Origin, Composition, and Processing: A Review. *Journal of Natural Gas Science and Engineering, 34*, 34–54. doi:10.1016/j.jngse.2016.06.030

Fires Classification. (2019). http://global-intrima.com/media/news/FIRE_CLASSIFICATION.pdf

Furton, K. G. (2000). Application of Solid-Phase Microextraction to the Recovery of Explosives and Ignitable Liquid Residues from Forensic Specimens. *Journal of Chromatography A, 885*, 419–432. papers2://publication/uuid/F55C29AC-08CD-4F01-A104-6F9164B92A72

Furton, K. G., Bruna, J., & Almirall, J. R. (1995). A Simple, Inexpensive, Rapid, Sensitive and Solventless Technique for the Analysis of Accelerants in Fire Debris Based on SPME. *Journal of High Resolution Chromatography*, *18*(10), 625–629.

Gorbett, G. E., Hicks, W., Hopkins, R., & Kennedy, P. M. (2010). Fire Patterns Analysis with Low Heat Release Rate Initial Fuels. *International Symposium on Fire Investigation Science and Technology*: 269–94.

Gupta, G., Lalita Jawale, M., & Bhattacharya, B. (2015). Various Methods for the Determination of the Burning Rates of Solid Propellants - an Overview. *Central European Journal of Energetic Materials*, *12*(3), 593–620.

Hall, J. R. (2006). *Home Cooking Fire Patterns and Trends*. https://iflss.net/CookingFires-NFPA report2006.pdf

Hine, G. (2004). Fire scene investigation: an introduction for chemists. In J. Almirall & K. Furton (Eds.), *Analysis and interpretation of fire scene evidence* (pp. 71–72). CRC Press.

Holleyhead, R. (1996). Ignition of Flammable Gases and Liquids by Cigarettes: A Review. *Science and Justice - Journal - Forensic Science Society*, *36*(4), 257–266.

Howard, R., & Wilson, I. B. (1948). The effect of surface on melting point. *Journal of Colloid Science*, *3*(6), 551–561.

International Association of Arson Investigators. (1999). *A pocket guide to accelerant evidence collection (Massachusetts chapter)* (2nd ed.). IAAI.

Lawrence, A. H. (1983). Chromatography Discussion Group Offers Reduced Subscriptions. *Trends in Analytical Chemistry*, *2*(12), V.

Lentini, J. J. (2013). Evidence Collection at Fire Scenes. *Encyclopedia of Forensic Sciences: Second Edition, 3*(December), 387–91. doi:10.1016/B978-0-12-382165-2.00211-7

Lilley, D. G. (1996). Fire Causes and Ignition. *Proceedings of the Intersociety Energy Conversion Engineering Conference, 1*, 85–90.

Lilley, D. G. (1997). Structural Fire Development. *35th Aerospace Sciences Meeting and Exhibit*. https://arc.aiaa.org/doi/abs/10.2514/6.1997-265

Lim, P. W., Ahmad, F. L. A., & Chang, K. H. (2018). Forensic Gas Chromatography Analysis of Time Elapsed Gasoline in Fire Scene Investigation. *The Malaysian Journal of Analytical Sciences*, *22*(1), 72–79.

Lu, Y., Chen, P., & Harrington, P. B. (2009). Comparison of Differential Mobility Spectrometry and Mass Spectrometry for Gas Chromatographic Detection of Ignitable Liquids from Fire Debris Using Projected Difference Resolution. *Analytical and Bioanalytical Chemistry*, *394*(8), 2061–2067.

Lu, Y., & Harrington, P. B. (2007). Forensic Application of Gas Chromatography-Differential Mobility Spectrometry with Two-Way Classification of Ignitable Liquids from Fire Debris. *Analytical Chemistry*, *79*(17), 6752–6759.

Ludwig, E. E. (Ed.). (1999). Process Safety and Pressure-Relieving Devices. In *Applied Process Design for Chemical and Petrochemical Plants*. Gulf Professional Publishing. https://www.sciencedirect.com/science/article/pii/S1874863599800089

Malmgren, A., & Riley, G. (2012). 5 Comprehensive Renewable Energy. In *Biomass Power Generation*. Elsevier Ltd. doi:10.1016/B978-0-08-087872-0.00505-9

NFPA 921- Guide for Fire and Explosion Investigations, 2017 Edition, Chapter 17.

NIJ (National Institute of Justice). (2009). A Guide for Investigating Fire and Arson | National Institute of Justice. *Office of Justice Programs*. https://www.nij.gov/topics/law-enforcement/investigations/crime-scene/guides/fire-arson/pages/welcome.aspx

Pawliszyn & Pawliszyn. (1997). Solid-Phase Microextraction. The Chemical Educator, 1–7.

Poole, C. F. (2005). Gas Chromatography - Detectors. Encyclopedia of Analytical Science: Second Edition, 95–105.

Redsicker, D. R., & O'Connor, J. J. (1996). *Practical Fire and Arson Investigation (2nd ed.)*. CRC Press.

Reno, J., Marcus, D., Leary, M. L., & Samuels, J. E. (2000). *Fire and Arson Scene Evidence: A Guide for Public Safety Personnel*. https://www.crime-scene-investigator.net/fireandarsonsceneevidenceNIJ.pdf

Schmidt-Rohr, K. (2015). Why Combustions Are Always Exothermic, Yielding about 418 KJ per Mole of O2. *Journal of Chemical Education*, *92*(12), 2094–2099.

Sherman. (1933). *Radiation From Luminous and Non-Luminous Natural Gas Flames.* The American Society of Mechanical Engineers.

Siegel, J., Knupfer, G., & Saukko, P. (2000). Methods. *Encyclopedia of Forensic Science*.

Simeoni, A., Reszka P., Colella F., & Torero J.L. (2011). *Flaming Ignition of Solid Fuels*. Academic Press.

Stauffer, E., Dolan, J. A., & Newman, R. (2008). Chemistry and Physics of Fire and Liquid Fuels. In *Fire Debris Analysis* (pp. 85–129). Elsevier.

Williams, M. R., & Sigman, M. (2007, May). Performance testing of commercial containers for collection and storage of fire debris evidence. *Journal of Forensic Sciences*, *52*(3), 579–585.

Chapter 7
Large Feature Mining With Ensemble Learning for Image Forgery Detection

Qingzhong Liu
Sam Houston State University, USA

Tze-Li Hsu
Sam Houston State University, USA

ABSTRACT

The detection of different types of forgery manipulation including seam-carving in JPEG images is a hot spot in image forensics. Seam carving was originally designed for content-aware image resizing. It is also being used for forgery manipulation. It is still very challenging to effectively identify the seam carving forgery under recompression. To address the highly challenging detection problems, this chapter introduces an effective approach with large feature mining. Ensemble learning is used to deal with the high dimensionality and to avoid overfitting that may occur with some traditional learning classifier for the detection. The experimental results validate the efficacy of proposed approach to detecting JPEG double compression and exposing the seam-carving forgery while the JPEG recompression is proceeded at the same quality and a lower quality, which is generally much harder for traditional detection methods. The methodology introduced in this chapter provides a strategy and realistic approach to resolve the highly challenging problems in image forensics.

DOI: 10.4018/978-1-7998-8386-9.ch007

INTRODUCTION

JPEG is a commonly used method of lossy compression for digital images. The detection of the forgery manipulations in JPEG images has been well studied (Amerini et al. 2011; Bianchi and Piva 2012; Liu and Chen 2015; Pun, Yuan and Bi 2015). For example, double JPEG compression may be involved on JPEG image tampering, the detection of double JPEG compression was actively conducted (Chen and Hsu 2011). While the detection is very effective if the second compression quality is higher than the first compression quality, it is not so strong if the second compression quality is lower than the first compression quality. A forgery maker may exploit the weakness of the current detection arts, manipulate the images, and produce the forgery at the same or at a lower quality level, to evade from being detected.

Seam carving, known as liquid rescaling, was originally designed for content-aware image resizing (Avidan and Shamir 2007). It establishes several seams (paths of least importance) in an image and then automatically removes seams to reduce image size or inserts seams to extend it. Seam carving allows manually defining areas in which pixels may not be modified and features the ability to remove whole objects from photographs. Seam carving has been implemented in Adobe Photoshop and other popular computer graphic applications including GIMP, digiKam, ImageMagic, and iResizer. While it is widely used in computer vision and multimedia processing for legitimate applications, it has been adopted for forgery manipulation. The proliferation of seam carving raises a serious challenge in image forensics.

In detecting seam carving–based image forgery, Sarkar et al. (2009) utilized a steganalysis detector, which was originally developed to detect JPEG-based steganograms. Fillion and Sharma (2010) detected seam-carved images including benign image reduction, benign image enlargement, and deliberate image reduction. They tested their method over a set of images consisting of 1484 uncompressed images. Unfortunately, the approach was not validated in JPEG images with seam carving manipulation. Liu and Chen (2015) made use of neighboring joint density that was designed in steganalysis to detect seam carving-based forgery in JPEG images. Chang et al. (2013) proposed a method to detect seam carving in JPEG images based on the symmetrical property of blocking artifact characteristics matrix (BCAM) and the extension. Wei et al. (2014) developed a patch analysis method to detect seam carving. Ryu and Lee (2014) discriminated seam carving from intact by energy bias and noise features. Wattanachote et al. (2015) presented a method to detect the JPEG seam-carving image based on the extended Markov feature. Yin et al. (2015) exposed seam carving forgery by combining half-seam features, energy bias and noise-based feature together from the local binary pattern. Sheng and Gao (2016) introduced a method based on Benford law. Zhang et al. (2017) implemented a blind detection approach for seam carved image with low scaling ratio by exploiting spatial and spectral entropies on multi-scale images. It has been observed that when a few seams are deleted from an original image, its SSE distribution is greatly changed. Taspinar et al. (2015) proposed to use the camera PRNU to detect the seam-carving tampered image. Han et al. (2018) exposed the location of removed objects by using 10-dimensional block artifact grid mispairing characteristic features. Although many methods have been proposed so far, most are only effective in exposing the seam carving while the seam-carved images are saved in uncompressed format. Especially if an original image was compressed in JPEG format and manipulated image after seam carving is saved at uncompressed format, the detection is much easier. While the doctored images are recompressed in JPEG format, especially on the same or a lower image quality, most seam-carving forgery detection methods are ineffective since the re-compression after seam carving generally compromises the existing forgery traces.

This chapter will introduce an effective approach to detecting seam-carving forgery in JPEG images by presenting a new set of large feature mining to which ensemble learning is applied for the model learning and detection (Liu 2019), which may shed light on solving other highly challenging problems in image forensics.

BACKGROUND

JPEG Re-Compression

In JPEG compression, it shows that the marginal density of DCT coefficients at each frequency approximately follows the generalized Gaussian distribution (GGD) (Sharifi and Leon-Garcia 1995), given by

$$\rho\left(x;\alpha,\beta\right)=\frac{\beta}{2\alpha\Gamma\left(1/\beta\right)}\exp\left\{-\left(\left|x\right|/\alpha\right)^{\beta}\right\} \tag{1}$$

Where $\Gamma(\cdot)$ is the Gamma function, scale parameter α models the width of the PDF peak, and shape parameter β models the shape of the distribution.

JPEG double compression may change the marginal density and the modification depends on the first compression quality and the second compression quality. Fig. 1 demonstrates a singly compressed JPEG image with quality '75' (a), doubly compressed JPEG images of the first compression quality '55' (b) and '90' (c) respectively, followed by the second compression quality factor '75' at the first row. The marginal densities are plotted on the frequency coordinate (1,0) at the second row, on the frequency coordinate (1,1) at the third row, and on the frequency coordinate

Figure 1. Marginal densities of a single compression (left) and double compressions (middle and right). X-axis and y-axis give the values of the DCT coefficients and the occurrences.
(Liu 2017)

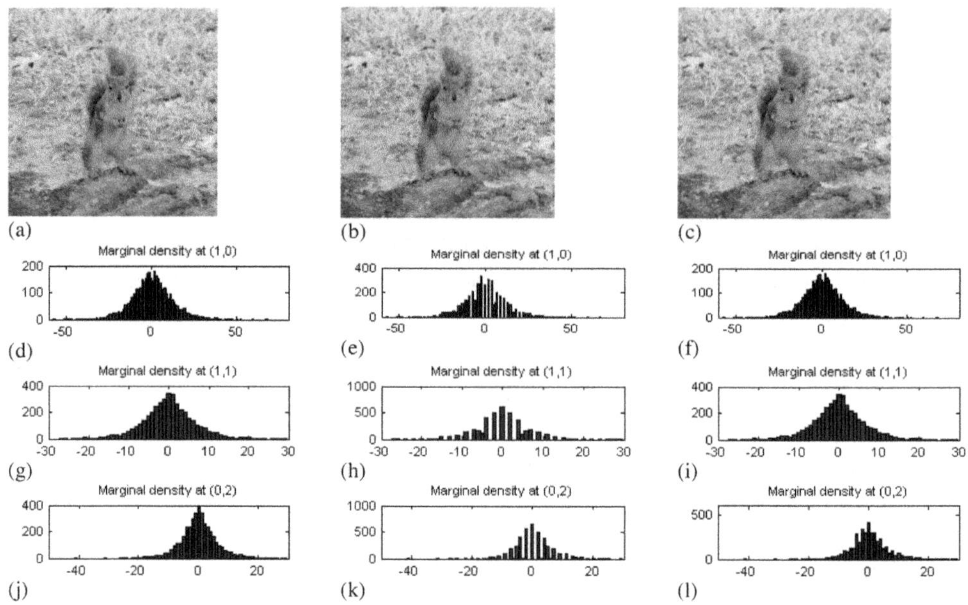

(0,2) at the fourth row. The marginal densities on the double compression from a lower first quality factor '55', shown in Fig. 1(e), (h), and (k), clearly demonstrate the periodical gaps. However, the marginal density on the double compression from a higher first quality factor '90', as shown by Fig. 1(f), (i), and (l), is not such noticeable.

In the detection of JPEG double compression, while the first compression quality is lower than the second compression quality, most detection methods are very effective. However, these methods are not well while the first compression quality is higher than the second compression quality. That is, the detection of JPEG down-recompression are not effective, especially while the first compression quality is considerably higher than the second compression quality (Chen and Hsu 2011).

To detect double JPEG compression with the same quantization matrix, Huang et al. (2010) designed a method based on an observation, that is, in the process of recompressing a JPEG image with the same quantization matrix repeatedly, the number of different JPEG coefficients between the two consecutive versions will monotonically decrease in general. There are, however, serious flaws with the method. Firstly, the observation is far from being always true. As indicated by Liu (2017), there are about 20% JPEG images without any modification of the quantized DCT coefficients between the first compression and the second compression by using the same quantization matrix. In other words, the foundation of the detection method designed in (Huang et al. 2010) may not be statistically sound. To resolve the issue, a method based on shift-recompression was developed to detect the misaligned forgery on the same quantization in JPEG images (Liu 2011). The method significantly improved the detection accuracy by about 32% to 45% (Liu 2011). However, the method was not examined for the detection of JPEG down-recompression (Liu 2011).

Considering the weakness of existing image forgery methods and powerless in detecting JPEG down-recompression, a forgery producer may manipulate a JPEG image in spatial domain, and then save the doctored JPEG image at a lower image quality. In such a case, the image tampering followed by JPEG down-recompression becomes very hard from being detected.

Introduction to Seam Carving

In seam carving, to achieve content-aware scaling, unnoticeable pixels on the least important seams that blend with their surroundings are removed or inserted. Formally, let **I** be a $p \times q$ image and a vertical seam is defined by:

$$S^X = \left\{ s_i^x \right\}_{i=1}^{p} = \left\{ \left(x\left(i\right), i \right) \right\}_{i=1}^{p}, \tag{2}$$

s.t. $\forall i, \left| x\left(i\right) - x\left(i-1\right) \right| \leq 1$

where x is a mapping $x : \left[1, \ldots, p\right] \rightarrow \left[1, \ldots, q\right]$.

Similarly, a horizontal seam is defined by:

$$S^Y = \left\{ s_j^y \right\}_{j=1}^{q} = \left\{ \left(j, y\left(j\right) \right) \right\}_{j=1}^{q}, \tag{3}$$

s.t. $\forall j, \left| y\left(j\right) - y\left(j-1\right) \right| \leq 1$

where y is a mapping $y : \left[1, \ldots, q\right] \rightarrow \left[1, \ldots, p\right]$.

The pixels of the path of seam S is denoted as $\boldsymbol{I}_s$. Given an energy function e, the cost of a seam is calculated by

$$E(S) = E(I_S) = \sum_{i=1}^{p} e(I(S_i)).$$

The optimal seam S^* generally minimizes the seam cost:

$$S^* = \min_S E(S) = \min_S \sum_{i=1}^{p} e(I(S_i)) \tag{4}$$

In the reference (Avidan and Shamir 2007), several image importance measures are examined as the energy function. Although no single energy function performs well across all images but in general the following two measures e_1 and e_{HoG} work quite well.

$$e_1(I) = \left|\frac{\partial}{\partial x} I\right| + \left|\frac{\partial}{\partial y} I\right| \tag{5}$$

$$e_{HoG}(I) = \frac{\left|\frac{\partial}{\partial x} I\right| + \left|\frac{\partial}{\partial y} I\right|}{\max\left(HoG\left(I(x,y)\right)\right)} \tag{6}$$

Where $HoG\left(I(x,y)\right)$ is a histogram of oriented gradients at each pixel.

While seam carving has been widely used for content aware image resizing, it is also used for tampering illusion such as removing or inserting some objects. Figure 2 shows one untouched image on the left and doctored image by seam carving on the right.

LARGE-FEATURE MINING-BASED APPROACH

Problem and Analysis

As discussed by Liu (2017), after seam carving, it is very easy for a forgery maker to save manipulated images in JPEG format. Given original JPEG image quality

Figure 2. An example of image tampering by seam carving (Liu 2017)

QF1 (before tampering) and the manipulated image quality QF2 (after seam carving in spatial domain and then saving in JPEG format), while the quality QF2 is lower than QF1, the following low-quality JPEG recompression may significantly destroy or compromise the traces left by seam carving. In image forensics, while the manipulated images are saved at a lower-level JPEG quality, the detection of such a forgery is very difficult.

In the circumstance that JPEG compression is performed to a never compressed image, the DCT coefficients before quantization and after quantization are denoted pc_{mn} and c_{mn} respectively at the frequency coordinate (m, n) (m=0, 1…7; n=0, 1…7), the quantization step is denoted Q_{mn}, and $R(\bullet)$ denotes the round function.

$$c_{mn} = R\left(\frac{pc_{mn}}{Q_{mn}}\right) \tag{7}$$

then

$$-0.5 \leq \frac{pc_{mn}}{Q_{mn}} - R\left(\frac{pc_{mn}}{Q_{mn}}\right) < 0.5, \text{ or } \left(c_{mn} - 0.5\right)Q_{mn} \leq pc_{mn} < \left(c_{mn} + 0.5\right)Q_{mn}$$

As for down-recompression, the quantization steps at the first and the second compression are denoted by $Q_{mn}^{(1)}$ and $Q_{mn}^{(2)}$ ($Q_{mn}^{(1)} < Q_{mn}^{(2)}$, $Q_{mn}^{(2)} Q_{mn}^{(2)} = Q_{mn}$). Without losing a generality, suppose c_{mn} is positive, the quantized DCT coefficient after down-recompression is denoted by $c_{mn}^{(2)}$, here the truncating error is negligible,

$$c_{mn}^{(2)} = R\left(R\left(\frac{pc_{mn}}{Q_{mn}^{(1)}}\right) \bullet \frac{Q_{mn}^{(1)}}{Q_{mn}^{(2)}}\right) \tag{8}$$

Let

$$c_{mn}^{(1)} = R\left(\frac{pc_{mn}}{Q_{mn}^{(1)}}\right) \tag{9}$$

Then,

$$c_{mn}^{(2)} = R\left(\frac{c_{mn}^{(1)} Q_{mn}^{(1)}}{Q_{mn}^{(2)}}\right) \tag{10}$$

$$c_{mn}^{(2)} = \begin{cases} c_{mn} - 1, & c_{mn}^{(1)} Q_{mn}^{(1)} < \left(c_{mn} - 0.5\right) Q_{mn}^{(2)} \leq pc_{mn} < \left(c_{mn}^{(1)} + 0.5\right) Q_{mn}^{(1)} \\ c_{mn}, & \left(c_{mn} - 0.5\right) Q_{mn}^{(2)} \leq c_{mn}^{(1)} Q_{mn}^{(1)} < \left(c_{mn} + 0.5\right) Q_{mn}^{(2)} \\ c_{mn} + 1, & \left(c_{mn}^{(1)} - 0.5\right) Q_{mn}^{(1)} \leq pc_{mn} < \left(c_{mn} + 0.5\right) Q_{mn}^{(2)} \leq c_{mn}^{(1)} Q_{mn}^{(1)} \end{cases} \tag{11}$$

As indicated by equation (11), compared to the single compression, the JPEG down-recompression may change the value of the DCT coefficient by 1, although the modification ratio is small. As a result, the marginal density of the DCT coefficients at each individual frequency and the neighboring joint density will be modified.

Generally, forgery manipulation operations including seam carving change the pixel values of original image and change the original DCT coefficients. As a result, seam carving also modifies the individual frequency-based marginal density and joint density.

Large Feature Mining

Directional Derivative-Based Joint Density Features

Image derivatives play very important roles in image analysis and computer vision. Image filters especially derivative filters can be used for the detection of discontinuities in an image, and they play a central role in sharpening an image (i.e., enhancing fine detail). The extraction of our first type of features is mostly based on directional derivative filters. An examination image is denoted by I and a directional derivative impulse response is denoted by H_i, after filtering, the output F_i is the convolution of I with the impulse response H_i. The first and higher order derivatives are given by

$$F_i^0 = I \tag{12}$$

$$F_i^j = F_i^{j-1} * H_i \ (\text{i=1, 2..., 14; j=1, 2, 3}) \tag{13}$$

In equation (13), H_i is set to the following directional derivative impulses:

$$h_1=\begin{bmatrix}1 & -1\end{bmatrix};\ h_2=\begin{bmatrix}1\\-1\end{bmatrix};\ h_3=\begin{bmatrix}1 & 0\\0 & -1\end{bmatrix};\ h_4=\begin{bmatrix}0 & 1\\-1 & 0\end{bmatrix};$$

$$h_5=\begin{bmatrix}1 & 0 & 0\\0 & 0 & -1\end{bmatrix};\ h_6=\begin{bmatrix}1 & 0\\0 & 0\\0 & -1\end{bmatrix};\ h_7=\begin{bmatrix}0 & 0 & 1\\-1 & 0 & 0\end{bmatrix};\ h_8=\begin{bmatrix}0 & 1\\0 & 0\\-1 & 0\end{bmatrix};$$

$$h_9=\begin{bmatrix}1 & 0 & 0 & 0\\0 & 0 & 0 & -1\end{bmatrix};\ h_{10}=\begin{bmatrix}1 & 0\\0 & 0\\0 & 0\\0 & -1\end{bmatrix};\ h_{11}=\begin{bmatrix}0 & 0 & 0 & 1\\-1 & 0 & 0 & 0\end{bmatrix};$$

$$h_{12}=\begin{bmatrix}0 & 1\\0 & 0\\0 & 0\\-1 & 0\end{bmatrix};\ h_{13}=\begin{bmatrix}0 & 1 & 0\\1 & -4 & 1\\0 & 1 & 0\end{bmatrix}\ \text{and}\ h_{14}=\begin{bmatrix}1 & 1 & 1\\1 & -8 & 1\\1 & 1 & 1\end{bmatrix}$$

Let m×n be the size of F_i^j, the following sub-arrays $S_d\left(F_i^j\right)$ (d=0, 1 ...20) with regard to F_i^j are given by

$$S_0\left(F_i^j\right) = F_i^j\left(1:m-4;4:n-4\right);S_1\left(F_i^j\right) = F_i^j\left(1:m-4;5:n-3\right);$$

$$S_2\left(F_i^j\right)=F_i^j\left(1:m-4;6:n-2\right);S_3\left(F_i^j\right)=F_i^j\left(1:m-4;7:n-1\right);$$

$$S_4\left(F_i^j\right)=F_i^j\left(1:m-4;8:n\right);S_5\left(F_i^j\right)=F_i^j\left(2:m-3;1:n-7\right);$$

$$S_6\left(F_i^j\right)=F_i^j\left(2:m-3;2:n-6\right);S_7\left(F_i^j\right)=F_i^j\left(2:m-3;3:n-5\right);$$

$$S_8\left(F_i^j\right)=F_i^j\left(2:m-3;4:n-4\right);S_9\left(F_i^j\right)=F_i^j\left(2:m-3;5:n-3\right);$$

$$S_{10}\left(F_i^j\right)=F_i^j\left(2:m-3;6:n-2\right);S_{11}\left(F_i^j\right)=F_i^j\left(2:m-3;7:n-1\right);$$

$$S_{12}\left(F_i^j\right)=F_i^j\left(3:m-2;2:n-6\right);S_{13}\left(F_i^j\right)=F_i^j\left(3:m-2;3:n-5\right);$$

$$S_{14}\left(F_i^j\right)=F_i^j\left(3:m-2;4:n-4\right);S_{15}\left(F_i^j\right)=F_i^j\left(3:m-2;5:n-3\right);$$

$$S_{16}\left(F_i^j\right)=F_i^j\left(3:m-2;6:n-2\right);S_{17}\left(F_i^j\right)=F_i^j\left(4:m-1;3:n-5\right);$$

$$S_{18}\left(F_i^j\right)=F_i^j\left(4:m-1;4:n-4\right);S_{19}\left(F_i^j\right)=F_i^j\left(4:m-1;5:n-3\right);$$

$$S_{20}\left(F_i^j\right)=F_i^j\left(5:m;4:n-4\right); \tag{14}$$

Thereafter $S_d\left(F_i^j\right)$ is denoted by $S_{i,d}^j$. The joint density between $S_0\left(F_i^j\right)$, denoted by $S_{i,0}^j$, and $S_0\left(F_i^{j+1}\right)$, denoted by $S_{i,0}^{j+1}$, is calculated by

$$J_i^j\left(k,l\right)=\frac{\sum_{(x,y)}\delta\left(\left|S_{i,0}^j\left(x,y\right)\right|=k,\left|S_{i,0}^{j+1}\left(x,y\right)\right|=l\right)}{\left(m-4\right)\bullet\left(n-7\right)} \tag{15}$$

The joint density between $S_{i,0}^j$ and $S_{i,d}^j$ is calculated by

$$J^{j}_{i,d}\left(k,l\right)=\frac{\sum_{(x,y)}\delta\left(\left|S^{j}_{i,0}\left(x,y\right)\right|=k,\left|S^{j}_{i,d}\left(x,y\right)\right|=l\right)}{2\bullet\left(m-4\right)\bullet\left(n-7\right)}+\frac{\sum_{(x,y)}\delta\left(\left|S^{j}_{i,0}\left(x,y\right)\right|=l,\left|S^{j}_{i,d}\left(x,y\right)\right|=k\right)}{2\bullet\left(m-4\right)\bullet\left(n-7\right)} \tag{16}$$

In equations (15) and (16), δ=1 if its arguments are satisfied, otherwise δ=0 (the same in the following equations if it appears). The features defined in equations (15) and (16) are grouped as type I.

Additionally, the following sub-arrays are obtained.

The following joint density features are calculated by

$$J^{j}_{i,ab}\left(k,l\right)=\frac{\sum_{(x,y)}\delta\left(\left|S^{j}_{i,a}\left(x,y\right)\right|=k,\left|S^{j}_{i,b}\left(x,y\right)\right|=l\right)}{2\bullet\left(m-8\right)\bullet\left(n-8\right)}+\frac{\sum_{(x,y)}\delta\left(\left|S^{j}_{i,a}\left(x,y\right)\right|=l,\left|S^{j}_{i,b}\left(x,y\right)\right|=k\right)}{2\bullet\left(m-8\right)\bullet\left(n-8\right)} \tag{17}$$

$$J^{j}_{i,ac}\left(k,l\right)=\frac{\sum_{(x,y)}\delta\left(\left|S^{j}_{i,a}\left(x,y\right)\right|=k,\left|S^{j}_{i,c}\left(x,y\right)\right|=l\right)}{2\bullet\left(m-8\right)\bullet\left(n-8\right)}+\frac{\sum_{(x,y)}\delta\left(\left|S^{j}_{i,a}\left(x,y\right)\right|=l,\left|S^{j}_{i,c}\left(x,y\right)\right|=k\right)}{2\bullet\left(m-8\right)\bullet\left(n-8\right)} \tag{18}$$

$$J^{j}_{i,ad}\left(k,l\right)=\frac{\sum_{(x,y)}\delta\left(\left|S^{j}_{i,a}\left(x,y\right)\right|=k,\left|S^{j}_{i,d}\left(x,y\right)\right|=l\right)}{2\bullet\left(m-8\right)\bullet\left(n-8\right)}+\frac{\sum_{(x,y)}\delta\left(\left|S^{j}_{i,a}\left(x,y\right)\right|=l,\left|S^{j}_{i,d}\left(x,y\right)\right|=k\right)}{2\bullet\left(m-8\right)\bullet\left(n-8\right)} \tag{19}$$

$$J^{j}_{i,bd}\left(k,l\right)=\frac{\sum_{(x,y)}\delta\left(\left|S^{j}_{i,b}\left(x,y\right)\right|=k,\left|S^{j}_{i,d}\left(x,y\right)\right|=l\right)}{2\bullet\left(m-8\right)\bullet\left(n-8\right)}+\frac{\sum_{(x,y)}\delta\left(\left|S^{j}_{i,b}\left(x,y\right)\right|=l,\left|S^{j}_{i,d}\left(x,y\right)\right|=k\right)}{2\bullet\left(m-8\right)\bullet\left(n-8\right)} \tag{20}$$

The features in equations (17) to (20) are grouped as type II.

Minimal Neighboring Difference-based Joint Density Features

We firstly apply image filtering to the examination image with the directional first derivative impulse response H_i, the output F_i is the convolution of I with the impulse response H_i, given by:

$$D_i^1 = I * H_i^1 \text{(i=1, 2..., 8)} \tag{21}$$

$$H_1^1 = \begin{bmatrix} 1 & 0 & 0 \\ 0 & -1 & 0 \\ 0 & 0 & 0 \end{bmatrix};\ H_2^1 = \begin{bmatrix} 0 & 1 & 0 \\ 0 & -1 & 0 \\ 0 & 0 & 0 \end{bmatrix};\ H_3^1 = \begin{bmatrix} 0 & 0 & 1 \\ 0 & -1 & 0 \\ 0 & 0 & 0 \end{bmatrix};\ H_4^1 = \begin{bmatrix} 0 & 0 & 0 \\ 0 & -1 & 1 \\ 0 & 0 & 0 \end{bmatrix};$$

$$H_5^1 = \begin{bmatrix} 0 & 0 & 0 \\ 0 & -1 & 0 \\ 0 & 0 & 1 \end{bmatrix};\ H_6^1 = \begin{bmatrix} 0 & 0 & 0 \\ 0 & -1 & 0 \\ 0 & 1 & 0 \end{bmatrix};\ H_7^1 = \begin{bmatrix} 0 & 0 & 0 \\ 0 & -1 & 0 \\ 1 & 0 & 0 \end{bmatrix};\ H_8^1 = \begin{bmatrix} 0 & 0 & 0 \\ 1 & -1 & 0 \\ 0 & 0 & 0 \end{bmatrix}.$$

The minimal neighboring difference arrays are given by:

$$\min D_1\left(x,y\right) = \min\left\{\left|D_1^1\left(x,y\right)\right|,\ \left|D_2^1\left(x,y\right)\right|,\ \left|D_3^1\left(x,y\right)\right|\right\} \tag{22}$$

$$\min D_2\left(x,y\right) = \min\left\{\left|D_3^1\left(x,y\right)\right|,\ \left|D_4^1\left(x,y\right)\right|,\ \left|D_5^1\left(x,y\right)\right|\right\} \tag{23}$$

$$\min D_3\left(x,y\right) = \min\left\{\left|D_5^1\left(x,y\right)\right|,\ \left|D_6^1\left(x,y\right)\right|,\ \left|D_7^1\left(x,y\right)\right|\right\} \tag{24}$$

$$\min D_4\left(x,y\right) = \min\left\{\left|D_7^1\left(x,y\right)\right|,\ \left|D_8^1\left(x,y\right)\right|,\ \left|D_1^1\left(x,y\right)\right|\right\} \tag{25}$$

Let m×n be the size of $minD_i$, the joint density features are calculated by

$$\min\_NJ_1(k,l)=\frac{\sum_{(x,y)}\delta\left(\min D_1(x,y)=k,\min D_3(x,y)=l\right)}{m\bullet n}\qquad(26)$$

$$\min\_NJ_2(k,l)=\frac{\sum_{(x,y)}\delta\left(\min D_4(x,y)=k,\min D_2(x,y)=l\right)}{m\bullet n}\qquad(27)$$

the sub-arrays $S_d(\min D_i)$ is denoted by $S_{i,d}^{\min D}$ (i=1, 2, 3, 4 and d=0, 1 ...20) with regard to $minD_1$, $minD_2$, $minD_3$, and $minD_4$, respectively, the joint density between $S_{i,0}^{\min D}$ and $S_{i,d}^{\min D}$ (d=1, 2…20) is given by

$$J_{i,d}^{\min D}(k,l)=\frac{\sum_{(x,y)}\delta\left(\left|S_{i,0}^{\min D}(x,y)\right|=k,\left|S_{i,d}^{\min D}(x,y)\right|=l\right)}{2\bullet(m-4)\bullet(n-7)}+\frac{\sum_{(x,y)}\delta\left(\left|S_{i,0}^{\min D}(x,y)\right|=l,\left|S_{i,d}^{\min D}(x,y)\right|=k\right)}{2\bullet(m-4)\bullet(n-7)}\qquad(28)$$

In equations (28), δ=1 if its arguments are satisfied, otherwise δ=0; *i* is set to 1, 2, 3, and 4; and *d* is set to 1 to 20 respectively. The features in equations (26) to (28) are grouped into type III.

High Order Joint Density Features on DCT Coefficients

JPEG image compression truncates the DCT coefficients at high frequency sub-bands to zero values during the coefficient quantization stage, and the non-zero DCT coefficients are concentrated at the low-frequency sub-bands. The quantized DCT coefficients are reordered in the zig-zag sequence.

We assume that the quantized DCT coefficient array of the image under examination contains $B_1 \times B_2$ blocks. The (*p*,*q*)-th block ($p = 1, 2, \ldots, B_1$; $q = 1, 2, \ldots, B_2$) consists of 8×8 quantized DCT coefficients. The coefficient at the ith (i=0,1, …63) zig-zag sequence order in the (*p*,*q*) block is denoted by $c_i(p,q)$. By making use of the index of the zig-zag sequence order, we design the following joint density features. Fig. 3 shows the coefficient positions.

Figure 3. Zig-zag index values in equations (29) and (30)

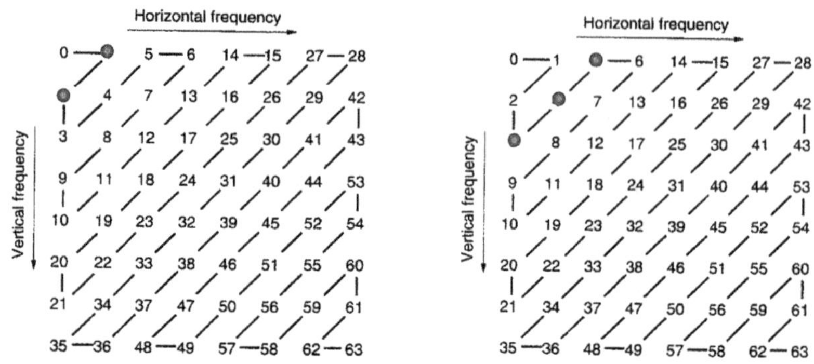

$$P_1(k_1,k_2)=\frac{\sum_{p=1}^{B_1}\sum_{q=1}^{B_2}\delta\left(\left|c_1(p,q)\right|=k_1;\left|c_2(p,q)\right|=k_2\right)}{B_1\bullet B_2} \quad (29)$$

$$P_2(k_1,k_2,k_3)=\frac{\sum_{p=1}^{B_1}\sum_{q=1}^{B_2}\delta\left(\left|c_3(p,q)\right|=k_1;\left|c_4(p,q)\right|=k_2;\left|c_5(p,q)\right|=k_3\right)}{B_1\bullet B_2} \quad (30)$$

The intra-block joint density features are given below.

$$P_{z_1z_2z_3}(k_1,k_2,k_3)=\frac{\sum_{p=1}^{B_1}\sum_{q=1}^{B_2}\delta\left(\left|c_{z_1}(p,q)\right|=k_1;\left|c_{z_2}(p,q)\right|=k_2;\left|c_{z_3}(p,q)\right|=k_3\right)}{B_1\bullet B_2} \quad (31)$$

In equation (31), the values of (z_1, z_2, z_3) are set to the following 41 sets: (1,4,8), (1,4,5), (1,5,6), (1,6,8), (2,3,9), (2,3,4), (2,4,7), (2,9,7), (3,9,10), (3,9,8), (3,8,12), (3,10,12), (4,8,11), (4,8,7), (4,7,13), (4,11,13), (5,7,12), (5,7,6), (5,6,14), (5,12,14), (6,13,17), (6,13,14), (6,14,15), (6,17,15), (7,12,18), (7,12,13), (7,13,16), (7,18,16), (8,11,19), (8,11,12), (8,12,17), (8,19,17), (9,10,20), (9,10,11), (9,11,18), (9,20,18), (10,20,19), (11,19,18), (12,18,17), (13,17,16), and (14,16,15).

Fig. 4 shows the four examples of zig-zig index sets: (1,4,5), (5,12,14), (1,4,8) and (4,11,13).

The features in (29), (30) and (31) are grouped type IV.

Additionally, the following joint density features on the inter-blocks are extracted, grouped into type V.

Figure 4. Four examples of zig-zag index values in equations (31)

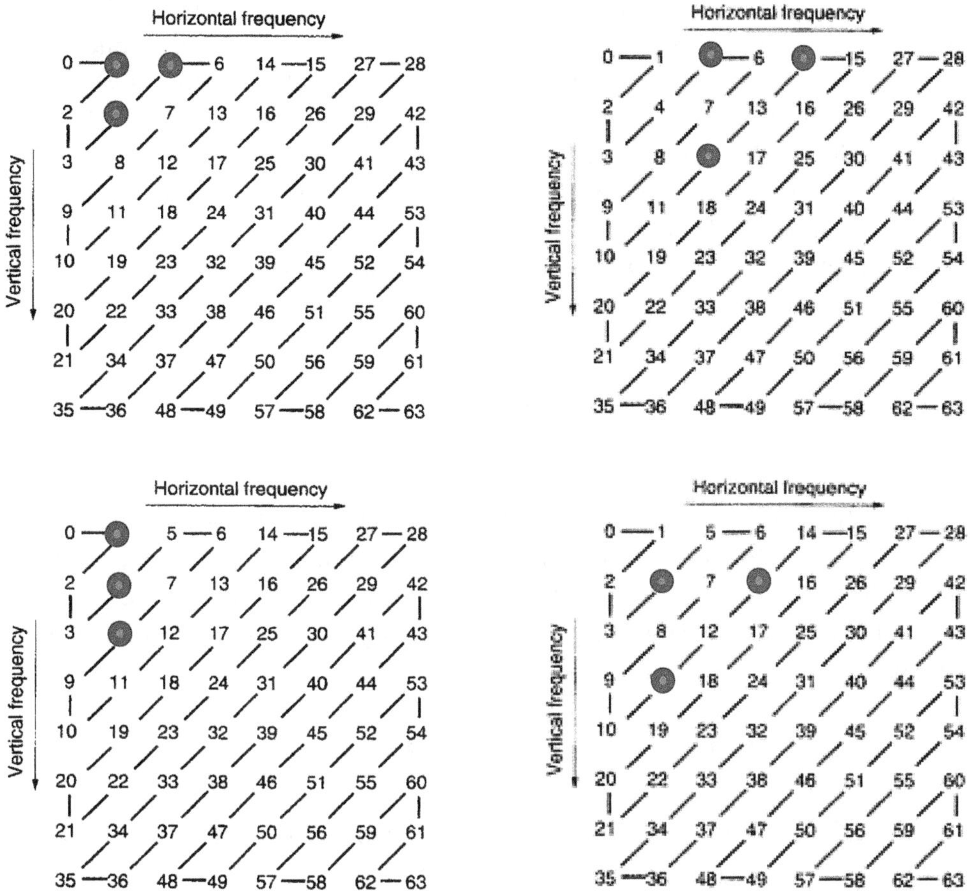

$$P_z\left(k_1,k_2,k_3\right)=\frac{\sum_{p=1}^{B_1-1}\sum_{q=1}^{B_2-1}\delta\left(\left|c_z\left(p,q\right)\right|=k_1;\left|c_z\left(p+1,q\right)\right|=k_2;\left|c_z\left(p,q+1\right)\right|=k_3\right)}{\left(B_{1-1}\right)\bullet\left(B_{2-1}\right)} \tag{32}$$

Gabor Filtering-Based Feature Design

In image analysis, Gabor filters have received considerable attention because the characteristics of certain cells in the visual cortex of some mammals can be approximated by these filters (Fogel and Sagi 1989). In addition, these filters have been shown to possess optimal localization properties in both spatial and frequency domain and thus are well suited for texture segmentation problems (Grigorescu, Petkov and Kruizinga 2002). Gabor filters have been used in many applications, such as texture segmentation, target detection, fractal dimension management, document analysis, edge detection, retina identification, image coding and image representation, and image steganalysis (Song et al. 2017). Inspired by these works, Gabor residual-based features are designed for seam-carving image forgery detection (Liu 2019), introduced below.

In the spatial domain, a 2D Gabor filter is a Gaussian kernel function modulated by a sinusoidal plane wave, given below.

$$g\left(x,y;\lambda,\theta,\varphi,\sigma,\gamma\right)=\exp\left(-\frac{x'^2+\gamma^2 y'^2}{2\delta^2}\right)\cos\left(2\pi\frac{x'}{\lambda}+\varphi\right) \tag{33}$$

In this equation, λ represents the wavelength of the sinusoidal factor, θ represents the orientation of the normal to the parallel stripes of a Gabor function, φ is the phase offset, σ is the sigma/standard deviation of the Gaussian envelope and γ is the spatial aspect ratio, and specifies the ellipticity of the support of the Gabor function, and

$$x' = x\bullet\cos\theta + y\bullet\sin\theta\,;\; y' = -x\bullet\sin\theta + y\bullet\cos\theta\,.$$

For the image I(x,y), the Gabor residual is denoted by u(x,y), given by the convolution of I and the Gabor function in (33).

$$u\left(x,y\right)=\iint_R I\left(\alpha,\beta\right)g\left(x-\alpha,y-\beta\right)d\alpha d\beta \tag{34}$$

Where the set of image points of I(x,y) is R and the integral variables are α and β.

A quantization is applied and rounded below:

$$v\left(x,y\right)=\text{round}\left(u\left(x,y\right)/\,w\right) \tag{35}$$

The matrix v is an m×n array on the same size of the image. According to the image JPEG compression blocks with the block size $B_1 \times B_2$, the matrix v on the (p,q)-*th* block (p = 1, 2, …, B_1; q = 1, 2, …, B_2) compression block is denoted by $v_{p,q}$. The following marginal density features are extracted:

$$m_v\left(k;i,j\right)=\frac{\sum_{p=1}^{B_1}\sum_{q=1}^{B_2}\delta\left(\left|v_{p,q}\left(i,j\right)\right|=k\right)}{B_1B_2} \tag{36}$$

Where, k is set from 0 to T by the step of 1, and the pair (i,j) enumerates the 64 positions in a JPEG compression block, from (1,1) to (8,8).

Additionally, the following neighboring joint density features are extracted by

$$nj_v^1\left(k,l\right)=\frac{\sum_{(i,j)=(1,1)}^{(8,8)}\sum_{p=1}^{B_1}\sum_{q=1}^{B_2-1}\delta\left(\left|v_{p,q}\left(i,j\right)\right|=k;\left|v_{p,q+1}\left(i,j\right)\right|=l\right)}{64B_1(B_2-1)} \tag{37}$$

$$nj_v^2\left(k,l\right)=\frac{\sum_{(i,j)=(1,1)}^{(8,8)}\sum_{p=1}^{B_1-1}\sum_{q=1}^{B_2}\delta\left(\left|v_{p,q}\left(i,j\right)\right|=k;\left|v_{p+1,q}\left(i,j\right)\right|=l\right)}{64(B_1-1)B_2} \tag{38}$$

$$nj_v^3\left(k,l\right)=\frac{\sum_{(i,j)=(1,1)}^{(8,7)}\sum_{p=1}^{B_1}\sum_{q=1}^{B_2}\delta\left(\left|v_{p,q}\left(i,j\right)\right|=k;\left|v_{p,q}\left(i,j+1\right)\right|=l\right)}{56B_1B_2} \tag{39}$$

$$nj_v^4\left(k,l\right)=\frac{\sum_{(i,j)=(1,1)}^{(7,8)}\sum_{p=1}^{B_1}\sum_{q=1}^{B_2}\delta\left(\left|v_{p,q}\left(i,j\right)\right|=k;\left|v_{p,q}\left(i+1,j\right)\right|=l\right)}{56B_1B_2} \tag{40}$$

Parameter Setting and Feature Extraction.

We empirically set the parameters, described below.

Type I: in equation (4), k =0, 1, …,10; l = 0, 1, … 10; i = 1, … 14; and j =1 and 2. In equation (5), k =0, 1, …,10; l = 0, … k; i = 1, … 14; d= 1, 2, … 20, and j =1, 2, 3. A sub-total of 58,828 features are extracted.

Type II: in equations (6), (7), (8), and (9), k =0, 1, …,10; l = 0, … k; i = 1, … 14; and j =1, 2, 3. A sub-total of 11,088 features are extracted.

Type III: In equations (15) and (16), k =0, 1… 10; 1 = 0, 1 … 10. In equation (17), k =0, 1, …,10; l = 0, … k; i = 1, 2, 3, 4; d= 1, 2, … 20. A sub-total of 5,522 features are extracted.

Type IV: in equation (18), the values of k_1 and k_2 are set from 0 to 10 step by 1, in a total of 121 features. In equation (19), the values of k_1, k_2, and k_3 are set from 0 to 8 step by 1, in a total of 729 features. In equation (20), the scope of k_1 depends on the zig-zag position z_1, meanwhile, the value of k_2 and k_3 depend on the value of k_1 and the zig-zag position z_2 and z_3.

M_z =10; (z=1, 2)

M_z = [4, 4, 4, 4, 5, 6, 7, 8, 8, 8, 8]; (z=3, 4, 5)

M_z = [4, 4, 4, 5, 6, 7, 7, 7, 7, 7, 7]; (z=6, 7, 8, 9)

M_z = [4, 4, 4, 4, 5, 6, 6, 6, 6]; (z=10, 11, 12, 13, 14)

M_z = [3, 3, 3, 4, 4, 4, 4, 4, 4]; (z=15, 16, 17, 18, 19, 20)

The feature extraction is given by:

```
for k1=0: max(Mz1)
          for k2=0:Mz2(k1+1)
                    for k3=0:Mz3(k1+1)
```

$$P_{z_1 z_2 z_3}\left(k_1, k_2, k_3\right) = \frac{\sum_{p=1}^{B_1} \sum_{q=1}^{B_2} \delta\left(\left|c_{z_1}(p,q)\right| = k_1; \left|c_{z_2}(p,q)\right| = k_2; \left|c_{z_3}(p,q)\right| = k_3\right)}{B_1 \bullet B_2}$$

```
                    end
          end
end
```

In Type IV, a total of 14,098 features are extracted.

***Type V*:** in equation (21), the value of z is set to 1, 2, 3 … 9. Correspondingly, the values of k1, k2 and k3 are set from 0 to m, m=8,8,7,7,7,6,6,6,6, corresponding to the value z from 1 to 9. A total of 4,366 features in Type V.

All these five types of feature are integrated together, denoted by **D³HNJ** (**D**irectional **D**erivative and minimal **D**ifference-based, and **H**igh order **N**eighboring **J**oint density features), in a total of 93,902 features.

Type VI: The features extracted from equations (36) to (40) are Gabor residual-based Rich features, denoted by GR. In our study, we examined different parameters and made the following setting: in (22), θ is set to the orientations from 0 to 175 by the step 5; φ is set to 0 and π/2, respectively; σ is set to 0.5 and 0.75 respectively; $\lambda = \frac{\sigma}{0.56}$; and γ is set to 0.5. *k* and/or *l* is set from 0 to T (T=8 in our study). In summary, all Gabor residual-based Rich feature set (GR) contains 129,600 features.

All six types of features are merged, denoted by **GR−D³HNJ**, in a total of 223,502 features.

Experiments

A. A Comparison Study in Detecting Recompressed Seam Carved JPEG Images on the Size 512×512

Two types of untouched and seam carved JPEG images are generated under recompression, described in Table I. Each type of data contains 7,838 untouched JPEG images under recompression and 7,838 seam carved JPEG images under recompression. Fig. 5 shows some untouched and doctored JPEG images. All untouched and seam carved JPEG images are cropped and aligned according to the JPEG compression blocks on the size of 512×512.

In the study (Liu 2019), a comparison study was carried out among the seam carving detectors LF-DCT (Liu et al. 2016), LF-DEDCT (Liu 2017), and rich modeling-based detector SRMQ1 that was originally designed in steganalysis (Fridrich and Kodovsky 2012), which was applied for seam carving detection (Liu and Chen 2015). Additionally, low-dimensional detectors cc-absNJ (Liu and Chen 2015), Markov-YCrCb (Wang et al. 2010), and other detectors in (Chang, Shih and Hsu 2013; Ryu and Lee 2014; Wei et al. 2014; Yin et al. 2015) were compared.

Liu (2019) utilized a well-established ensemble learning with fisher linear discriminant weak learner that was originally developed in steganalysis by Kodovsky,

Table 1. Untouched and seam carved JPEG images (512×512) under same and lower recompression qualities

Data Set	Untouched JPEG Images	Doctored JPEG Images by Seam Carving
1	Original JPEG images are resaved in JPEG at the same quality '75', cropped and aligned on the size of 512×512, according to JPEG compression block alignment.	After manipulating original JPEG images (quality '75') by seam carving, the doctored images are saved in JPEG at the quality '75', cropped and aligned on the size of 512×512, according to JPEG compression block alignment.
2	Original JPEG images at the quality '75' are resaved in JPEG at the quality '75', and then resaved in JPEG at the quality '40', cropped and aligned on the size of 512×512, according to JPEG compression block alignment.	Firstly applying seam carving to the original JPEG images (quality '75'), secondly the doctored images are saved in JPEG at the quality '75', and finally these images are resaved in JPEG at the quality '40', cropped and aligned on the size of 512×512, according to JPEG compression block alignment.

Fridrich, and Holub (2012). Table II shows the mean detection accuracy over 30 experiments. In each experiment, we randomly selected 70% images for training. The remaining 30% are used for testing. The outcomes were classified as True Positive (TP), False Positive (FP), False Negative (FN), and True Negative (TN). The detection accuracy is calculated by 0.5*TP/(TP+FN) + 0.5*TN/(TN+FP).

The results show that the large feature mining-based approaches are significantly superior to other low dimensional detectors. In comparison to the large feature detectors LF-DCT (Liu et al. 2016), LF-DEDCT (Liu 2017) and the rich modeling-based detector SRMQ1(Fridrich and Kodovsky 2012), the new detector D^3HNJ, Gabor residual-based detector GR, and the combination GR−D^3HNJ have achieved better detection accuracy values.

GR−D^3HNJ contains over 200,000 features, some features may be useless for the detection, and therefore, Liu (2019) applied a feature selection method by Roffo, Melzi, and Cristani (2015) to GR−D^3HNJ. The selected feature set is denoted by S-GR/D^3HNJ that only retains 10% to 17% features. The last row in Table II gives the best detection accuracy by S-GR/D^3HNJ. Compared to GR−D^3HNJ, with much smaller number of features, S-GR/D^3HNJ even obtained a higher detection accuracy values on data set 1.

B. A Comparison Study in Detecting Recompressed Seam Carved JPEG Images on the Size 256×256

On the size of 256×256, we generated two types of untouched and seam carved JPEG images under recompression, described in Table III. Each type of data contains 10,558 untouched JPEG images under recompression and 10,558 seam carved JPEG

Figure 5. Untouched samples on the odd rows and tampered on the even (Liu 2019)

images under recompression. Fig. 6 shows some untouched and doctored JPEG images in data set 3.

In this study, LF-DCT (Liu et al. 2016), LF-DEDCT (Liu 2017), SRMQ1 (Fridrich and Kodovsky 2012), D^{3}HNJ (Liu 2019), GR (Liu 2019), GR−D^{3}HNJ (Liu 2019) and reduced GR−D^{3}HNJ feature set by applying the feature selection method in (Roffo, Melzi, and Cristani 2015), denoted by S-GR/D^{3}HNJ in (Liu 2019) were compared, and the ensemble learning (Kodovsky, Fridrich, and Holub 2012) was applied to these feature detectors.

Table IV shows the mean detection accuracy over 30 experiments. In each experiment, we randomly selected 80% images from untouched and 80% from tampering for training. The remaining 20% images are used for testing. The results show that all compared detectors are effective, and SRMQ1 is slightly better than our proposed approach. The detection performance on data set 4 is less accurate than the detection on data set 3. The last row in Table II gives the best detection

Table 2. Mean detection accuracy (%) on seam carving under the same and a lower recompression quality on the size 512×512

Detector/Method	Detection Accuracy (%)	
	Data Set 1	Data Set 2
Markov-YCrCb (Wang et al. 2010)	62.2	54.4
Chang, Shih and Hsu (2013)	63.9	53.5
Yin et al. (2015)	55.4	51.9
Wei et al. (2014)	51.9	51.8
Ryu and Lee (2014)	64.4	56.4
cc-absNJ (Liu and Chen 2015)	93.0	74.5
LF-DCT (Liu et al. 2016)	92.3	89.8
LF-DEDCT (Liu 2017)	98.7	94.0
SRMQ1 (Fridrich and Kodovsky 2012)	99.3	87.0
D^3HNJ (Types I-V) (Liu 2019)	99.4	94.4
GR (Type VI) (Liu 2019)	99.4	94.2
GR–D^3HNJ (all types) (Liu 2019)	99.5	94.5
S-GR/D^3HNJ (Liu 2019)	99.6	92.2

accuracy by S-GR/D^3HNJ. Original GR–D^3HNJ contains over 200,000 features, after feature selection, S-GR/D^3HNJ only retains 10% to 15% features, but higher detection accuracy values are obtained by using the smaller number of features.

It is worth noting that the parameters in the large feature mining (Liu 2019) were not optimized yet, to validate the discriminative capability of the proposed features,

Table 3. Untouched and seam carved JPEG images (256×256) under same and lower recompression qualities

Data Set	Untouched JPEG Images	Doctored JPEG Images by Seam Carving
3	Original JPEG images are resaved in JPEG at the same quality '75', cropped and aligned on the size of 256×256, according to JPEG compression block alignment.	After manipulating original JPEG images (quality '75') by seam carving, the doctored images are saved in JPEG at the quality '75', cropped and aligned on the size of 256×256, according to JPEG compression block alignment.
4	Original JPEG images at the quality '75' are resaved in JPEG at the quality '75', and then resaved in JPEG at the quality '40', cropped and aligned on the size of 256×256, according to JPEG compression block alignment.	Firstly, applying seam carving to the original JPEG images (quality '75'), secondly the doctored images are saved in JPEG at the quality '75', and finally these images are resaved in JPEG at the quality '40', cropped and aligned on the size of 256×256, according to JPEG compression block alignment.

Figure 6. Some examples in data set 3 (untouched on the odd rows and touched on the even rows)
(Liu, 2019)

Table 4. Mean detection accuracy (%) on seam carving under the same and a lower recompression quality on the size 256×256

Detector/Method	Detection Accuracy (%)	
	Data Set 3	**Data Set 4**
LF-DCT (Liu 2017)	84.2	57.1
LF-DEDCT (Liu 2017)	88.1	59.8
SRMQ1 (Fridrich and Kodovsky 2012)	94.3	67.1
D³HNJ (Liu 2019)	88.3	58.8
GR (Liu 2019)	92.8	64.5
GR–D³HNJ (Liu 2019)	93.1	64.7
S-GR/D³HNJ (Liu 2019)	93.5	66.3

Figure 7. Histograms of p-values in t-test on the four data sets
(Liu 2019)

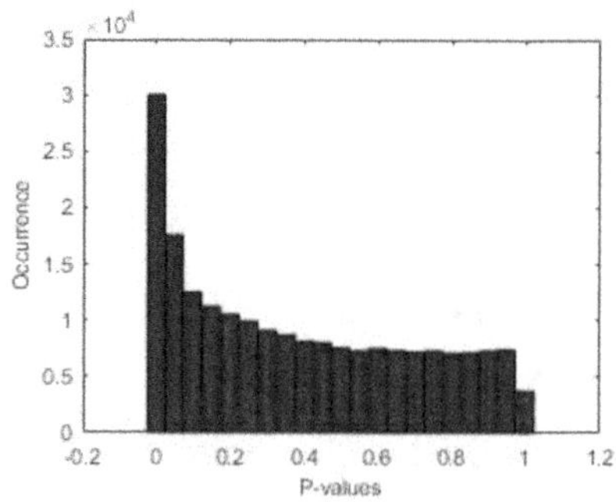

(a) Data set 1

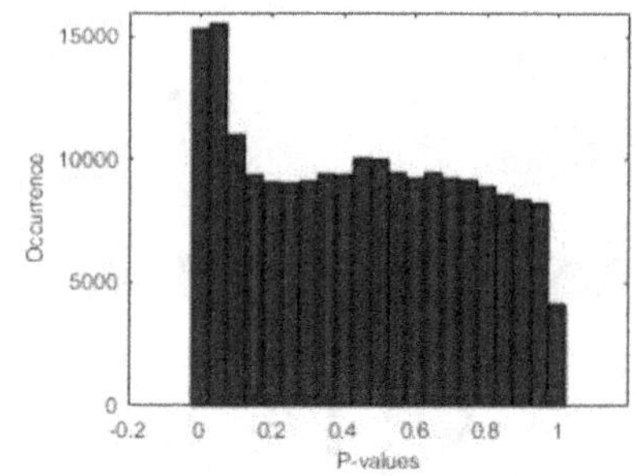

(b) Data set 2

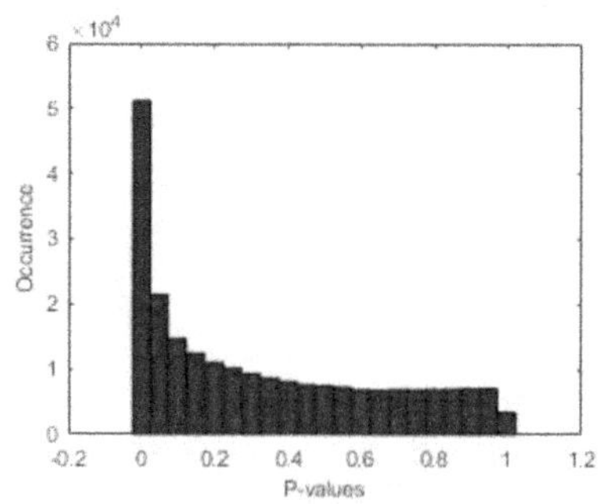

(c) Data set 3

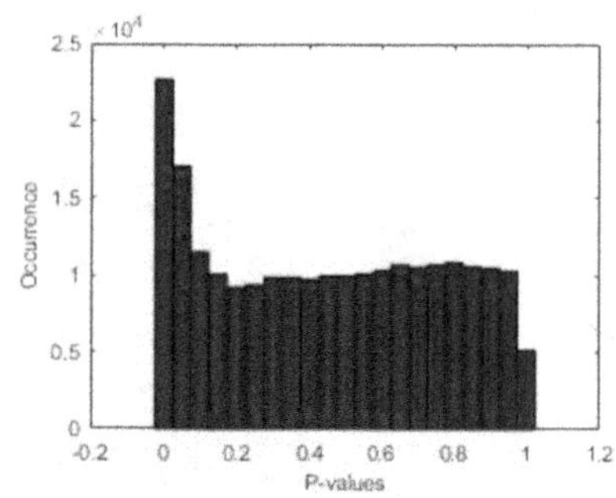

(d) Data set 4

Liu (2019) performed a t-test of the hypothesis that two independent samples come from distributions with equal means with the result of the test in H. H=0 indicates that the null hypothesis ("means are equal") cannot be rejected at the 5% significance level. Fig. 7 shows the p-value occurrences of all proposed features by using the t-test on data sets 1, 2, 3, and 4. Most p-values are aggregated at zero, indicating the good discriminative capability of proposed features. It also implies that some features could be useless for the detection.

FUTURE RESEARCH DIRECTIONS

Based on our previous study, we proposed a large-scale feature mining-based approach to detecting JPEG down-recompression and seam carving forgery under recompression anti-forensics. Experimental results show that proposed approach significantly improved the detection accuracy.

Enormous digital multimedia data are daily created and widely spread around the world in today's digital era. Because of society uses and adopts digital multimedia data increasingly, the digital dependence also continues to grow recently. As digital multimedia data can be easily manipulated with a variety of malicious or criminal intents, investigators confront unprecedented threats and new challenges in digital forensics and cybersecurity. Cybercrime and cyber-enabled crime are pivotal issue in modern society, and the cost of these cybercrimes is increasing. In the United States, the cost to react these cybersecurity issues is estimated to be around $120 billion annually, with the impact to over 500,000 jobs every year (Lewis 2018; Demirdjian and Mokatsian 2015). Different age population is also suffering from different cybersecurity threats based on digital multimedia data on social media.

Nowadays, deep learning methods are widely used in many areas including image forgery detection (Agarwal et al. 2021; Barad and Goswami 2020). The image forgery detection is also pivotal to have impact on cybersecurity among different population, especially older adults. Yang et al. (2020) summarized the visual perception enabled industry is needed to consider the forgery detection as an important application to detect visual images, and visual images have been applied in many different industries. Liu et al. (2020) also mentioned that image sharing on social media has become a pivotal part in our daily social activities by sharing pictures or posting visual images on online social network. Morrison et al. (2020) discussed the implications for cybersecurity and vulnerability to affect older retirees, and they mention aging population is vulnerable as victims of cybercrime after retirement as major life transition to anyone in modern societies. In our future study, we will adopt and modify existing deep learning models that were originally developed in computer vision and investigate image forgery localization with deep learning methods.

CONCLUSION

To detect seam carving-based forgery under JPEG recompression anti-forensics, we first analyze the statistical modification to the DCT coefficients by JPEG down-recompression, and design large-scale feature mining-based approach to the detection. Ensemble classifier is applied to the rich features for the detection. Experimental results show that our approach redefines the state of detecting JPEG down-recompression and it also effectively detects the seam carving forgery under recompression anti-forensics, which was not well explored before.

Our following study indicates that by integrating feature mining with deep learning-based approaches together, which may also be expanded to other image forensics problems that were not well exposed before. These combined approaches may be employed to reduce the computational cost and the cost to react for cybercrimes or improve cybersecurity to vulnerable population including older adults and children using digital multimedia data increasingly. Texture analysis will be also considered according to each individual examination image, especially the increasing digital dependence posting images on social media as increasing daily social activities. Furthermore, blind detection will be investigated, and its performance will be evaluated. These techniques are pivotal to understand the implication of forgery detection for digital multimedia data in modern society with increasing digital dependence.

REFERENCES

Agarwal, R., Verma, O. P., Saini, A., Shaw, A., & Patel, R. (2021). The Advent of Deep Learning-Based Image Forgery Detection Techniques. In J. S. Raj, A. M. Iliyasu, R. Bestak, & Z. A. Baig (Eds.), *Innovative Data Communication Technologies and Application. Lecture Notes on Data Engineering and Communications Technologies* (Vol. 59). Springer. doi:10.1007/978-981-15-9651-3_56

Amerini, I., Ballan, L., Caldelli, R., Del Bimbo, A., & Serra, G. (2011). A SIFT-based forensic method for copy–move attack detection and transformation recovery. *IEEE Transactions on Information Forensics and Security*, *6*(3), 1099–1110. doi:10.1109/TIFS.2011.2129512

Avidan, S., & Shamir, A. (2007). Seam carving for content-aware image resizing. *ACM Transactions on Graphics*, *26*(3), 10. doi:10.1145/1276377.1276390

Barad, Z., & Goswami, M. (2020). Image Forgery Detection using Deep Learning: A Survey. *Proceedings of 2020 6th International Conference on Advanced Computing and Communication Systems (ICACCS)*, 571-576. doi: 10.1109/ICACCS48705.2020.9074408

Bianchi, T., & Piva, A. (2012). Image forgery localization via block-grained analysis of JPEG artifacts. *IEEE Transactions on Information Forensics and Security*, *7*(3), 1003–1017. doi:10.1109/TIFS.2012.2187516

Chang, W., Shih, T. K., & Hsu, H. (2013). Detection of seam carving in JPEG images. *Proceedings of the 2013 International Joint Conference on Awareness Science and Technology and Ubi-media Computing*, 632-638. 10.1109/ICAwST.2013.6765516

Chen, Y. L., & Hsu, C. T. (2011). Detecting recompression of JPEG images via periodicity analysis of compression artifacts for tampering detection. *IEEE Transactions on Information Forensics and Security*, *6*(2), 396–406. doi:10.1109/TIFS.2011.2106121

Demirdjian, Z. S., & Mokatsian, Z. (2015). The cost of cyber crimes to business and society. *ASBBS Proceedings, 22*(1), 104.

Fillion, C., & Sharma, G. (2010). Detecting content adaptive scaling of images for forensic applications. *Proceedings of SPIE, Media Forensics and Security II, 7541*, 888–896. doi:. 2010.10.1117/12.838647

Fogel, I., & Sagi, D. (1989). Gabor filters as texture discriminator. *Biological Cybernetics, 61*(2). doi:10.1007/BF00204594

Fridrich, J., & Kodovsky, J. (2012). Rich models for steganalysis of digital images. *IEEE Transactions on Information Forensics and Security*, *7*(3), 868–882. doi:10.1109/TIFS.2012.2190402

Grigorescu, S. E., Petkov, N., & Kruizinga, P. (2002). Comparison of texture features based on Gabor filters. *IEEE Transactions on Image Processing, 11*(10), 1160–1167. doi:10.1109/TIP.2002.804262 PMID:18249688

Han, R., Ke, Y., Du, L., Qin, F., & Guo, J. (2018). Exploring the location of object deleted by seam-carving. *Expert Systems with Applications*, *95*, 162–171. doi:10.1016/j.eswa.2017.11.023

Huang, F. J., Huang, J., & Shi, Y. (2010). Detecting double JPEG compression with the same quantization matrix. *IEEE Transactions on Information Forensics and Security*, *5*(4), 848–856. doi:10.1109/TIFS.2010.2072921

Kodovsky, J., Fridrich, J., & Holub, V. (2012). Ensemble classifiers for steganalysis of digital media. *IEEE Transactions on Information Forensics and Security*, *7*(2), 432–444. doi:10.1109/TIFS.2011.2175919

Lewis, J. A. (2018). *Economic impact of cybercrime-no slowing down.* Available: http://csis-website-prod.s3.amazonaws.com/s3fs-public/publication/economic-impact-cybercrime.pdf

Liu, C., Zhu, T., Zhang, J., & Zhou, W. (2020). *Privacy Intelligence: A Survey on Image Sharing on Online Social Networks.* arXiv preprint arXiv:2008.12199.

Liu, Q. (2011). Detection of misaligned cropping and recompression with the same quantization matrix and relevant forgery. *Proc. 3rd International ACM workshop on Multimedia in Forensics and Intelligence*, 25-30. 10.1145/2072521.2072528

Liu, Q. (2017). An approach to detecting JPEG down-recompression and seam carving forgery under recompression anti-forensics. *Pattern Recognition*, *65*, 35–46. doi:10.1016/j.patcog.2016.12.010

Liu, Q. (2019). An Improved Approach to Exposing JPEG Seam Carving under Recompression. *IEEE Transactions on Circuits and Systems for Video Technology*, *29*(7), 1907–1918. doi:10.1109/TCSVT.2018.2859633

Liu, Q., & Chen, Z. (2015). Improved approaches with calibrated neighboring joint density to steganalysis and seam-carved forgery detection in JPEG images. *ACM Transactions on Intelligent Systems and Technology*, *5*(4), 63. doi:10.1145/2560365

Liu, Q., Sung, A. H., Zhou, B., & Qiao, M. (2016). Exposing inpainting forgery in JPEG images under recompression attacks. *Proc. 15th IEEE International Conference on Machine Learning and Applications*, 164-169. 10.1109/ICMLA.2016.0035

Morrison, B. A., Coventry, L., & Briggs, P. (2020). Technological Change in the Retirement Transition and the Implications for Cybersecurity Vulnerability in Older Adults. *Frontiers in Psychology*, *11*, 623. doi:10.3389/fpsyg.2020.00623 PMID:32425841

Pun, C., Yuan, X., & Bi, X. (2015). Image forgery detection using adaptive oversegmentation and feature point matching. *IEEE Transactions on Information Forensics and Security*, *10*(8), 1705–1716. doi:10.1109/TIFS.2015.2423261

Roffo, G., Melzi, S., & Cristani, M. (2015). Infinite feature selection. *Proceedings of the 2015 IEEE International Conference on Computer Vision*, 4202-4210.

Ryu, S., Lee, H., & Lee, H.-K. (2014). Detecting trace of seam carving for forensics analysis. *IEICE Transactions on Information and Systems*, *97*(5), 1304–1311. doi:10.1587/transinf.E97.D.1304

Sarkar, A., Nataraj, L., & Manjunath, B. (2009). Detection of seam carving and localization of seam insertions in digital images. *Proceedings of the 11th ACM workshop on Multimedia and security*, 107–116. 10.1145/1597817.1597837

Sharifi, K., & Leon-Garcia, A. (1995). Estimation of shape parameter for generalized Gaussian distributions in subband decompositions of video. *IEEE Transactions on Circuits and Systems for Video Technology*, *5*(1), 52–56. doi:10.1109/76.350779

Sheng, G., & Gao, T. (2016). Detection of seam-carving image based on benford's law for forensic applications. *International Journal of Digital Crime and Forensics*, *8*(1), 51–61. doi:10.4018/IJDCF.2016010104

Song, X., Liu, F., Zhang, Z., Yang, C., Luo, X., & Chen, L. (2017). 2D Gabor filters-based steganalysis of content-adaptive JPEG steganography. *Multimedia Tools and Applications*, *76*(24), 26391–26419. doi:10.100711042-016-4157-9

Taspinar, S., Mohanty, M., & Memon, N. (2017). PRNU-based camera attribution from multiple seam-carved images. *IEEE Transactions on Information Forensics and Security*, *12*(12), 3065–3080. doi:10.1109/TIFS.2017.2737961

Wang, W., & Dong, J. (2010). Image tampering detection based on stationary distribution of Markov chain. *Proceedings of the 17th International Conference on Image Processing*, 2101-2104. 10.1109/ICIP.2010.5652660

Wattanachote, K., Shih, T. K., Chang, W.-L., & Chang, H.-H. (2015). Tamper detection of jpeg image due to seam modifications. *IEEE Transactions on Information Forensics and Security*, *10*(12), 2477–2491. doi:10.1109/TIFS.2015.2464776

Wei, J., Lin, Y., & Wu, Y. (2014). A patch analysis method to detect seam carved images. *Pattern Recognition Letters*, *36*, 100–106. doi:10.1016/j.patrec.2013.09.026

Yang, J., Wang, C., Jiang, B., Song, H., & Meng, Q. (2020). Visual perception enabled industry intelligence: State of the art, challenges and prospects. *IEEE Transactions on Industrial Informatics*, *17*(3), 2204–2219. doi:10.1109/TII.2020.2998818

Yin, T., Yang, G., Li, L., Zhang, D., & Sun, X. (2015). Detecting seam carving based image resizing using local binary patterns. *Computers & Security*, *55*, 130–141. doi:10.1016/j.cose.2015.09.003

Zhang, D., Yin, T., Yang, G., Xia, M., Li, L., & Sun, X. (2017). Detecting image seam carving with low scaling ratio using multi-scale spatial and spectral entropies. *Journal of Visual Communication and Image Representation*, *48*(C), 281–291. doi:10.1016/j.jvcir.2017.07.006

Chapter 8
Forensic Camera Identification in Social Networks via Camera Fingerprint

Tzuhuan Lin
New Taipei City Police Department, Taiwan

Yu-Ru Wang
Yilan County Government Police Bureau, Taiwan

ABSTRACT

Image-related crimes cause the urgent demand for tracing the origin of digital images. The breakthrough is a passive detection method via photo response non-uniformity (PRNU) analysis proposed by Lukáš et al. Recently, digital images are often shot with handheld devices (such as smartphones) and transmitted using social media (such as LINE). Most of the images are distorted (such as compressed and resized) during transmission. Previous studies are less focused on the impact of transmission compression through social networks. Thirty-one different Apple mobile phones were used to capture digital images in the experiment. Images were uploaded to the photo album via LINE software and then downloaded. The modified signed peak correlation energy (MSPCE) statistics is used to evaluate the correlation between the PRNU values of the disputed images and the pattern noise of the experimental devices. Experimental results show that the PRNU analysis method can effectively trace the source of the shot device using the distorted images which are compressed and resized during the transmission in LINE.

DOI: 10.4018/978-1-7998-8386-9.ch008

INTRODUCTION

Due to the wide use of mobile phones, the acquisition of an image is simple. In addition, the rise of convenient social media has made communication using images easier. Consequently, some illegal information such as obscene pornography, personal privacy, and national security-related content has been released to the public by images or videos.

To incriminate the criminal behavior of these suspects who created illegal images, it is important to identify that these images were taken from devices owned by these suspects. Kurosawa *et al.* (1999) were the first investigators to propose the source camera identification (SCI) technique. From the study, the non-uniformity noise or defects can be retained in the charge-coupled devices (CCDs) in the photosensitive element manufacturing process. It can be used as a fingerprint as bullet scratches to match a bullet identity in forensic science. They performed SCI experiments with 9 camcorders in 4 types by using dark current on CCD chips and obtained 8 fixed pattern noises from 9 camcorders. However, they used the dark current noise which is a signal collected from the sensor when it is not exposed to light. Dark current is only extracted from dark frames. This limits the method because camera identification is not possible from non-dark frames. The mainstream of traceability technology for digital imaging equipment is the passive detection and traceability analysis of photo-response non-uniformity noise (PRNU) published by Lukas *et al.* (2005a, 2005b, 2006). The PRNU is caused by different sensitivity of sensor pixels to light. They (2005a, 2005b) used the Daubechies 8 wavelet (db8) filter to estimate PRNU and used the correlation coefficient for digital camera sources to make a comparison of 9 types of different digital cameras and 2 kinds of file formats (TIFF and JPEG). In Lukas *et al.* (2006), they updated the wavelet filter (Mihcak *et al.*, 1999) as the denoise filter, and used different Gamma correction coefficients and compression ratios to analyze the sensor pattern noise in 9 different digital cameras. This technology has been used in child rape convictions in Scotland (Spy Blog, 2009) and applied to various international technical specifications (Scientific Working Group on Digital Evidence, 2018). Yang *et al.* (2021) considered the camera rolling problem and reduced the false positive rate as compared to existing methods used in the field of forensic examination.

Recently, the disputed images obtained from criminal investigations are often digital images transmitted through social media. It is worth exploring whether the PRNU analysis technology can still be efficiently used to trace the source of the shot device.

METHOD

The proposed method that extracts the PRNU factor of the device, and evaluates the similarity between the PRNU factors of the disputed images and those of the device is proposed. Fig. 1 illustrates the flow chart of our proposed SCI method. The adaptive de-noising filter is first applied to images to extract the rough device artifacts. Then, the enhancement algorithm is applied to remove the non-unique artifacts. Next, to reduce the effect of rolling[1], the obtained PRNU factors are rotated to the adaptive direction. For training images, the refined noise residues are averaged and then stored in the database. For disputed images, the modified signed peak correlation energy (MSPCE) statistics and a theoretical threshold are used to determine the source camera. The following subsections will describe each step of the proposed method in detail.

Figure 1. The flow chart of our proposed method

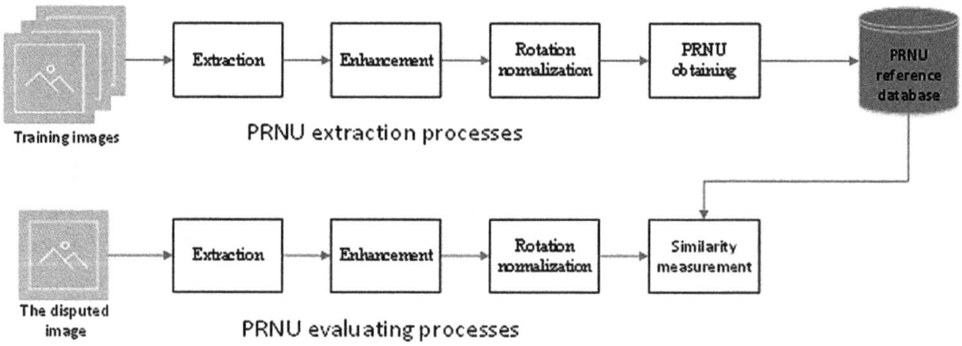

PRNU Extraction Processes

The PRNU factor of each device must be obtained before comparing it to the PRNUs of the disputed images. The main steps are discussed below.

Step 1: Artifact Extraction

According to Yang *et al.* (2021), the Mıhçak *et al.*'s (1999) wavelet-based de-noising filter is selected in the proposed method. An image I is obtained as follows:

$$I = I^o + I^o K + \Omega, \quad (1)$$

where I^o denotes the sensor output without noises, $I^o\mathrm{K}$ denotes the PRNU of the sensor, and Ω denotes a complex of independent random noises. The artifacts are extracted as Equations (2), (3), and (4). PRNU factor, K, is a two-dimension matrix obtained from the following equations:

$$\mathrm{K} = \left(I - I^o\right) / I^o + \Omega / I^o, \tag{2}$$

I^o could be estimated by $\hat{I}^o$,

$$I^o \approx \hat{I}^o = f\left(I\right), \tag{3}$$

where $f\left(.\right)$ denotes Mıhçak *et al.*'s (1999) wavelet-based de-noising function.

From Equations (2) and (3), K could be estimated by $\hat{\mathrm{K}}$

$$\mathrm{K} \approx \hat{\mathrm{K}} = \left(I - f\left(I\right)\right) / f\left(I\right) + \Omega / f\left(I\right). \tag{4}$$

Step 2: Enhancement

Zero-Mean operation in the spatial domain and Wiener filter operation in the Fourier domain (Chen *et al.*, 2008) are used to remove non-unique noises from the obtained artifacts.

Step 3: Rotation Normalization

In order to decrease the rolling effect of the imaging, Equation (5) is used to obtain refined artifacts with a consistent rotation.

$$K' = \begin{cases} R\left(\mathrm{K}, 90\right), & if\, m > n, \\ K, & otherwise. \end{cases}, \tag{5}$$

where each estimated PRNU factor with a resolution of *m* rows by *n* columns and *R*(*A*, *B*) is a rotation operation that rotates *A* by *B* degrees.

Step 4: PRNU Obtaining

The averaging operation is used to remove some independent and identically distributed noises to obtain the device PRNU factor, SPN, for each corresponding device.

$$SPN = \sum_{i=1}^{N} K' / N$$

$$= \sum_{i=1}^{N} \left[f(I_i) / \left(I_i - f(I_i)\right)\right] / N + \sum_{i=1}^{N} \left[\Omega / \left(I_i - f(I_i)\right)\right] / N$$

$$= \sum_{i=1}^{N} \left[f(I_i) / \left(I_i - f(I_i)\right)\right]. \tag{6}$$

PRNUs Evaluating Processes

The disputed images are analyzed through the following steps. The first 3 steps are the same as those in steps 1-3 of the PRNU extraction process.

Step 4: Similarity Measurement

The modified signed peak correlation energy (MSPCE) statistics and the suggestive threshold value of 41.82 as described in (Yang *et al.*, 2021) are used to determine the source device of the disputed image. The detailed operations of MSPCE statistics are shown in Equations (7), (8), and (9).

$$MSPCE\left(SPN, \mathrm{K}'\right) = \max\left(SPCE\left(SPN, \mathrm{K}'\right), SPCE\left(SPN, R(\mathrm{K}', 180)\right)\right), \tag{7}$$

$$SPCE\left(SPN, \mathrm{K}'\right) = \frac{\rho^2\left(SPN, \mathrm{K}'; 0, 0\right) \times sign\left(\rho\left(SPN, \mathrm{K}'; 0, 0\right)\right)}{\frac{1}{mn - \left|N_{\max}\right|} \sum_{(x_s, y_s) \notin N_{\max}} \rho^2\left(SPN, \mathrm{K}'; x_s, y_s\right)}, \tag{8}$$

$$\rho\left(U,V;x_s,y_s\right)=\frac{\sum_{x,y}\left(U_{x,y}-\bar{U}\right)\left(V_{x+x_s,y+y_s}-\bar{V}\right)}{\sqrt{\sum_{x,y}\left(U_{x,y}-\bar{U}\right)^2}\sqrt{\sum_{x,y}\left(V_{x+x_s,y+y_s}-\bar{V}\right)^2}}, \tag{9}$$

where max(,) denotes a function that returns maximum value among the given numbers, $\rho\left(U,V;x_s,y_s\right)$ stands for a cross-correlation function standing for the range of indices, x and $x_s \in 1, 2, \ldots, m$, y and $y_s \in 1, 2, \ldots, n$. $\bar{U}$ and $\bar{V}$ respectively denotes the sample mean of $U_{x,y}$ and $V_{x,y}$. *sign*(.) denotes a signum function. N_{max} stands for a small neighborhood around the origin.

EXPERIMENT

In this experiment, 31 different Apple mobile phones (see Table I) were used to create the dataset. 200 photos of random scenes were acquired by each mobile phone. Then, half of the photos were randomly selected to be the training set and the others were the testing set (as the disputed images). Each mobile phone has 100 photos on each set. Our algorithms are developed in Matlab R2020 and the experimental platform is Microsoft Windows 10, Intel Core i7-9750H 2.60 GHz with 32 GB RAM.

As the foregoing description, the digital images are usually distorted (by resize operation, recompression operation, etc.) during the transmission in the social network. In practice, the PRNU based camera identification method is significantly influenced by the different sizes between the testing image and the device PRNU factors. Therefore, three different scenarios based on the previous forensic examination experience are designed to assure the same size between the device's PRNU factors and those of the distorted images, and to verify the effectiveness of the proposed method.

In scenario 1 (see Fig. 2), the training images are transmitted through LINE software and then downloaded to extract the device's PRNU factors. The sizes of the device's PRNU factors and those of the distorted images are the same.

Scenario 2 (see Fig. 3) is the traditional PRNU analysis method. The distorted images are resized to fit the image of suspect devices and then extracted the PRNU factors to do PRNU analysis.

In scenario 3 (see Fig. 4), the training images are resized to fit the distorted images before the device PRNU factors extraction.

Table 1. Mobile phones used in the experiment

No.	Model	Serial Number
1	iphone 5	DQGLX7BSDTWH
2	iphone 6	FDNQV2HRG5MR
3	iphone 6 plus	F2MNPV6FG5QW
4	iphone 7	F17VW0QHHG7P
5	iphone 8 plus	C39VDX9WJCM4
6	iphone SE	FFMCJB4WPLJQ
7	iphone SE	FFMCJETDPLJQ
8	iphone SE	FFWCJ411PLJT
9	iphone X	G6WVP81DJCL8
10	iphone XR	F2LZH069KXK2
11	iphone XR	F4GV8392KXK2
12	iphone 11	C6KZ62G2N735
13	iphone 11	C6KZ92R2N736
14	iphone 11	C6KZ8TLJN735
15	iphone 11	C7CZ9AEHN736
16	iphone 11	DNPZN076N73P
17	iphone 11	F4GZ41FTN737
18	iphone 11	F4GZ8ALWN738
19	iphone 11	F4GZ9NL3N739
20	iphone 11	F4GZ9PKVN739
21	iphone 11	F4GZXFK3N73F
22	iphone 11 pro	C39Z6GV0N6Y2
23	iphone 11 pro	C39Z71VHN6Y3
24	iphone 11 pro	C39Z8606N6Y4
25	iphone 11 pro	C39ZC8JYN6Y5
26	iphone 11 pro	C39ZJN1MN6Y5
27	iphone 11 pro max	G6TZ9TGQN70W
28	iphone 11 pro max	G6TZ60S8N70R
29	iphone 11 pro max	G6TZ809YN70R
30	iphone 11 pro max	G6TZ9J2ZN70W
31	iphone 11 pro max	G6TZ9U3GN70W

Figure 2. The flow chart of scenario 1

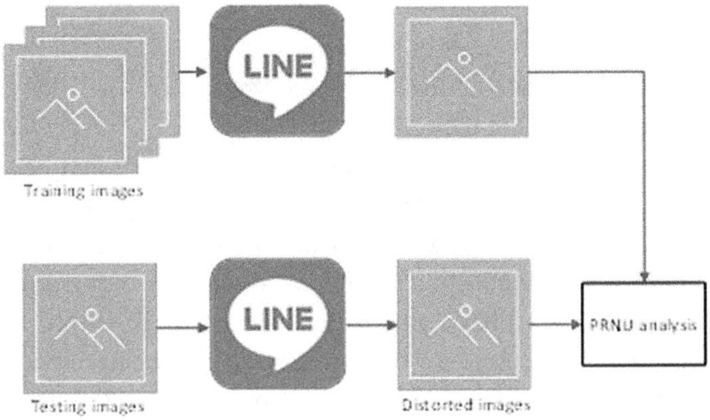

Figure 3. The flow chart of scenario 2

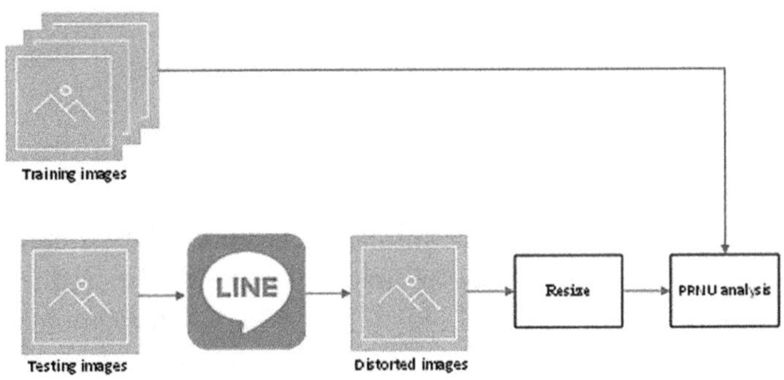

Figure 4. The flow chart of scenario 3

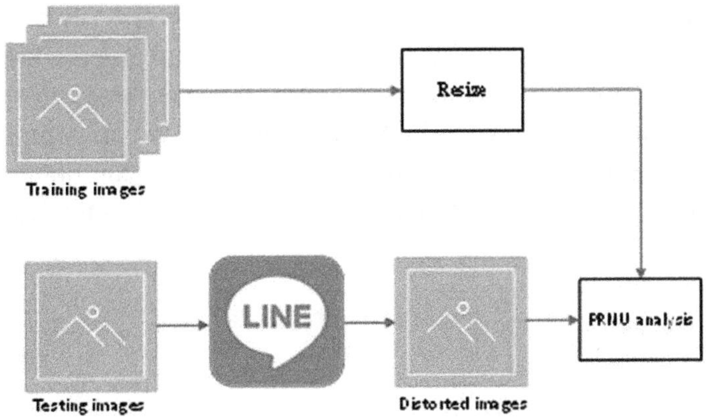

Analysis of Scenarios

In this subsection, the effectiveness of the proposed method used in the LINE social network will determine by the analysis of three scenarios. The MSPCE values of three scenarios are discussed in the following. Tables 2 and 3 respectively show the average MSPCE values from the source device and non-original devices in three scenarios. The source device and non-original devices respectively denote if the disputed images were shot by the device or not. From Table 2 and 3, the average MSPCE values are high if and only if the shot device of the disputed images is the same as it of the device PRNU factor. It means that the PRNU based camera identification method is effective even the disputed images have been transmitted through LINE software. Moreover, although the average MSPCE value of scenario 1 is the largest, scenario 3 is our suggestive method using in forensic practice. Because scenario 1 has a risk to break the chain of custody during the uploading and downloading of the training images, we suggest that scenario 3 is the better SCI method used in LINE social network.

The Error Rate in Scenarios

The false-positive rate (FPR), well known as the rate of the type I error, means the probability of the source of the disputed image is identified as an irrelative device by mistake. That means it might cause an innocent person to be arrested. Yang *et al.* (2021) pointed out the theoretic threshold of the MSPCE value should be larger than 41.82 to verify two different images shot by the same device when the population size of devices is about 10 billion. Here, the false-positive (FP) error is defined as the MSPCE value of the disputed image is higher than a theoretic threshold 41.8 when the source of the disputed image is identified to an irrelative device by mistake. In contrast, the false-negative rate (FNR), the rate of type II error, means the probability of the disputed image is not identified to its real source device. This means the evidence might be useless. The false-negative (FN) error is defined as the MSPCE value of the disputed image is lower than a theoretic threshold 41.82 when the disputed image is identified to its real source device. Table 4 shows the FPR and FNR values in each scenario. From Table 4, the experimental results with zero FPR and very low FNR show the proposed method can obtain the perfect accuracy when the disputed images have been transmitted through LINE software.

Table 2. The average MSPCE values from the source devices

No.	Scenario 1			Scenario 2			Scenario 3		
	Maximum	Minimum	Average	Maximum	Minimum	Average	Maximum	Minimum	Average
1	26648.16	13056.73	20959.81	11479.11	8351.17	9965.92	23691.74	12860.63	19044.70
2	3481.26	1809.18	2467.89	1964.66	987.94	1406.34	3736.34	1961.20	2625.05
3	53573.53	31912.66	41020.39	26943.42	14148.90	21178.10	46544.45	27849.70	36061.55
4	3828.68	1430.86	1791.76	2965.65	1495.25	1974.42	3926.86	1576.21	1968.91
5	45305.21	37545.46	41571.20	7790.85	5188.19	6853.68	24233.61	18390.51	21240.76
6	9762.49	3402.14	4828.31	4556.80	1707.66	2429.31	8079.31	2613.92	3936.85
7	10891.61	3874.11	6006.22	7505.44	2601.50	4083.91	11878.39	3798.60	6252.42
8	11396.95	3481.57	5477.08	5049.97	1737.95	2484.73	10896.97	3385.62	5058.07
9	1800.37	161.95	775.14	1170.40	143.19	395.91	1910.47	193.53	674.86
10	3549.56	57.52	477.61	1905.78	**40.96**	205.09	2862.66	**25.11**	326.52
11	85398.58	8133.79	23535.84	27558.05	4337.05	8411.24	75243.00	8251.72	21006.39
12	11840.03	1819.47	4218.59	5386.66	1055.26	2007.35	12106.92	1953.84	4298.68
13	8407.74	1475.63	3175.60	4640.95	702.35	1572.42	9017.60	1219.68	3174.97
14	16450.99	1843.31	3456.26	7692.73	888.13	1749.96	17180.99	1574.98	3325.24
15	11649.02	1645.39	4375.84	6797.95	1154.03	2262.65	13544.72	1799.47	4632.84
16	1237.94	136.92	509.35	725.53	89.83	270.59	1276.71	123.91	461.88
17	13471.14	2193.19	4637.12	5708.01	1182.28	2059.75	14607.56	2246.09	4870.24
18	6391.38	1511.80	2821.40	2561.17	412.42	1035.44	5967.29	852.60	2270.69
19	8752.02	2307.52	3515.14	3809.45	956.16	1562.65	10698.91	2415.65	3843.71
20	27677.46	2158.44	6490.01	12412.61	1163.60	2995.60	26423.33	2653.86	5829.28
21	3379.98	907.95	1635.95	1380.33	354.69	683.56	2530.54	635.12	1144.66
22	29139.77	14237.04	18717.50	12408.33	5947.16	8077.64	25365.15	11747.41	16180.90
23	15448.70	5220.92	7687.06	8115.22	2621.81	3903.56	17222.09	5407.97	8160.93
24	14599.05	2602.92	4805.10	8880.78	1389.83	2663.27	17691.62	2747.37	5337.13
25	15252.96	801.31	4946.28	8782.18	401.16	2597.45	18798.95	506.58	5521.41
26	6953.58	1300.59	2466.55	2388.55	360.11	811.08	4925.94	683.16	1687.55
27	7440.94	1359.78	3522.04	3020.51	467.25	1467.01	6630.67	707.29	3183.46
28	20318.94	1015.13	10840.14	8477.76	376.30	4613.55	15980.88	594.36	8679.21
29	30007.51	4287.97	7319.36	16293.49	2742.81	3858.83	31564.08	5110.31	7811.46
30	10604.08	3229.41	4597.95	5365.35	1665.66	2462.27	11289.33	3469.96	4912.79
31	81197.54	19643.98	29289.53	37284.71	10089.19	14119.73	70002.45	16719.81	25596.09

Table 3. The average MSPCE values from non-original devices

No.	Scenario 1			Scenario 2			Scenario 3		
	Maximum	Minimum	Average	Maximum	Minimum	Average	Maximum	Minimum	Average
1	21.64	-4.03	0.99	19.45	-5.13	0.85	11.91	-5.87	0.79
2	17.87	-7.51	1.05	19.08	-6.43	0.96	14.29	-6.21	0.90
3	19.73	-6.89	1.18	15.52	-3.61	0.88	13.81	-4.76	0.82
4	21.02	-5.93	1.43	17.34	-9.79	0.95	12.15	-4.14	0.90
5	15.98	-4.86	1.21	19.04	-4.70	0.91	15.47	-3.38	0.94
6	26.20	-5.30	2.46	15.06	-5.53	0.96	15.22	-5.78	0.87
7	21.21	-3.86	2.46	14.08	-4.73	0.96	11.32	-4.81	0.87
8	20.94	-3.81	2.48	15.11	-6.26	0.87	14.60	-8.79	0.84
9	15.82	-4.77	1.09	15.81	-5.89	0.93	14.54	-5.19	0.83
10	34.16	-4.09	2.67	14.97	-4.96	0.84	12.87	-4.58	0.81
11	30.19	-4.65	3.32	15.19	-5.07	0.88	14.30	-5.64	0.84
12	34.68	-3.91	3.64	15.51	-5.03	0.95	13.49	-3.04	0.89
13	24.44	-4.96	2.69	14.72	-4.21	1.03	14.87	-4.50	0.95
14	36.00	-3.33	3.40	11.49	-4.94	0.89	14.97	-3.80	0.89
15	32.24	-3.10	3.87	13.71	-5.13	0.97	13.80	-3.24	0.98
16	26.76	-3.01	2.97	16.50	-3.95	0.96	18.29	-5.96	0.89
17	34.80	-2.61	3.86	17.72	-5.80	1.01	19.78	-4.94	0.95
18	27.09	-3.35	2.35	14.75	-5.46	0.96	14.13	-4.06	0.90
19	31.99	-3.92	2.71	15.34	-6.86	0.96	12.95	-5.56	0.95
20	37.13	-3.40	3.82	25.77	-3.64	1.05	12.09	-3.69	1.02
21	27.70	-4.23	3.28	25.97	-4.99	0.97	15.66	-4.52	0.94
22	25.02	-3.79	2.56	16.43	-10.27	0.96	13.98	-3.69	0.98
23	34.10	-2.23	3.70	17.50	-4.42	1.03	18.78	-5.80	1.02
24	32.29	-4.68	3.53	12.95	-3.81	0.99	18.21	-4.15	0.99
25	31.51	-2.55	2.88	20.94	-4.65	0.93	18.83	-3.45	0.90
26	25.33	-7.04	2.26	18.53	-5.29	1.05	14.01	-4.10	0.99
27	23.42	-3.63	2.40	19.28	-5.87	0.99	13.82	-5.85	0.96
28	25.11	-3.05	2.62	13.57	-7.78	1.02	13.46	-5.87	0.95
29	32.20	-4.97	3.81	15.02	-4.97	1.05	17.11	-4.63	1.00
30	40.98	-2.55	3.81	16.89	-4.34	1.01	13.97	-5.47	0.96
31	33.88	-2.49	3.89	12.59	-4.09	1.00	12.27	-3.81	1.01

Table 4. The error rate of the experimental results in each scenario

Scenario	Error Rate	
	FPR	FNR
1	0	0
2	0	0.00001
3	0	0.00002

CONCLUSION

Traditional PRNU based camera identification methods are passive detection and propose excellent performance to trace the source of digital images. Recently, the disputed images obtained from criminal investigations are often digital images transmitted through social media. The disputed images usually are distorted (by resize operation, recompression operation, etc.) during the transmission in social networks. It is worth notice that the PRNU based camera identification method is significantly influenced by the different sizes between the disputed image and the PRNU factors of the suspect device. It causes the urgent demand for tracing the origin of digital images transmitted through social media.

An effective source camera identification and verification method based on PRNU analysis for use in LINE social networks had been proposed. The main contribution of this chapter is we consider the effects of LINE social network in possible SCI methods. Experimental results demonstrated our method cannot only provide effectiveness but with the zero FPR and low FNR for use in LINE social network.

ACKNOWLEDGMENT

This study was supported by the National Science Council, Taiwan, Republic of China [MOST 109-2221-E-015-002-].

REFERENCES

Kurosawa, K., Kuroki, K., & Saitoh, N. (1999). CCD Fingerprint Method-Identification of a Video Camera from videotaped Images. *Proceedings of IEEE International Conference on Image Processing*. 10.1109/ICIP.1999.817172

Lukas, J., Fridrich, J., & Goljan, M. (2005a) Determining Digital Image Origin Using Sensor Imperfections. In *Proceedings SPIE Electronic Imaging* (pp. 249-260). 10.1117/12.587105

Lukas, J., Fridrich, J., & Goljan, M. (2005b) Digital Bullet Scratches for Images. *Proceedings of IEEE International Conference on Image Processing.*

Lukas, J., Fridrich, J., & Goljan, M. (2006). Digital Camera Identification from Sensor Noise. *IEEE Transactions on Information Forensics and Security, 1*(2), 205–214. doi:10.1109/TIFS.2006.873602

Mihcak, M. K., Kozintsev, I., & Ramchandran, K. (1999) Spatially adaptive statistical modeling of wavelet image coefficients and its application to denoising. In Proceedings IEEE International Conference Acoustics, Speech, Signal Processing (vol. 6, pp. 3253-3256). IEEE.

Scientific Working Group on Digital Evidence. (2018). *SWGDE Best Practices for Image Authentication*. https://drive.google.com/file/d/1Z0DsJMa6aDZlFJ9kRfOL8cow5VjhVT0t/view

Spy Blog - Watching Them Watching Us. (2009). *Operation Algebra child rape convictions in Scotland: open WiFi tracking, digital camera image forensics*. https://spyblog.org.uk/ssl/spyblog/2009/05/09/operation-algebra-child-rape-convictions-in-scotland-open-wifi-tracking-digi.html

Yang, W. C., Jiang, J., & Chen, C. H. (2021). A Fast Source Camera Identification and Verification Method Based on PRNU Analysis for Use in Video Forensic Investigations. *Multimedia Tools and Applications, 80*(5), 6617–6638. doi:10.100711042-020-09763-z

ENDNOTE

1 Rolling is defined as the rotation around the optical axis of the camera.

Chapter 9
Vehicle License Plate Recognition With Deep Learning

Chi-Hsuan Huang
National Taiwan University, Taiwan

Yu Sun
National Taiwan University, Taiwan

Chiou-Shana Fuh
National Taiwan University, Taiwan

ABSTRACT

In this chapter, an AI (artificial intelligence) solution for LPR (license plate recognition) on moving vehicles is proposed. The license plates in images captured with cameras on moving vehicles have unpredictable distortion and various illumination which make traditional machine vision algorithms unable to recognize the numbers correctly. Therefore, deep learning is leveraged to recognize license plate in such challenging conditions for better recognition accuracy. Additionally, lightweight neural networks are chosen since the power supply of scooter is quite limited. A two-stage method is presented to recognize license plate. First, the license plates in captured images are detected using CNN (convolutional neural network) model and the rotation of the detected license plates are corrected. Subsequently, the characters are recognized as upper-case format (A-Z) and digits (0-9) with second CNN model. Experimental results show that the system achieves 95.7% precision and 95% recall at high speed during the daytime.

DOI: 10.4018/978-1-7998-8386-9.ch009

Figure 1. The illustration of a typical LPR system

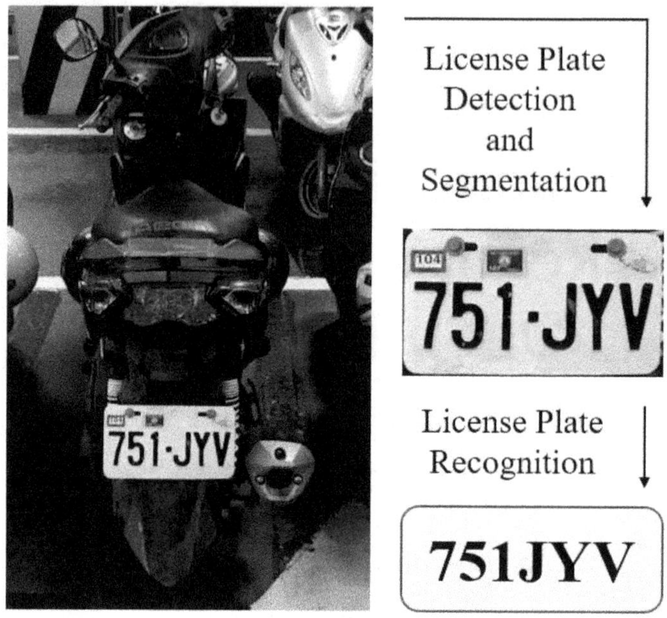

1. INTRODUCTION

1.1. Overview

The objective of LPR is to recognize the vehicle registration numbers from images and identify the owner or the origin of the vehicle. The acquired information can be used for stolen vehicle tracking and traffic management, such as ETC (Electronic Toll Collection), access control, traffic law enforcement, and parking lot management. A typical LPR system consists of license plate detection, segmentation, and recognition as shown in Fig. 1-1.

With the growth of surveillance cameras in Taiwan, there has been a steady decline in theft rate, e.g., 82.94% and 77.22% in motorcycle and car theft rate, respectively (Peng 2015). However, the video tracking of theft vehicles is one of, if not the most labor-intensive and time-consuming task, which turns LPR system into an indispensable part in the whole process, various LPR systems were invented consequently.

Due to the variations in luminance, contrast, noise, and geometry of the license plates captured in natural scene, traditional machine vision methods are unable to solve the problem completely. In recent years, many computer vision tasks achieved state-of-the-art results leveraging deep learning, particularly, CNN (Convolutional

Figure 2. License plate with green background and white characters

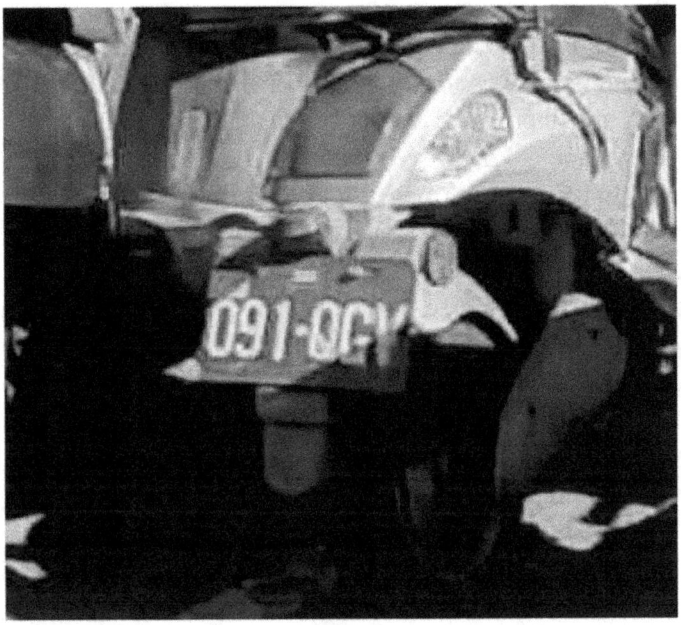

Neural Network) with exceptional performance in image feature extraction. By integrating different low-level features with the deep network structure, the network can detect complicated high-level features to conquer diverse environments. Despite the remarkable performance of deep learning, there still exist some challenges to be tackled.

To begin with, training deep networks with insufficient data could cause the networks to memorize all the training data, massive amounts of data are required to prevent overfitting. A network is said to be overfitted when it is much more accurate in predicting known data than new data. Due to rare old license plates such as green background with white characters in Fig. 1-2 or some seldom used characters such as "I" and "4" (same pronunciation as death in Mandarin Chinese), overfitting may occur.

Furthermore, class imbalance problem is common in classification problem where training with disproportionate ratio of observations in each class. Unfortunately, character distributions of Taiwanese license plate are severely biased, while most machine learning algorithms work best when the training samples are equally distributed.

Finally, with the high computational complexity of deep learning, high-end GPUs (Graphics Processing Units) and similar processing units are used to ensure better efficiency and less time consumption. These processing units are delicate, costly, and most of all—power-consuming.

Since a deep learning-based LPR solution is to be deployed on moving vehicles, scooters in particularly, power consumption is a critical problem. Far from having household power, it is impractical to supply adequate processing power for general high-end GPUs on scooters. However, most state-of-the-art deep networks are too complex to have real-time performance without high-end GPUs, it is a three-way trade-off between inference speed, power consumption, and accuracy.

This chapter focuses on recognizing license plate from cameras on moving scooters. Accordingly, the main complexities faced by the LPR system are varying illumination especially low light at night and ensuing long exposure and thus blur, motion blur, scaling due to varying camera and license plate distances, skewing due to viewing angle and perspective projection, and translations from the continuously moving viewpoint. Lightweight neural networks and less power-consuming VPUs (Vision Processing Units) are chosen to reduce computational intensity and power consumption while maintaining competitive accuracies and real-time performance. This chapter aims to solve the overfitting problem and the data imbalance problem caused by the shortage of some Taiwanese license plate characters such as letter "I" (easily confused with number "1"), letter "O" (easily confused with number "0"), and "4" (same pronunciation as death in Mandarin Chinese) through AI techniques. Accuracies with or without these techniques are compared and tabulated.

1.2. License Plates in Natural Scene

In this chapter, scene images have been captured without any perspective, distance, illumination, background or weather constraints, as long as the license plates in images are legible for human beings. Two cameras, which are front view and right-side view, are mounted at the front of the scooter to capture front vehicles and street parking vehicles respectively.

The cars and buses are in a majority in front view. As for right-side view, most are scooters and cars. An example of scene image captured with front view camera on a rainy day is shown in Fig. 3.

Due to various parking angles of the street parking vehicles and the way cameras are mounted, the license plates are usually more distorted in right-side view than in front view. Several examples of right-side view from various perspective are shown in Figs. 4 to 7.

Figure 3. Vehicles captured with front view camera on a rainy day

1.3 Chapter Organization

The related works about the license plate recognition are briefly introduced in Chapter 2. The techniques involved in our LPR system are introduced in Chapter 3. The details of our LPR system are described in Chapter 4. The experimental results are demonstrated in Chapter 5. Chapter 6 contains the summary of this chapter and some future works. The final chapter is the list of references.

2. RELATED WORKS

Due to the unpredictable variations in luminance, contrast, noise, and geometry of the license plates in unconstrained scenarios, most license plate recognition systems based on traditional machine vision methods are limited to specific scenarios, e.g., ETC and automatic parking management system proposed by M. M. Rashid et al. (Rashid 2012). Deep neural networks, with good generalization behavior, are used to deal with this problem.

Warped Planar Object Detection Network (WPOD-NET) proposed by S. M. Silva et al. (Silva 2018) searches for license plates and regresses one affine transformation per detection, allowing a rectification of the license plate area to a

Figure 4. Distorted far front scooter license plates captured with right-side view camera

rectangle resembling a frontal view. H. Li et al. (Li 2018) proposed an end-to-end deep neural network which can localize license plates in natural scenes and recognize the letters simultaneously in a single forward pass.

Despite outstanding recognition accuracies using complex networks, hardware requirements are demanding as well owing to the massive amount of computation, and thus some efficient approaches were introduced. R. Laroca et al. (Laroca 2019) proposed an end-to-end, efficient and layout-independent license plate recognition system based on modified YOLOv2 object detector, which achieves more than 30 Frames Per Second (FPS) when there are 4 vehicles in the scene on a NVIDIA Titan Xp GPU. Another YOLO-based model proposed by S. M. Silva et al. (Silva 2017) is able to detect and recognize a single license plate with 76 FPS on a NVIDIA Titan X, and 9 FPS on a NVIDIA GeForce GT 750M. Though the execution time is quite impressive, less than 65% of the license plates were correctly recognized in the SSIG-SegPlate dataset (Gonçalves 2016). According to the authors, the main accuracy bottleneck of their approach is character recognition since the characters in training dataset are highly unbalanced.

Figure 5. Less distorted scooter license plates captured with right-side view camera when camera optical axis is almost perpendicular to license plate

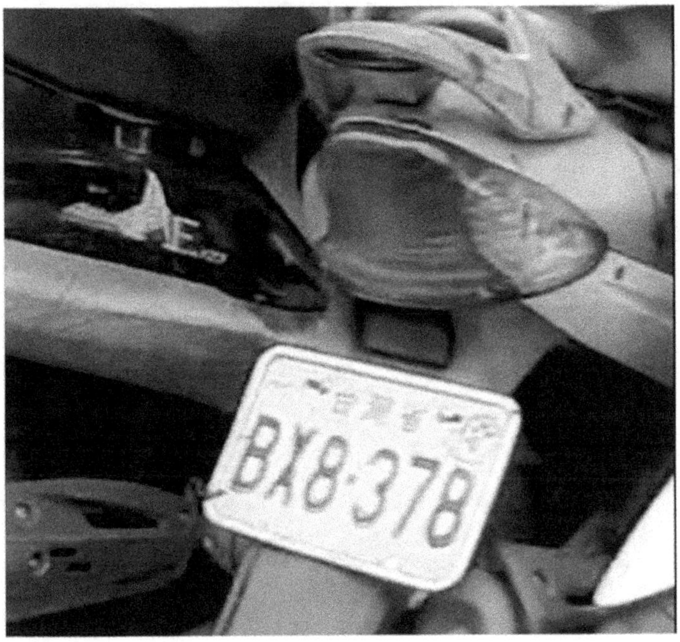

Figure 6. Less distorted car license plates captured with right-side view camera when camera optical axis is almost perpendicular to license plate

Figure 7. Distorted car license plates captured with right-side view camera when camera optical axis and license plate normal vector have difference of more than 45 degrees

3. BACKGROUND

3.1 MobileNetV2

MobileNetV2 proposed by M. Sandler et al. (Sandler 2018) is based on the inverted residual structure and depthwise separable convolutions. Unlike the typical residual structure which creates shortcut between the layers with higher channels as shown in Fig. 3-1, the inverted residual structure links the thin bottleneck layers as shown in Fig. 3-2.

To reduce computational complexity, the intermediate expanded layers in inverted residual structure use depthwise separable convolutions to extract features and employ ReLU6 as activation function as a source of non-linearity. *Regarding* the computational cost of the standard convolutions, it takes $h_i \times w_i \times d_i \times d_j \times k \times k$ multiply-adds to produce an $h_i \times w_i \times d_j$ output tensor with an $h_i \times w_i \times d_i$ input tensor and convolutional kernels $K \in \mathbb{R}^{k \times k \times d_i \times d_j}$. In contrast to the standard convolutions, the computational cost of the depthwise separable convolutions takes only $h_i \times w_i \times d_i \times \left(k^2 + d_j\right)$ multiply-adds. ReLU6 is the same as the usual ReLU except that the activated value has an upper limit, and can be defined as:

Figure 8. An illustration of typical residual structure where the layers with higher channels are connected

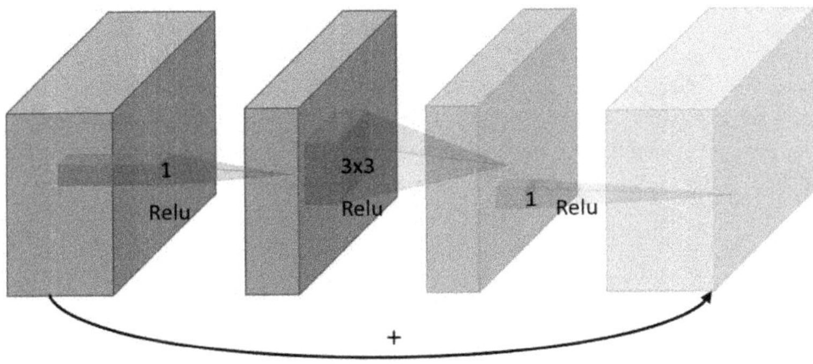

$$f(x) = \min\left(\max\left(0, x\right), 6\right)$$

where x is the input value and $f\left(x\right)$ is the activated value.

In depthwise separable convolutions, the original convolutional layer is separated into two subtasks: a depthwise convolution followed by a pointwise convolution. In contrast to a normal convolution, each kernel of a depthwise convolution is only responsible for a single channel of input data. For instance, a 7×7 pixel image with 3 channels will still have a resolution of 7×7 pixels and 3 channels after depthwise convolving as shown in Fig. 3-3. Each channel in the resultant feature maps is unrelated, hence the need for a pointwise convolution.

Pointwise convolution represents convolving the input data with 1×1 kernels, and which can be used to change the dimension of the feature maps or perform feature transformation across channels, connecting highly correlated features in the same

Figure 9. An illustration of inverted residual structure where the thin bottleneck layers are connected

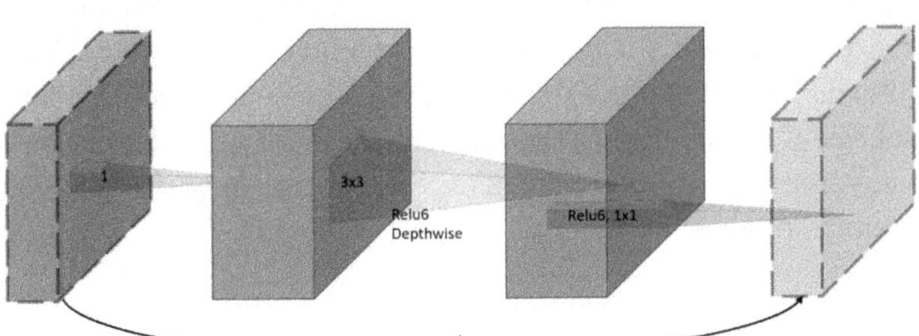

Figure 10. An illustration of depthwise convolution

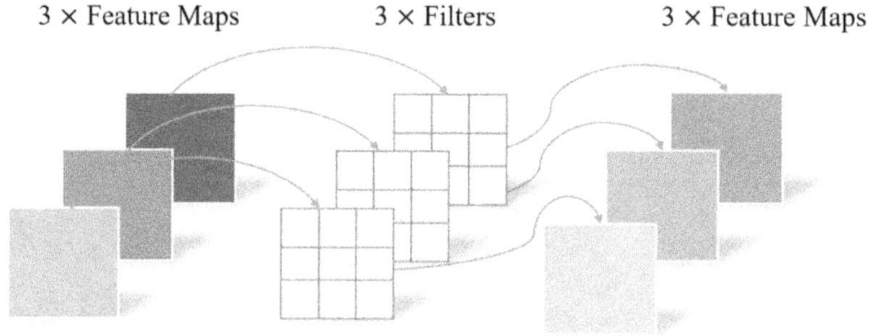

spatial location but different channels. An illustration of pointwise convolution is shown in Fig. 3-4.

Figure 11. An illustration of pointwise convolution

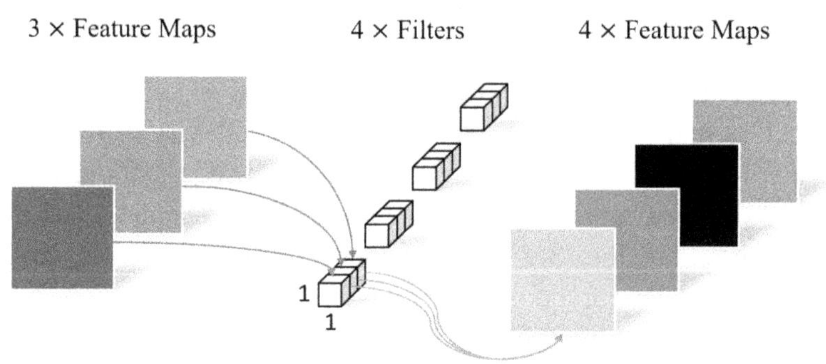

The basic block of MobileNetV2 is a bottleneck residual block as shown in Fig. 3-5. To preserve representational power, the nonlinear activation function in the thin bottleneck layer is removed.

The full MobileNetV2 network structure is shown in Table 1, where each line represents a sequence of 1 or more layers; T represents the expansion factor of bottleneck expansion layer; C represents the number of output channels; N represents the number of identical layers in each sequence; S represents the stride for the first layer of each sequence; and all spatial convolution kernels are 3×3. The whole network uses 3.4 million parameters and has a computational cost of 300 million multiply-adds.

Figure 12. The illustration of bottleneck residual block

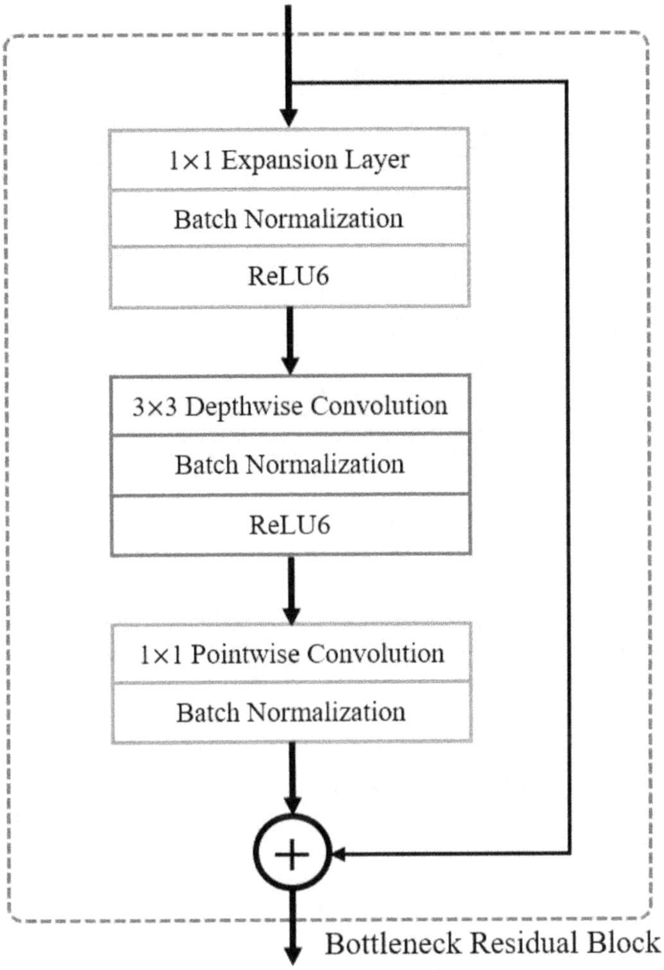

3.2 SSD: Single-Shot MultiBox Detector

SSD is a deep neural network using multi-level features to detect multi-scale objects in images proposed by W. Liu et al. (Liu 2016). The whole network structure can be separated into feature extraction layers and object detection layers as shown in Fig. 3-6. The original SSD uses the VGG16 proposed by K. Simonyan et al. (Simonyan 2014) as the backbone of feature extraction. Object detection layers consist of small convolution filters used to compute confidence of each class and object locations for each cell in the feature maps.

Table 1. The network structure of MobileNetV2

Input	Operator	*T*	*C*	*N*	*S*
224^2×3	Convolution 2D		32	1	2
112^2×32	Bottleneck	1	16	1	1
112^2×16	Bottleneck	6	24	2	2
56^2×24	Bottleneck	6	32	3	2
28^2×32	Bottleneck	6	64	4	2
14^2×64	Bottleneck	6	96	3	1
14^2×96	Bottleneck	6	160	3	2
7^2×160	Bottleneck	6	320	1	1
7^2×320	1×1 Convolution 2D		1280	1	1
7^2×1280	7×7 Average Pooling			1	
1×1×1280	1×1 Convolution 2D				

Figure 13. An illustration of SSD network structure which consists of feature extraction layers and object detection layers
(Liu 2016)

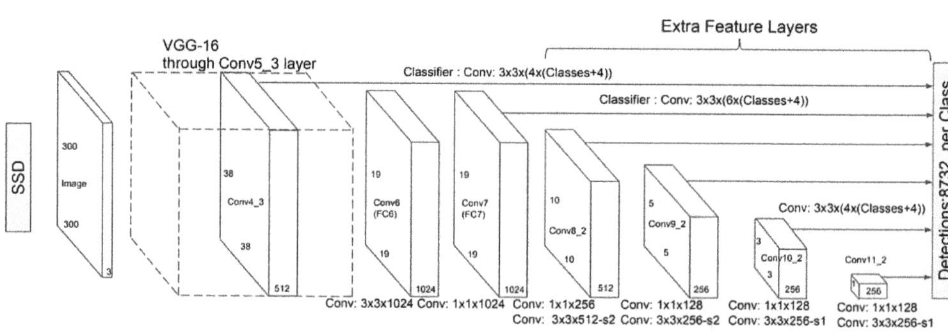

SSD uses multi-scale feature maps to detect objects independently; it detects smaller objects when the feature map has higher resolution and detects larger objects as the dimension of feature maps gradually reduces by CNN in Fig. 3-7.

To make the network stable, instead of predicting the absolute coordinates of the bounding boxes of the objects, the object centers and sizes are estimated relative to the predefined default boxes. Predictions made from the same feature map employ the same set of default boxes, and there are distinct sets of default boxes corresponding to each feature map. Each cell of the feature maps makes t predictions where t is

Figure 14. An illustration of detection in multi-scale feature maps. SSD detects small objects in feature maps with higher resolution (left) and detects large objects in feature maps with lower resolution (right)

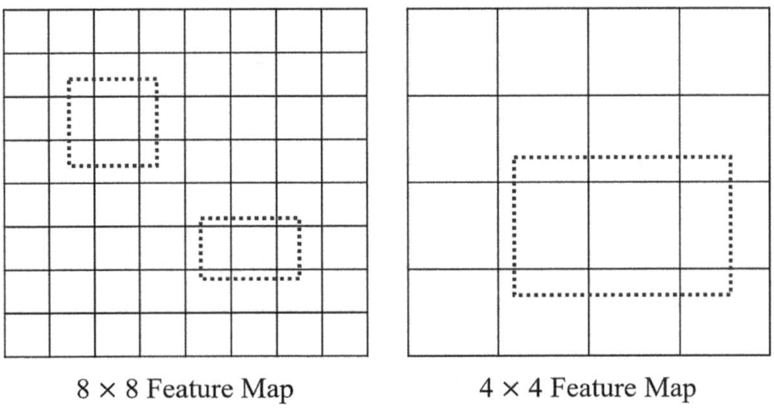

the number of default boxes responsible for that specific feature map. Each prediction produces the confidence of each class, and 4 values relative to the width, height, and center coordinates of the bounding box. Specifically, a $f_w \times f_h$ feature map would produce $f_w \times f_h \times f_t \times (P+4)$ predictions where P is the number of classes. The relationship between the actual predicted bounding box center coordinates b_{cx} and b_{cy}, the default boxes center coordinates d_{cx} and d_{cy}, and the network outputs n_{cx} and n_{cy} can be described as:

$$b_{cx} = n_{cx} \times d_w + d_{cx}$$

$$b_{cy} = n_{cy} \times d_h + d_{cy}$$

The relationship between the actual predicted bounding box width b_w and height b_h, the default boxes width d_w and height d_h, and the network outputs n_w and n_h can be described as:

$$b_w = \exp(n_w) \times d_w$$

$$b_h = \exp(n_h) \times d_h$$

The default boxes are count in positive matches if they have an IoU (Intersection over Union) greater than 0.5 with any ground truth box, negative matches otherwise. Let $x_{i,j}^{p} = \{1,0\}$ represent the matching result of i-th default box with j-th ground truth box of class p. The target loss function L of SSD is the weighted sum of localization loss L_{loc} and confidence loss L_{conf} defined as:

$$L(x,c,n,g) = \frac{1}{M}\left(\alpha L_{loc}(x,n,g) + L_{conf}(x,c)\right)$$

where c is the class confidence; g is the ground truth box; α is the weight of localization loss; and M is the number of positive matches. The total loss is set to 0 if $M = 0$.

The localization loss between the network output n and ground truth box g is a Smooth L_1 loss only calculated from positive matches, and can be defined as:

$$L_{loc}(x,n,g) = \sum_{i \in Positive}^{M} \sum_{v \in \{cx,cy,w,h\}} x_{i,j}^{p} Smooth_{L_1}\left(n_i^v - \hat{g}_j^v\right)$$

where

$$\hat{g}_j^{cx} = \left(g_j^{cx} - d_i^{cx}\right) / d_i^{w},$$

$$\hat{g}_j^{cy} = \left(g_j^{cy} - d_i^{cy}\right) / d_i^{h},$$

$$\hat{g}_j^{w} = \ln\left(\frac{g_j^{w}}{d_i^{w}}\right),$$

and

$$\hat{g}_j^{h} = \ln\left(\frac{g_j^{h}}{d_i^{h}}\right).$$

The confidence loss is the loss of classification which is computed according to the probability of each class. For positive matches, the losses are the sum of binary cross entropy to the ground truth of each class; for negative matches, the losses are the sum of binary cross entropy to the background class, which is labeled class 0, of each class, and can be defined as:

$$L_{conf}\left(x,c\right)=\sum_{i\in Positive}^{M} x_{i,j}^{p}\sum\nolimits_{p} BCE\left(c_{j}^{p},c_{i}^{p}\right)+\sum_{i\in Negative}\sum\nolimits_{p} BCE\left(c_{bg}^{p},c_{i}^{p}\right)$$

where

$$BCE\left(c_{j},c_{i}\right)=-\left(c_{j}\log\left(c_{i}\right)+\left(1-c_{j}\right)\log\left(1-c_{i}\right)\right).$$

3.3 Connectionist Temporal Classification

3.3.1. Overview

It is impractical to have every single horizontal position in handwriting character images labeled to the corresponding character for training a handwriting recognition network. For one thing, two distinct characters could exist in the same horizontal position, which makes the sequence labeling task ambiguous as shown in Fig. 3-8. For another, the predicted sequence from the network without further processing would possibly have many duplicates due to a single character to span multiple horizontal positions as shown in Fig. 3-9. The repeated character "o" in prediction makes the original word "to" become "too", which is an incorrect prediction.

Apparently, labeling every horizontal position or so-called time-step in sequence data is not feasible. CTC proposed by A. Graves et al. (Graves 2006) is a sequence-to-sequence mapping function used to solve the troublesome and ambiguous sequence labeling problem in many real-world sequence recognition tasks, such as handwriting recognition and speech recognition. With CTC, the only thing needed to do for training a sequence recognition network is to provide the ground truth text without any alignment.

Let S be the training dataset from a distribution $D_{Q\times W}$ where Q and W are input space and output space, respectively. The input space Q is the set of all real-valued vector sequences; the output space W is the set of all possible sequences over the characters in given character set, i.e., labels. Each datum in S is a pair of sequences (q, w). The sequence length of input q is always greater than or equal to the sequence length of output w for all sequence pairs in S. The goal of CTC is to train a mapping function $f: Q \rightarrow W$ with dataset S that minimizes the errors.

Figure 15. An illustration of an ambiguity in sequence labeling task for handwriting recognition where the horizontal position overlapped with two different characters are labeled as "?"

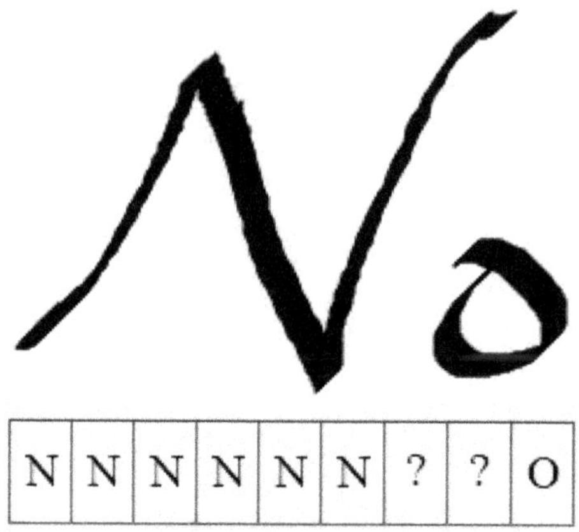

3.3.2. From Network Outputs to Sequences

During the prediction phase, the output sequence w must be decoded from the network output. A network well-trained with CTC outputs a matrix containing probabilities for each character at each time-step as shown in Fig.3-10.

The characters with the highest probability at each time-step are then selected and concatenated. During the merging phase, two characters are merged together if two consecutive time-steps predict the same character. To avoid the characters being erroneously combined, the blank character is introduced in CTC used to indicate

Figure 16. An illustration of an erroneous prediction which mistakenly predicts "to" into "too"

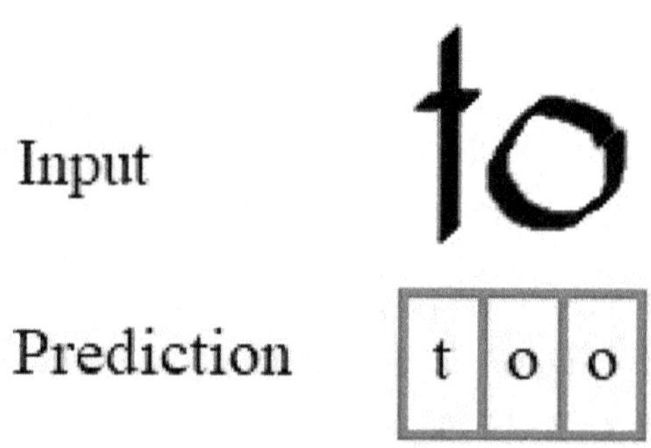

Figure 17. An illustration of network output matrix containing probabilities for each character at each time-step

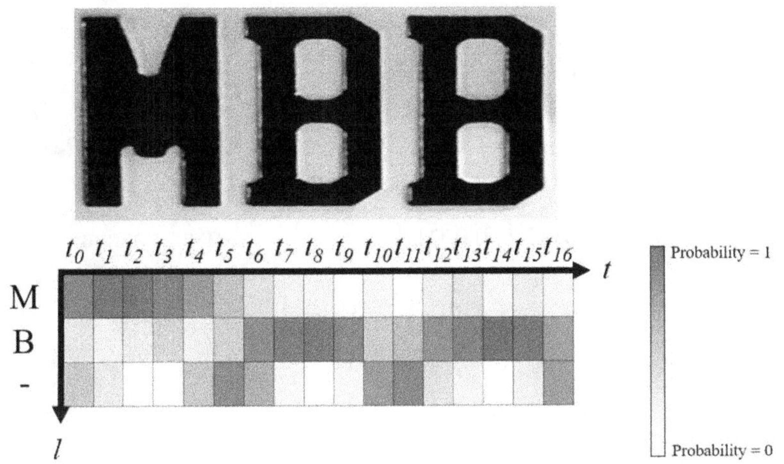

that there is no prediction at the given time-step. Therefore, the repeated characters can be merged together while the instances of the same character are separated with blank character, which is denoted as "-", as shown in Figs. 3-11 and 3-12.

3.3.3 CTC Loss Function

Since the similarity between output sequence and the ground truth is the only thing that matters in CTC, and it has no difference whether the output sequence is exactly aligned with the input sequence or not. A target output sequence could have many different alignments, e.g., a sequence input has length 3 could produce an output sequence "ab" with the following alignments: "aab", "-ab", "a-b", "ab-", and "abb". Let m_l^t be the probability of label l predicted at time t. Denote one possible alignment as π and maximum length of the sequence as T. The conditional probability of one possible alignment output π given input sequence q then can be defined as:

$$p(\pi \mid q) = \prod_{t=1}^{T} m_{\pi_t}^{t}$$

Together, the total conditional probability of the output sequence w given input sequence q can be acquired by summing the probabilities of all different alignments, which can be described as:

Figure 18. An illustration of incorrect merging result due to the lack of the separator between instances of the same character

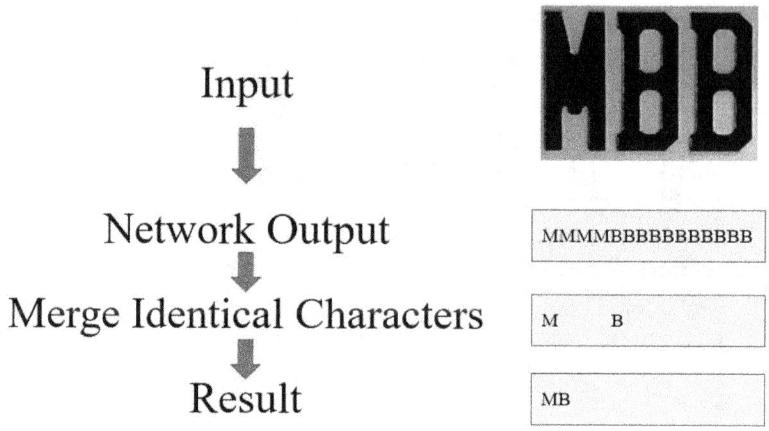

$$p(w \mid q) = \sum_{\pi} p(\pi \mid q)$$

and the CTC loss function is the negative natural log of it, which defined as:

$$-\ln(p(w \mid q))$$

Figure 19. An illustration of correct merging result benefit from the blank separator. Two instances of character "B" are separated by the blank separator

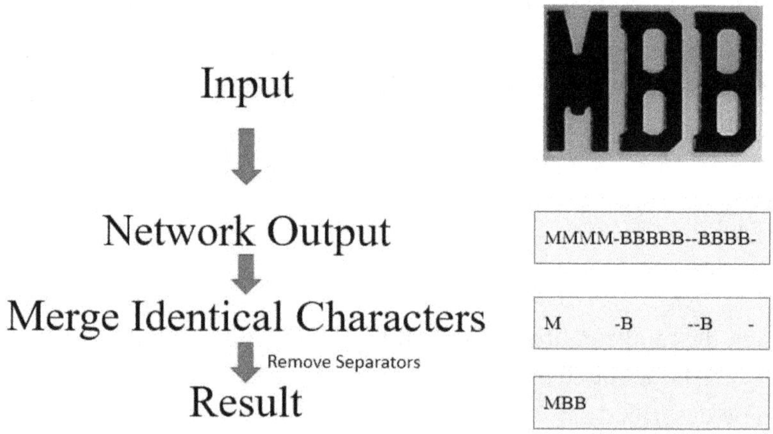

3.4 OpenVINO: Open Visual Inference and Neural Network Optimization

3.4.1. Overview

OpenVINO, developed by Intel, enables CNN-based inference on the Intel platforms, e.g., Intel CPU, Intel Integrated Graphics, Intel FPGA, Intel Movidius Neural Compute Stick, and Intel Vision Accelerator Design with Intel Movidius VPUs.

OpenVINO Toolkit (Intel 2020) is a free and cross-platform toolkit which aims to help users rapidly develop computer vision solutions based on CNNs and distributes workloads across Intel platforms through OpenVINO Inference Engine (IE) plugin library.

Deploying neural network models from the training environment to embedded platforms for inference could be troublesome sometimes. For instance, the typical training processes usually take place on cloud server or personal computer with high-end GPUs. However, inference processes might happen not only on server side but also on embedded platforms. Embedded systems could have limited power supply; therefore, the network optimizations are required for less power consumption. Furthermore, there are several commonly used deep learning frameworks, such as PyTorch, TensorFlow, MXNet, Caffe, and so on. The support of the original training framework on such platforms are uncertain. As a result, it is sometimes unreasonable to use original training framework for inference on embedded systems.

OpenVINO Model Optimizer (MO) is used to optimize a deep learning model and transfer the model to an Intermediate Representation (IR) format which contains the architecture of the trained network, weights, and biases. The IR format model can be loaded and inferred on Intel platforms with the IE. The MO optimizes the model through many mechanisms such as pruning unnecessary model components that are only required at the time of training, combining operations that can be merged to a single operation, and providing an option to convert float point precision from 32 bits to 16 bits.

3.4.2 Deployment Workflow

The illustration of typical workflow for deploying a trained neural network model is shown in Fig. 3-13. The steps to deploy a neural network model to work on Intel hardware are:

1. Train the model on supported deep learning framework and configure MO for that specific framework.

2. Use MO to generate an optimized IR of the model from the architecture of the trained network, weights, and biases.
3. Load the IR format model with IE and the target hardware plugin.
4. Integrate IE with user application.

Figure 20. An illustration of typical workflow for deploying a trained neural network model

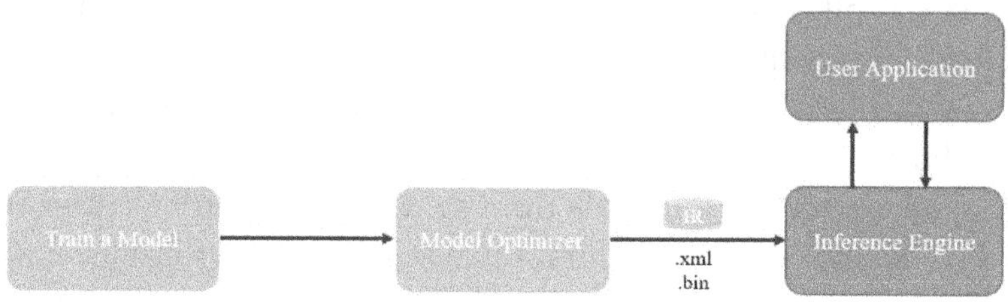

4. METHODOLOGY

4.1 Overview

OpenVINO, developed by Intel, enables CNN-based inference on the Intel platforms, e.g., Intel CPU, Intel Integrated Graphics, Intel FPGA, Intel Movidius Neural Compute Stick, and Intel Vision Accelerator Design with Intel Movidius VPUs.

The methods employed in our LPR system are introduced in this chapter. Our proposed method contains three stages. To start with, the license plates in input images are detected using MobileNetV2-based SSD. Next, crop the detected license plates and correct the rotation. Finally, recognize the characters using compact CNNs with CTC from the cropped license plates. The details of license plate detection and rotation correction are explained in Section 4.2. The details of license plate recognition are described in Section 4.3. Section 4.4 shows the augmentation techniques applied during neural network training. The workflow of our LPR system is shown in Fig. 4-1.

4.2 License Plate Detection

To lower power consumption while maintaining real-time performance for both cameras, MobileNetV2-based SSD is employed in license plate detection stage. The input size is set to 512×512 pixels, and the original backbone of feature extraction is replaced with MobileNetV2 in Section 3.1 where only the first 19 layers are

Figure 21. An illustration of complete workflow of our LPR system

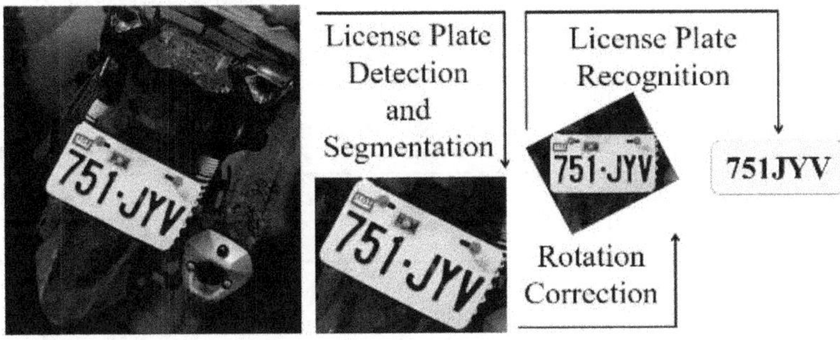

kept, and 4 extra feature layers are appended after Layer 19. Feature maps produced after Layer 15, Layer 19, and 4 appended layers are used to make predictions with corresponding default box set. The sizes of default boxes are calculated using *K*-means on training data where the distance function is relative to the IoU of two bounding boxes (Redmon 2017). The employed feature map resolutions and corresponding default boxes in our system are tabulated in Table 4-1.

Each feature map layer is followed by a convolutional layer to produce the predictions on class confidences and bounding box. The architecture of the Prediction Layer is shown in Fig. 4-2. The whole detection network structure is shown in Fig. 4-3.

The license plates in images could have different distortion since the images are captured from the cameras mounted on moving scooters. To ensure the robustness of the consecutive recognition stage, the license plates are classified according to the orientation of characters on them during the detection stage. The character angle between 30 degrees left and 30 degrees right are divided into 6 categories. In addition, another 2 categories are set for the extreme cases beyond this range. Examples of the license plate for each category are shown in Fig. 4-4.

Considering the number of bounding boxes made during detection, an NMS (Non-Maximum Suppression) is needed to preserve only the best. The predictions with confidences less than 0.03 are discarded for efficiency, and NMS with an IoU threshold of 0.45 applied on those retained afterwards. In contrast to the original NMS, the applied NMS deals with all classes at a time. Since the license plates with different orientations are the same object, only one detected bounding box with the highest confidence should be kept for a license plate instance. The remaining predictions are then cropped and rotated according to their categories, making the characters upwards before feeding into the recognition network. Several examples of rotation corrected license plate are shown in Fig. 4-5.

Table 2. The sizes of default boxes and corresponding feature map resolutions. The widths and heights of default boxes are relative to the side length of square network input

Feature Map Layer	Feature Map Resolution	Default Boxes	
		Width	Height
Layer 15	32×32	0.03	0.03
		0.045	0.042
Layer 19	16×16	0.061	0.053
		0.081	0.055
		0.085	0.079
Extra Feature Layer 1	8×8	0.12	0.075
		0.113	0.106
		0.152	0.098
Extra Feature Layer 2	4×4	0.178	0.124
Extra Feature Layer 3	2×2	0.215	0.143
Extra Feature Layer 4	1×1	0.255	0.17

Figure 22. The architecture of the Prediction Layer

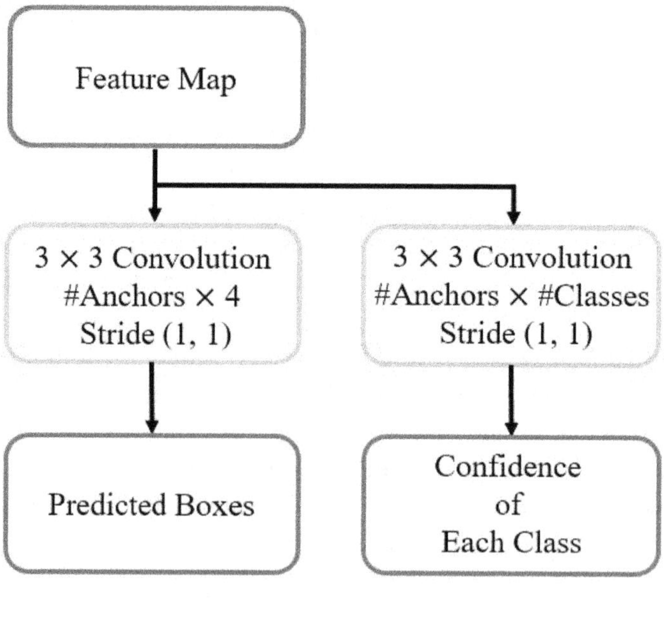

Figure 23. Network structure of MobileNetV2-based SSD

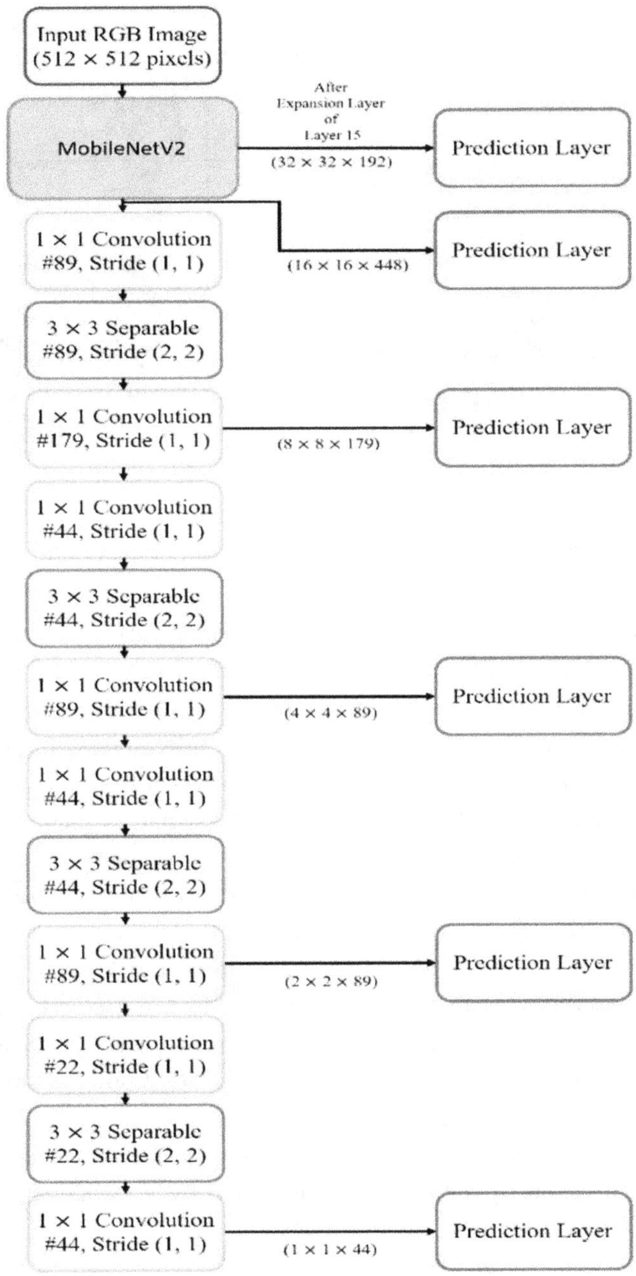

Figure 24. An illustration of 8 categories of character orientation with corresponding license plate images

Degree Left	License Plate Image	Degree Right	License Plate Image
0 – 10		0 – 10	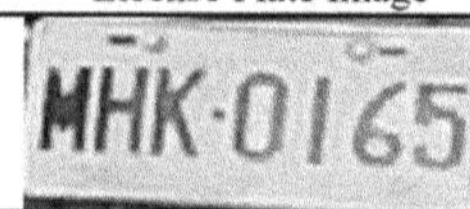
10 – 20		10 – 20	
20 – 30		20 – 30	
> 30		> 30	

4.3 License Plate Recognition

Once the license plates have been detected and rotation corrected according to their character orientations, a modified LPRNet (Zherzdev 2018) is employed to recognize the license plate. The network input size is set to 94×48 pixels, which is close to the aspect ratio of Taiwanese license plates. The architecture of the Basic Block is shown in Fig. 4-6, and the network structure is shown in Fig. 4-7.

The output sequences consist of upper-case format English alphabet (A-Z) and digits (0-9). For each input image, the network produces a matrix containing probabilities for each character at each time-step and generate resultant sequence from that matrix with CTC in Section 3.3. Some heuristic rules are used to get final prediction, e.g., license plates with 7 characters containing 3 leading English letters and 4 consecutive digits, and several frequently misclassified characters in Table 4-2 are replaced accordingly. Predicted numbers which rarely occur in consecutive frames are also filtered since the exact same license plate should be captured multiple times while the camera passes by.

Moreover, to overcome the imbalance problem in license plate characters, the focal CTC loss, inspired by the focal loss proposed by T. Y. Lin et al. (Lin 2017), is used during the training process.

Figure 25. An illustration of license plates with and without rotation corrected

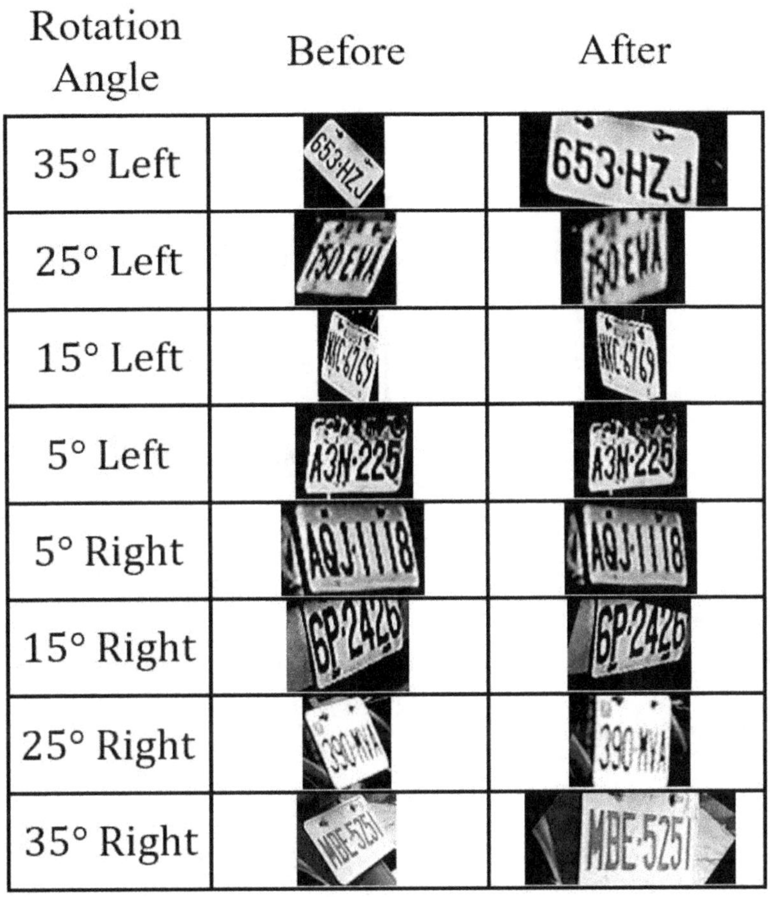

4.3.1 Focal CTC Loss

The character distribution of Taiwanese license plates is severely biased, and thus some techniques are required to achieve better accuracy in license plate recognition. The character distribution of our dataset is shown in Fig. 4-8.

As described in Section 3.4, the conditional probability of the output sequence w given input sequence q is $p\left(w|q\right)$, or just p for simplicity, and the CTC loss function is the negative natural log of it, which is:

$$-\ln\left(p\right)$$

Figure 26. The architecture of the Basic Block

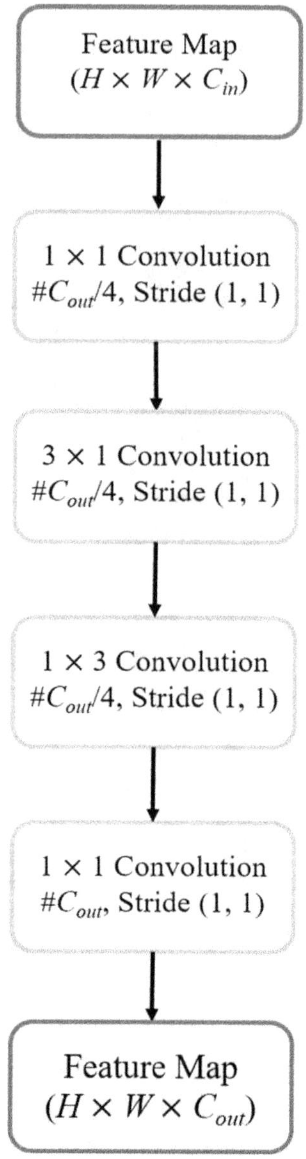

To make the network focus on samples which produce lower probabilities, i.e., those are hard to predict, the focal CTC loss can be defined as:

$$-(1-p)^{r}\ln\left(p\right)$$

Figure 27. The network structure of modified LPRNet employed in our system

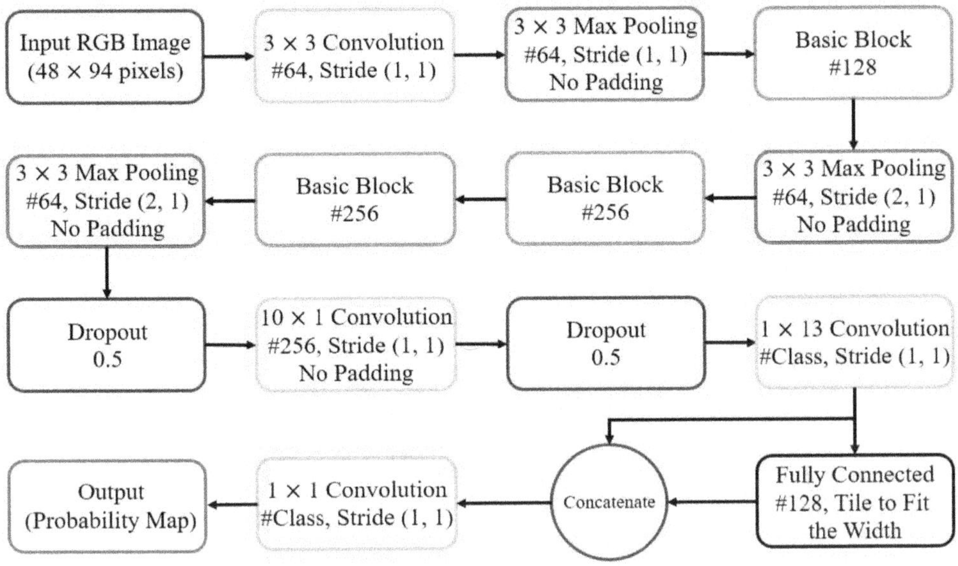

where r is a hyperparameter used to control the loss. The higher the probability p is, the lower $(1-p)^r$ will be, and thus makes the network pay less attention to those well-recognized samples. Setting larger r value does more impact on well-recognized samples relative than hard samples while setting r to 0 equals normal CTC loss. An illustration of the effect of setting different r values for focal CTC loss is shown in Fig. 4-9.

4.4 Data Augmentation

To reduce the possibility of overfitting and enhance the robustness of neural network models, several types of augmentation are applied. During the training process of license plate detection network, scene images are blurred, motion blurred, scaled, cropped, and horizontally flipped. In addition, the contrast, hue, saturation, and

Table 3. Confusing character pairs of Taiwanese license plate

English Letter	A	B	D	G	S	Z
Digit	4	8	0	6	5	7

Figure 28. An illustration of the character distribution of our dataset

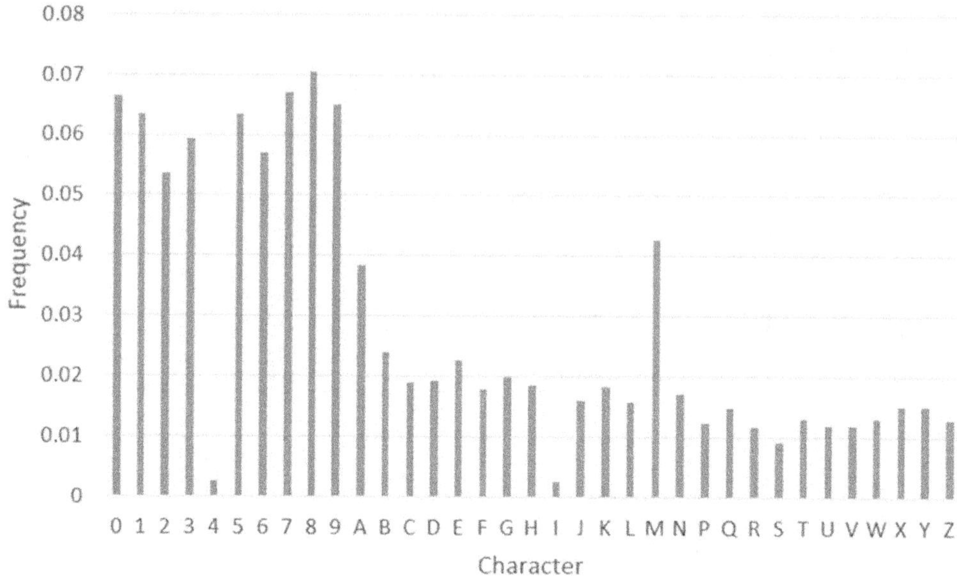

Figure 29. An illustration of the effect of setting different r *values for focal CTC loss*

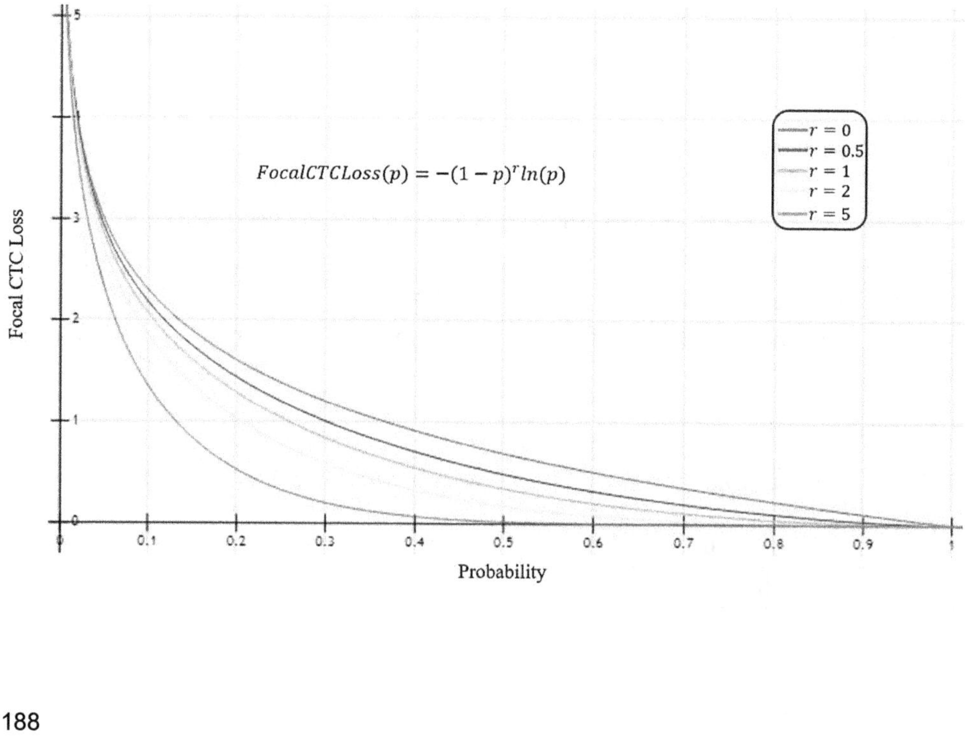

Figure 30. The augmented result of the blurred Fig. 3

Figure 31. The augmented result of the cropped Fig. 3

Figure 32. The augmented result of the horizontally flipped Fig. 3

Figure 33. The augmented result of the grayscale of Fig. 3

Figure 34. An illustration of license plates and augmented results

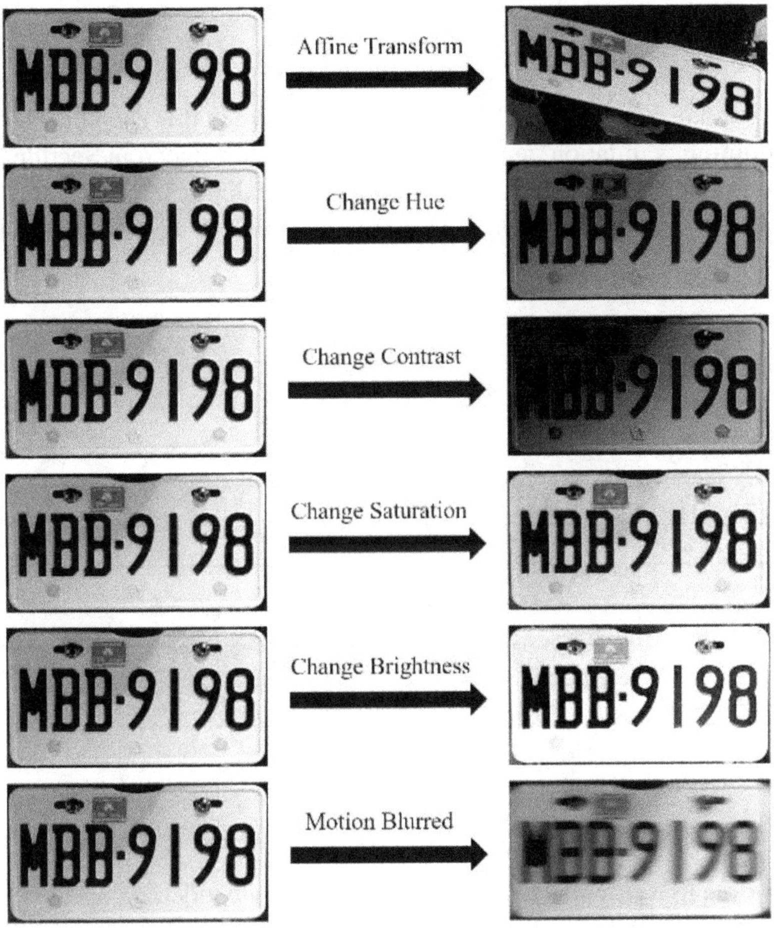

brightness of images are changed, and the grayscale images are generated. Several augmented results of Fig. 1-3 are shown in Figs. 4-10 to 4-13.

Apart from the photometric augmentation methods described above, affine transformations are applied on license plate images, imitating various perspectives on moving scooters. Several license plates and augmented results are shown in Fig. 4-14.

5. EXPERIMENTAL RESULTS

5.1 Overview

The experimental results of our LPR system are shown in this chapter. The datasets used for training, validation, and testing are briefly introduced in Section 5.2. The performance evaluation methods are explained in Section 5.3. Finally, the detection and recognition results are shown in Sections 5.4. The system environment is described in Table 5-1.

Table 4. System environment for our LPR system

System Environment	
Operating System	Windows 10
Central Processing Unit	Intel® Core™ i7-7700T
Random Access Memory	16.0 GB
OpenVINO Toolkit	2019 R1.1
Vision Accelerator Card	Mustang-V100-MX8-R11 with 8 × Movidius Myriad X
Deep Learning Framework	TensorFlow 1.13.1
Programming Language	Python 3.6.5

5.2 Datasets

In our experiments, self-collected datasets are used for network training and validation. Our datasets consist of scene dataset and license plate dataset.

5.2.1 Scene Dataset

There are 513 images and a total of 978 labeled license plates in our scene dataset. This dataset is divided into training data and validation data in Table 5-2.

Table 5. The numbers of images and labeled license plates in our scene dataset

Data	Number of Images	Number of License Plates
Training Data	410	763
Validation Data	103	215
Total	513	978

Figure 35. An illustration of the scene image with labeled license plates

Table 6. The number of license plates of each class in our scene dataset

Category	Training Data	Validation Data	Total
Greater than 30 Degrees Left	1	0	1
20 to 30 Degrees Left	19	6	25
10 to 20 Degrees Left	68	22	90
0 to 10 Degrees Left	227	56	283
0 to 10 Degrees Right	194	47	241
10 to 20 Degrees Right	172	59	231
20 to 30 Degrees Right	73	22	95
Greater than 30 Degrees Right	9	3	12
Total	763	215	978

Table 7. The numbers of images and characters in our license plate dataset

Data	Number of Images	Number of Characters
Training Data	582	3,660
Validation Data	167	1,059
Total	749	4,719

Figure 36. An illustration of the character distribution of our license plate training data

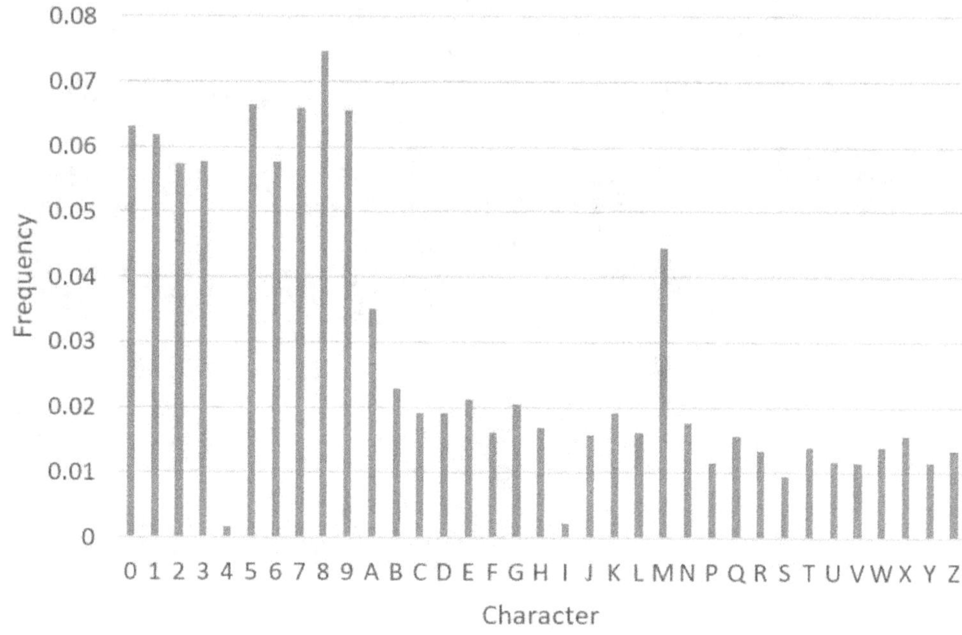

The license plates are labeled according to the orientation of characters on them. The character angle between 30 degrees left and 30 degrees right are divided into 6 classes, and another 2 classes are set for the cases beyond this range, i.e., angle greater than 30 degrees left or 30 degrees right. An illustration of the scene image with labeled license plates is shown in Fig. 5-1. The number of license plates of each class is in Table 5-3.

5.2.2 License Plate Dataset

There are 749 images and a total of 4,719 characters in our license plate dataset. This dataset is divided into training data and validation data in Table 5-4.

The character frequency of Taiwanese license plate is disproportionate, and thus our license plate dataset is imbalanced inevitably. The character distributions of our training data and validation data are shown in Figs. 5-2 and 5-3.

Figure 37. An illustration of the character distribution of our license plate validation data

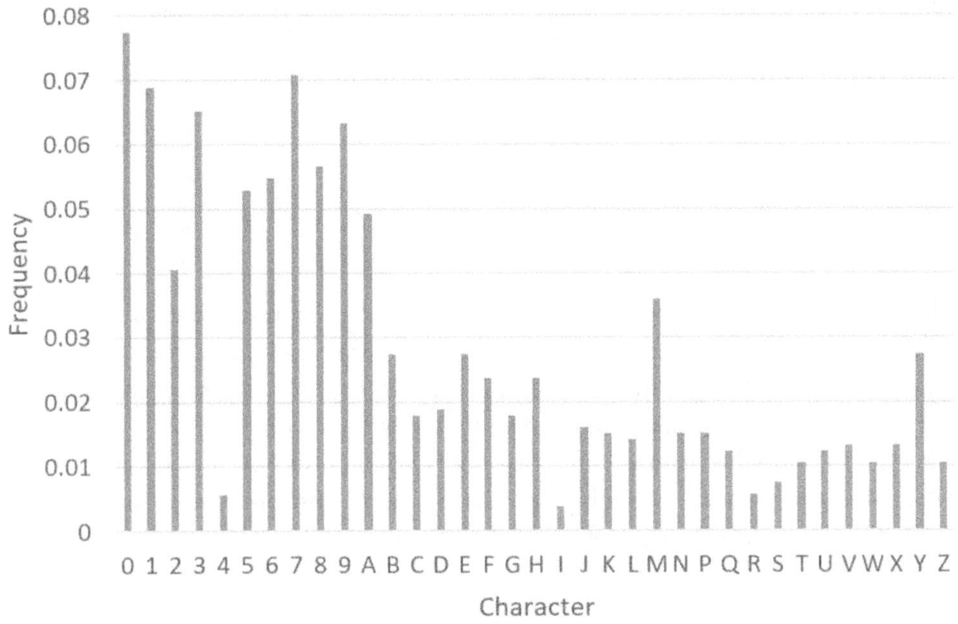

5.3 Performance Evaluation Methods

The performance of an LPR system is poor if it detects and recognizes 1 out of 100 license plates correctly, but unable to detect the remaining 99 license plates. In such condition, this system has 100% precision but only 1% recall due to the

Figure 38. Three conditions for recognition where N represents the nonexistent but incorrectly recognized license plates; M represents the missing license plates; and C stands for the existent and correctly recognized license plates

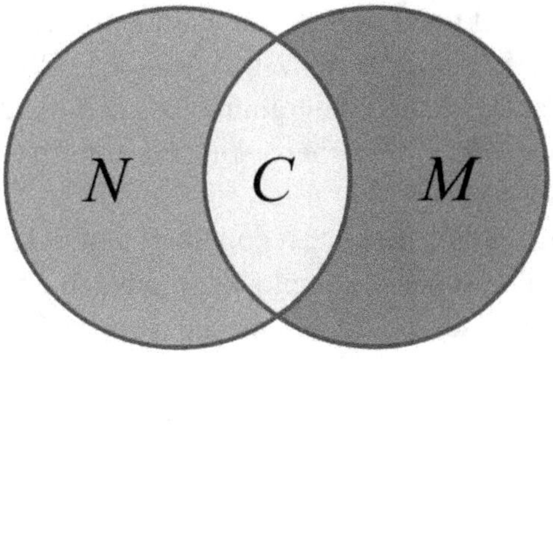

undetected ones are not correctly recognized. Therefore, it is inadequate to evaluate the performance with only precision. To better define the performance of our LPR system, three conditions for recognition are shown in Fig. 5-4 where *N* represents the nonexistent but incorrectly recognized license plates; *M* represents the missing license plates; and *C* stands for the existent and correctly recognized license plates.

The precision is then defined as:

$$\frac{C}{N+C},$$

which represents how accurate the system is. Since the same license plate usually appear in many consecutive frames in real street scene and the correct prediction only count once, if the system is not stable enough, many similar yet different license plate numbers would be count in *N*, making precision lower.

As the example described in the beginning of Section 5.3, it is insufficient to evaluate the performance with only precision. The proportion of how many human eye legible license plates are identified is also essential. The recall is defined as:

$$\frac{C}{M+C},$$

which describes the sensitivity of the system. The performance of our system is evaluated with precision and recall, and the results are demonstrated in Section 5.4. Several examples of human eye legible and illegible license plates are shown in Figs. 5-5 and 5-6.

5.4 Results and Comparisons

5.4.1Training Stage

In the training stage of license plate detection network, the Momentum optimizer is used; the batch size is set to 32 images; the number of iterations is set to 65,000 batches (approximately 858 epochs = 32 images × 65,000 ÷ 2,424 images); the learning rates are in Table 5-5; and the training loss and validation loss are shown in Fig. 5-7. The model weights for further testing are chosen from the iteration with the lowest validation loss.

For the training stage of license plate recognition network, the learning rates are in Table 5-6, the Adam optimizer is employed; the batch size is set to 32 images; the number of iterations is set to 250,000 batches (approximately 1,704 epochs =

Figure 39. Examples of human eye legible license plates

License Plate Image	License Plate Number
	GKT232
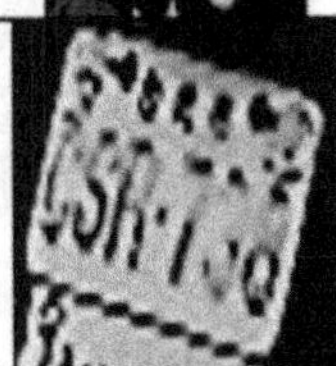	CSR738
	265BMA
	BLZ697
	AXC7671
	EWJ3330

32 images × 250,000 ÷ 4,695 images); the r value for focal CTC loss is set to 2; the training loss is shown in Fig. 5-8; and the validation accuracy is shown in Fig. 5-9. The comparison of losses between using focal CTC loss and ordinary CTC loss for training is shown in Fig. 5-10.

Figure 40. Examples of human eye illegible license plates

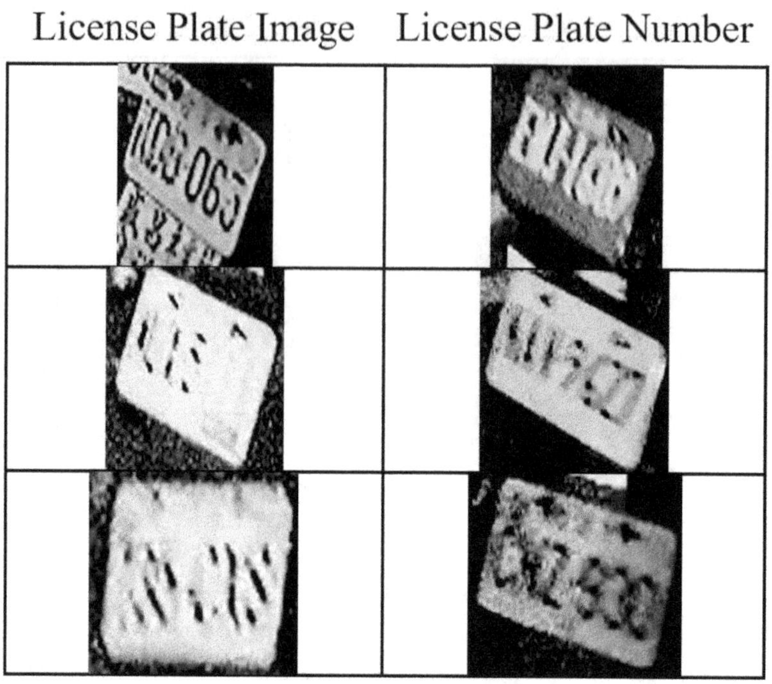

5.4.2 Testing Stage

In testing stage, several street scene videos are used to evaluate our model. There are 1,056 license plates in total captured with 2 different cameras, C_1 and C_2; C_1 works fine at lower speed during both day and night, while C_2 works better during the daytime even at high speed yet suffer from blurring at any speeds during the night. The results of the experiments with different speeds where camera C_1 is

Table 8. The learning rate table for license plate detection network

Iterations	Learning Rate
<1000	0.0004
<10000	0.01
<40000	0.005
<55000	0.0005
<65000	0.00005

Figure 41. Training loss and validation loss of license plate detection network

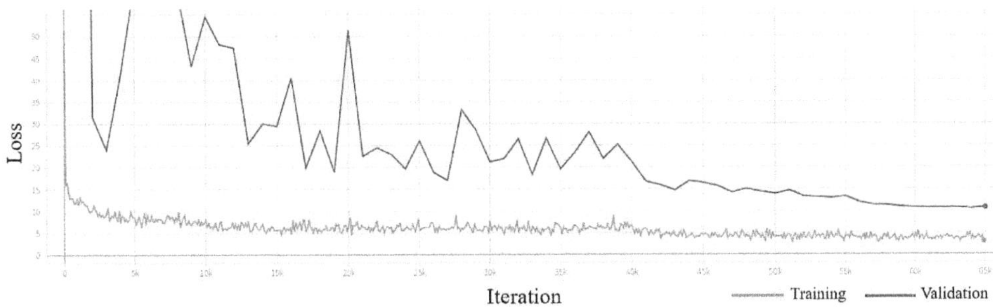

Table 9. The learning rate table for license plate recognition network

Iterations	Learning Rate
<150000	0.001
<200000	0.0001
<250000	0.00001

positioned 30, 45, and 60 degrees to the right in daytime are shown in Figs. 5-11 to 5-13.

Since our system has higher average recall when camera C_1 is positioned 45 degrees to the right in daytime because target motorcycles are parked on our right side and perpendicular to our forward motion direction. The experiments during the night are conducted with this setting, and the results are shown in Fig. 5-14.

There are no recognition results for over 30 km per hour since the license plates are mostly not human eye legible for those speeds using C_1. The average ratio of

Figure 42. Training loss of license plate recognition network

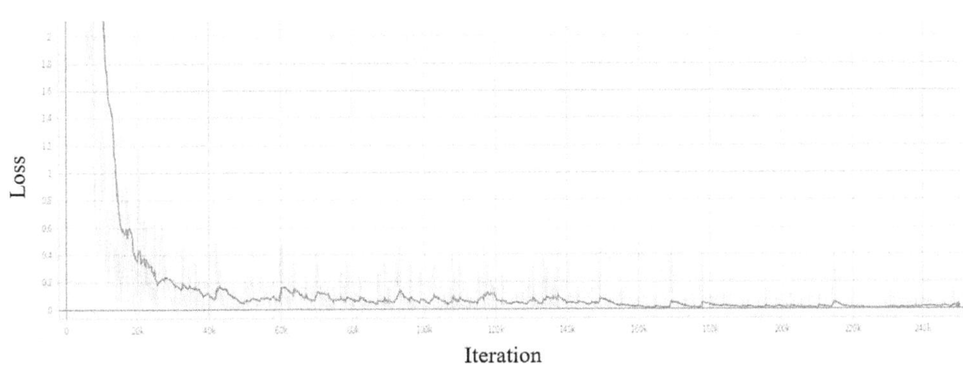

Figure 43. The validation accuracy of license plate recognition network

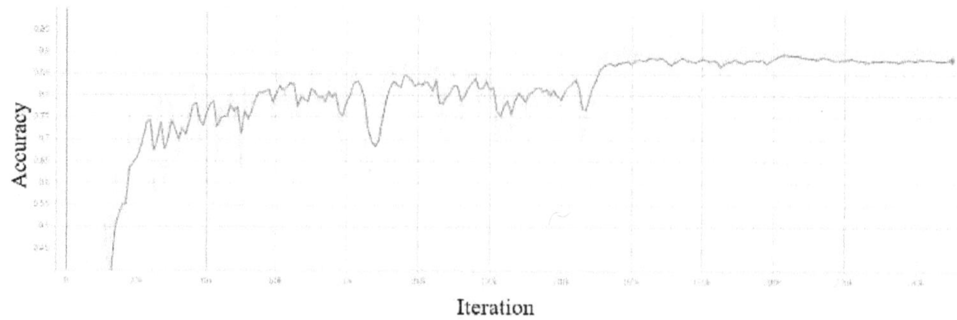

Figure 44. Comparison of CTC losses between using ordinary CTC loss (left) and focal CTC loss (right) for training

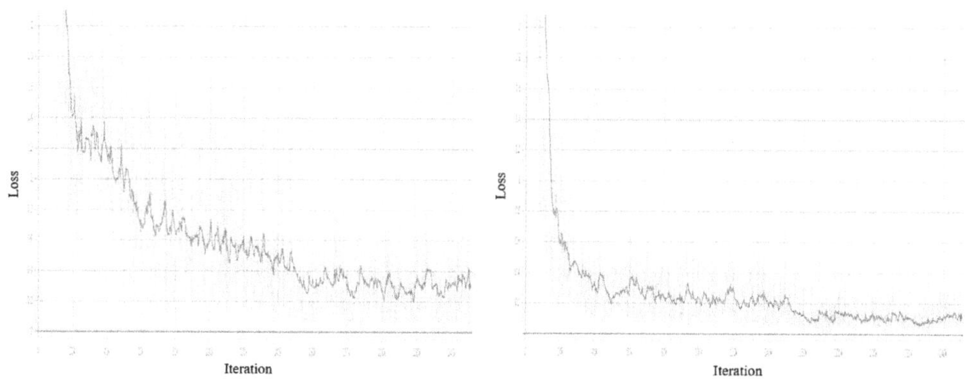

Figure 45. The results of the experiment with different speeds where camera is positioned 30 degrees to the right during the daytime

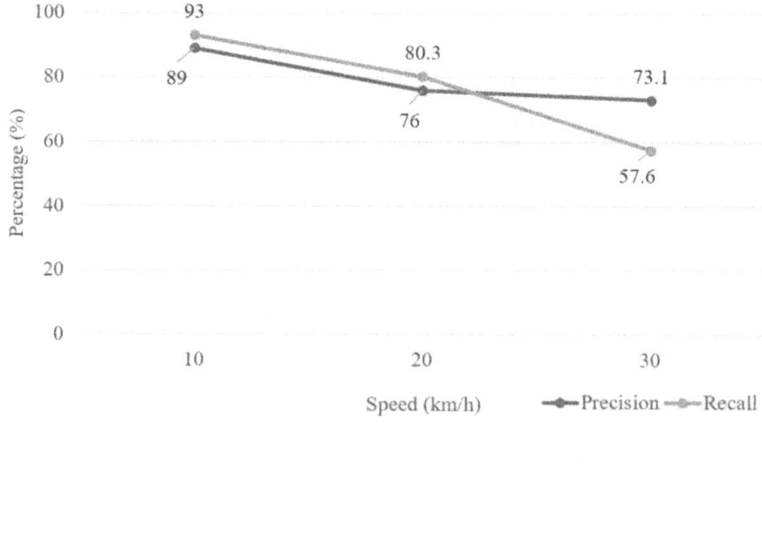

Figure 46. The results of the experiment with different speeds where camera is positioned 45 degrees to the right during the daytime

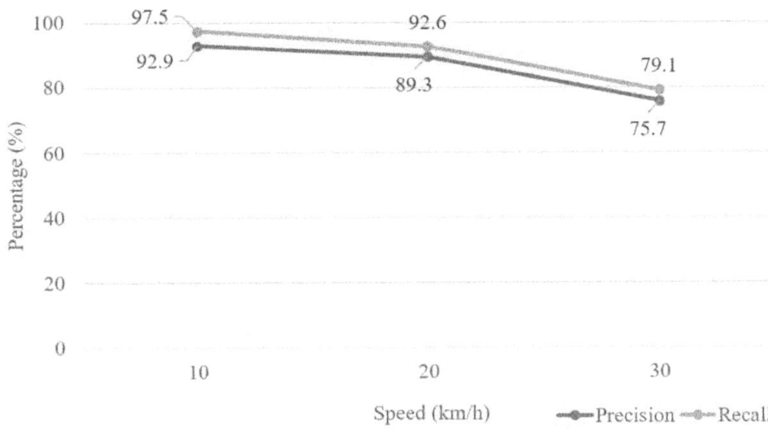

human eye legible license plates for each speed using C_1 during day and night are in Tables 5-7 and 5-8. As for using C_2 as image source, there are 99.3% of license plates are human eye legible at a speed of 60km/h during the daytime, and lower than 10% at a speed of 10km/h during the night.

Using C_2, our system can achieves 95.7% precision and 95% recall at high speed during the daytime. Several techniques are adopted in our system for better performance, the effects of these techniques on recognition results are tabulated in Table 5-9.

Figure 47. The results of the experiment with different speeds where camera is positioned 60 degrees to the right during the daytime

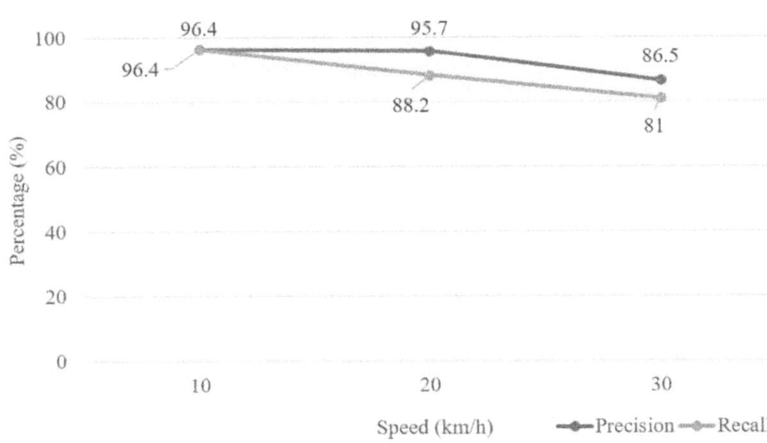

Figure 48. The results of the experiment with different speeds where camera is positioned 45 degrees to the right during the night

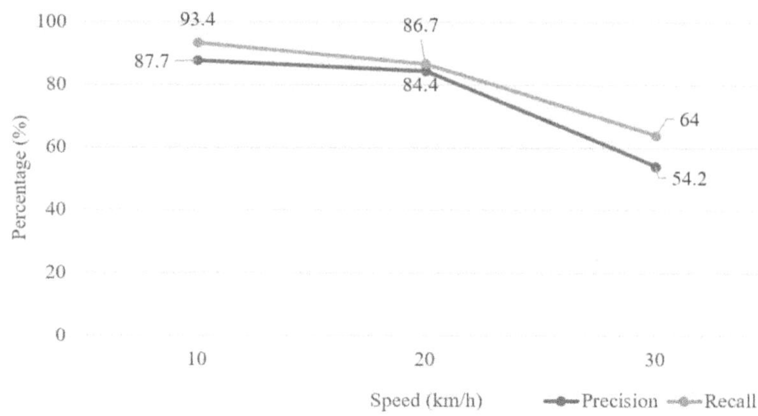

Among the recognition results of our testing data, some license plates our system correctly recognized are shown in Figs. 5-15 to 5-24, and missing license plates are shown in Figs. 5-25 to 5-34, and nonexistent but incorrectly recognized (false alarm) are shown in Figs. 5-35 to 5-44.

The inference speeds with certain number of license plates in given scene of our system is in Table 5-10. The input images are 640×480 pixels. According to the results, our system can achieve real-time performance for 2 cameras when there are 3 license plates are detected in the scene, and 7 license plates for 1 camera.

Figure 49. The average ratio of human eye legible license plates in video for each speed using C_1 during the daytime

Speed	**Ratio of Human Eye Legible License Plates in Video**
10 km/h (81 license plates/40 sec.)	0.965
20 km/h (76 license plates/27 sec.)	0.935
30 km/h (70 license plates/22 sec.)	0.85
40 km/h (31 license plates/17 sec.)	0.365

Figure 50. The average ratio of human eye legible license plates in video for each speed using C_1 during the night

Speed	Ratio of Human Eye Legible License Plates in Video
10 km/h (76 license plates/42 sec.)	0.938
20 km/h (75 license plates/28 sec.)	0.915
30 km/h (50 license plates/20 sec.)	0.61
40 km/h (23 license plates/16 sec.)	0.28

Last, since the power consumption is quite concerned as mentioned above, with the 8×Movidius Myriad X VPUs and Intel i7-7700T as our main processing units, our system consumes only 91.8W at full load, which is enough for scooters to supply.

Figure 51. The effects of different techniques employed in our system

Scene Augmentation		V	V	V	V
Rotation Correction				V	V
License Plate Augmentation		V	V	V	V
Focal CTC Loss			V		V
Precision (%)	40.0	77.8	88.7	94.1	95.7
Recall (%)	4.4	71.5	80.3	92.7	95

6. CONCLUSION AND FUTURE WORKS

In this chapter, an efficient AI solution for LPR on moving scooters is presented. Lightweight MobileNetV2-based SSD is chosen to detect license plates in the scene, and LPRNet is used to recognize the license plates. License plates are detected

Figure 52. License plate 1 our system correctly recognized during testing stage

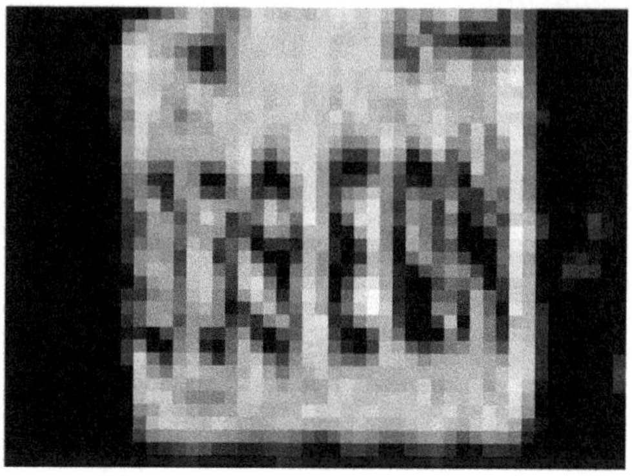

depending on their character orientations, and are rotation corrected accordingly for better recognition precision. Several data augmentation methods are employed to enhance the robustness of our models while preventing overfitting, and focal CTC loss is adopted to conquer the data imbalance problem. Experimental results on real street scenes show that our model achieves 95.7% precision and 95% recall at high speed during the daytime, and an average precision of 86.1% and an avergae recall of 90.1% are attained when the speed is lower than 20 km/h during the night. In addition, our system has real-time performance even if there are 7 license plates in the given scene, and with a maximum power consumption of only 91.8W.

Figure 53. License plate 2 our system correctly recognized during testing stage

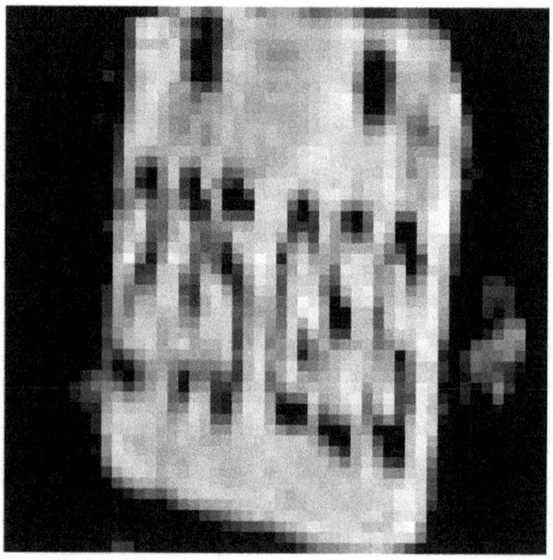

Figure 54. License plate 3 our system correctly recognized during testing stage

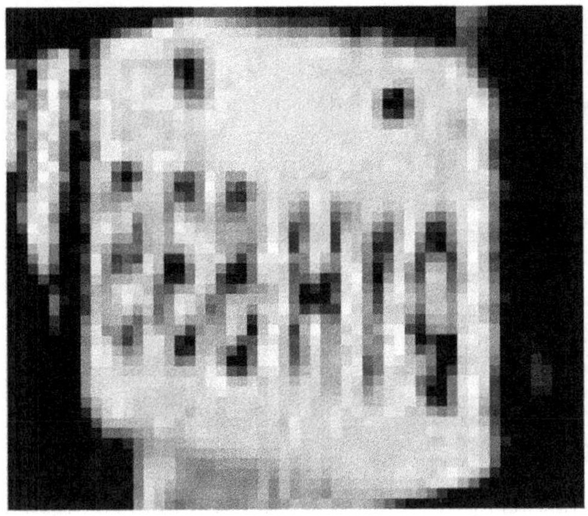

For future work, our system could have better performance with state-of-the-art object detection methods such as YOLOv4 (Bochkovskiy 2020) which are both fast and accurate at the same time. Our adopted MobileNetV2-based SSD is quite fast yet less accurate.

Furthermore, there are some different ways to solve the data imbalance problem. For instance, use Conditional Generative Adversarial Network (CGAN) proposed by P. Isola et al. (Isola 2017) to generate different valid license plates for training. CGAN is effective at synthesizing images from given labels, and thus it is quite convenient to control the character distribution of license plate dataset with CGAN owing to its controllability.

Figure 55. License plate 4 our system correctly recognized during testing stage

Figure 56. License plate 5 our system correctly recognized during testing stage

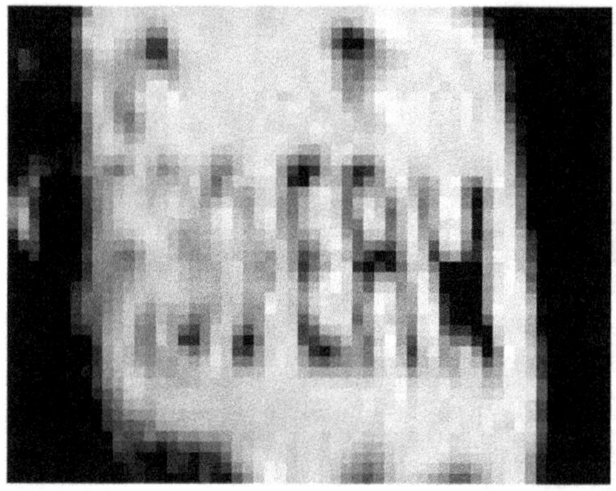

Figure 57. License plate 6 our system correctly recognized during testing stage

Figure 58. License plate 7 our system correctly recognized during testing stage

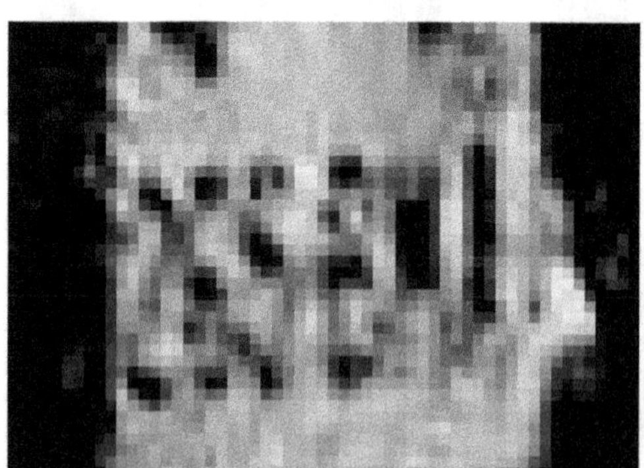

Figure 59. License plate 8 our system correctly recognized during testing stage

Figure 60. License plate 9 our system correctly recognized during testing stage

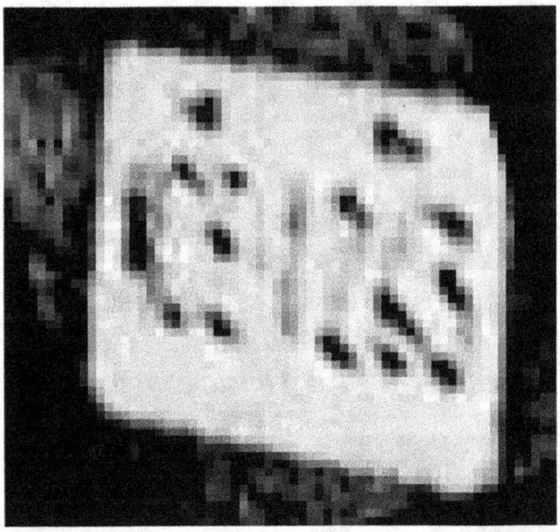

Figure 61. License plate 10 our system correctly recognized during testing stage

Figure 62. Missing license plate 1 during testing stage

Figure 63. Missing license plate 2 during testing stage

Figure 64. Missing license plate 3 during testing stage

Figure 65. Missing license plate 4 during testing stage

Figure 66. Missing license plate 5 during testing stage

Figure 67. Missing license plate 6 during testing stage

Figure 68. Missing license plate 7 during testing stage

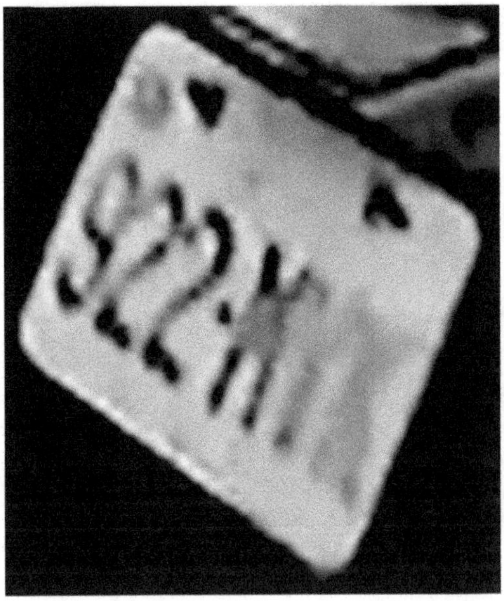

Figure 69. Missing license plate 8 during testing stage

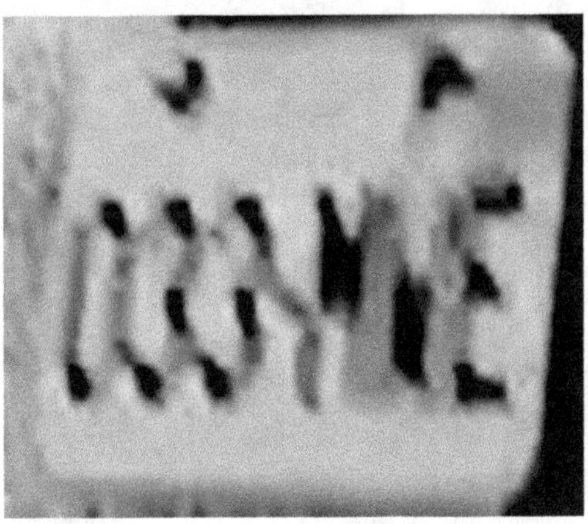

Figure 70. Missing license plate 9 during testing stage

Figure 71. Missing license plate 10 during testing stage

Figure 72. Nonexistent but incorrectly recognized (false alarm) license plate 1

Figure 73. Nonexistent but incorrectly recognized (false alarm) license plate 2

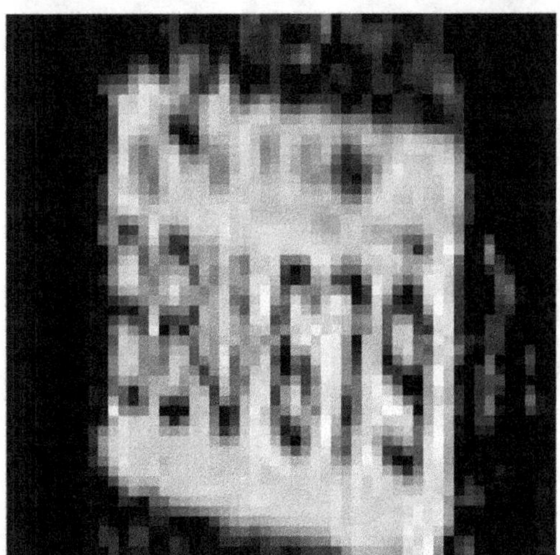

Figure 74. Nonexistent but incorrectly recognized (false alarm) license plate 3

Figure 75. Nonexistent but incorrectly recognized (false alarm) license plate 4

Figure 76. Nonexistent but incorrectly recognized (false alarm) license plate 5

Figure 77. Nonexistent but incorrectly recognized (false alarm) license plate 6

Figure 78. Nonexistent but incorrectly recognized (false alarm) license plate 7

Figure 79. Nonexistent but incorrectly recognized (false alarm) license plate 8

Figure 80. Nonexistent but incorrectly recognized (false alarm) license plate 9

Figure 81. Nonexistent but incorrectly recognized (false alarm) license plate 10

Figure 82. The number of license plates in the scene and the corresponding inference speed of our system

Number of License Plates	0	1	2	3	4	5	6	7
FPS	90	85	81	65	57	46	38	31

REFERENCES

Bochkovskiy, A., Wang, C. Y., & Liao, H. Y. M. (2020). *YOLOv4: Optimal Speed and Accuracy of Object Detection*. https://arxiv.org/pdf/2004.10934.pdf

Gonçalves, da Silva, Menotti, & Schwartz. (2016). Benchmark for License Plate Character Segmentation. *Journal of Electronic Imaging, 25*(5).

Graves, A., Fernandez, S., Gomez, F., & Schmidhuber, J. (2006). Connectionist Temporal Classification: Labelling Unsegmented Sequence Data with Recurrent Neural Networks. *Proceedings of International Conference on Machine Learning*, 369-376. 10.1145/1143844.1143891

Intel. (2020). *Intel Distribution of OpenVINO Toolkit*. https://software.intel.com/en us/openvino

Isola, P., Zhu, J. Y., Zhou, T., & Efros, A. A. (2017). Image to Image Translation with Conditional Adversarial Networks. *Proceedings of IEEE Conference on Computer Vision and Pattern Recognition*, 1125-1134. 10.1109/CVPR.2017.632

Laroca, R., Zanlorensi, L. A., Gonçalves, G. R., Todt, E., Schwartz, W. R., & Menotti, D. (2019). *An Efficient and Layout Independent Automatic License Plate Recognition System Based on the YOLO Detector*. https://arxiv.org/pdf/1909.01754.pdf

Li, H., Wang, P., & Shen, C. (2018). Toward End—to—End Car License Plate Detection and Recognition with Deep Neural Networks. *IEEE Transactions on Intelligent Transportation Systems*, *20*(3), 1126–1136. doi:10.1109/TITS.2018.2847291

Lin, T. Y., Goyal, P., Girshick, R., He, K., & Dollár, P. (2017). Focal Loss for Dense Object Detection. *Proceedings of IEEE International Conference on Computer Vision*, 2980-2988.

Liu, W., Anguelov, D., Erhan, D., Szegedy, C., Reed, S., Fu, C. Y., & Berg, A. C. (2016). SSD: Single Shot MultiBox Detector. *Proceedings of the European Conference on Computer Vision*, 21-37. 10.1007/978-3-319-46448-0_2

Peng, F. K. (2015). *The Spatial and Temporal Analysis of the Impacts of Video Surveillance System on the Theft Crime Rates*. Master Chapter, Department of Crime Prevention and Corrections, Central Police University. https://hdl.handle.net/11296/4swq89

Rashid, M. M., Musa, A., Rahman, M. A., Farahana, N., & Farhana, A. (2012). Automatic Parking Management System and Parking Fee Collection Based on Number Plate Recognition. *International Journal of Machine Learning and Computing*, *2*(2), 93–98. doi:10.7763/IJMLC.2012.V2.95

Redmon, J., & Farhadi, A. (2017). YOLO9000: Better, Faster, Stronger. *Proceedings of IEEE Conference on Computer Vision and Pattern Recognition*, 6517-6525.

Sandler, M., Howard, A., Zhu, M., Zhmoginov, A., & Chen, L. C. (2018). MobileNetV2: Inverted Residuals and Linear Bottlenecks. *Proceedings of IEEE Conference on Computer Vision and Pattern Recognition*, 4510-4520.

Silva, S. M., & Jung, C. R. (2017). Real—Time Brazilian License Plate Detection and Recognition Using Deep Convolutional Neural Networks. *Proceedings of Conference on Graphics, Patterns, and Images (SIBGRAPI)*, 55—62.

Silva, S. M., & Jung, C. R. (2018). License Plate Detection and Recognition in Unconstrained Scenarios. *Proceedings of the European Conference on Computer Vision*, 593-609. 10.1007/978-3-030-01258-8_36

Simonyan, K., & Zisserman, A. (2014). *Very Deep Convolutional Networks for Large—Scale Image Recognition*. https://arxiv.org/pdf/1409.1556.pdf

Zherzdev, S., & Gruzdev, A. (2018). *LPRNet: License Plate Recognition via Deep Neural Networks*. https://arxiv.org/pdf/1806.10447.pdf

Chapter 10
Leveraging Machine Learning in Financial Fraud Forensics in the Age of Cybersecurity

Md Ariful Haque
Old Dominion University, USA

Sachin Shetty
Old Dominion University, USA

ABSTRACT

Financial sectors are lucrative cyber-attack targets because of their immediate financial gain. As a result, financial institutions face challenges in developing systems that can automatically identify security breaches and separate fraudulent transactions from legitimate transactions. Today, organizations widely use machine learning techniques to identify any fraudulent behavior in customers' transactions. However, machine learning techniques are often challenging because of financial institutions' confidentiality policy, leading to not sharing the customer transaction data. This chapter discusses some crucial challenges of handling cybersecurity and fraud in the financial industry and building machine learning-based models to address those challenges. The authors utilize an open-source e-commerce transaction dataset to illustrate the forensic processes by creating a machine learning model to classify fraudulent transactions. Overall, the chapter focuses on how the machine learning models can help detect and prevent fraudulent activities in the financial sector in the age of cybersecurity.

DOI: 10.4018/978-1-7998-8386-9.ch010

1. INTRODUCTION

Cybersecurity is one of the biggest concerns because of the growing number of fraudulent activities both online and offline. Online fraud activities are disrupting and manipulating real-time data transmission and thus stealing the credentials which the hackers could use legitimate identity to make online transactions. For example, in e-commerce transactions, the exchange is happening in real-time. Therefore, any attempt to steal and use legitimate credentials to make purchases fraudulently is an online fraud activity. On the other hand, offline fraud activities can be planting malware or spyware in the target system and then collecting the credentials by monitoring the system's message communications. One such example is the recent fraud activity of the Bangladesh Central Bank's SWIFT payment systems (Zetter, 2016). The hackers had stolen the credentials by planting Spywares in the SWIFT system and then making transaction instructions during non-business hours. Thus, there are no questions on the need for developing cyber forensic capabilities in almost all sectors, specifically in the financial sectors, because of massive loss to the institutions and customers.

According to PwC's global economic crime and fraud survey, cybercrime increases 34 percent year over year (Global, 2020). Identity theft accounted for approximately 20.33 percent of the nearly 3.2 million fraud cases reported to the Federal Trade Commission (FTC) in 2019, according to the Ascent (Ascent, 2020). Credit card fraud is the most prevalent form of identity theft, accounting for 41.8 percent of all identity theft reports, according to Ascent's report, "Identity theft and credit card fraud statistics for 2020." In February 2016, the Bangladesh Central Bank was the target of a cyberattack. Hackers used the SWIFT network to send 35 fraudulent instructions to transfer nearly US$1 billion from a Bangladesh Bank account at the Federal Reserve Bank of New York. Five out of thirty-five fraudulent instructions were successful in stealing US$101 million. The investigation into this case is ongoing, and efforts to reclaim the stolen funds keep continuing.

The statistics above reflects the underlying reasons why the institutions are overly concerned to handle cybercrime issues and therefore, investing in developing preventive mechanisms for cybercrime. According to Mckinsey & Company (Hasham, Joshi, & Mikkelsen, 2019), private companies spent approximately $8.2 billion on anti–money laundering controls alone in 2017. Thus, there are always ongoing efforts to develop preventive measures and forensic capabilities to deal with cybercrimes in the financial sectors. In this age of cybersecurity and advanced analytics, machine-learning-based models and preventive mechanisms gain popularity because of the model's accuracy to predict and classify the potential anomalies in cyberspace. In this chapter, we would start with some background discussions on the cybercrime prevention techniques currently in place and then how we can leverage machine learning-based algorithms to assist in the development of financial crime forensic capabilities.

Machine learning-based algorithms also have their inherent challenges because of the quality of the data that is the core of developing and training the model. Often, there are security concerns of sharing the customer data, and thus the available datasets may not accurately reflect the actual system under consideration. Another big concern is the imbalanced datasets. The datasets become imbalanced because the fraction of the reported fraud activities in the datasets are tiny compared to the complete dataset and thus are highly skewed. In this chapter, we would use an open-source e-commerce dataset collected from a data mining contest. We then provide enough details on the data mining, data preprocessing, data field details, and algorithms which could be the potential candidates to handle the imbalance datasets problem. We would evaluate several machine learning algorithms such as Random Forest, Logistic Regression, K-nearest neighbors (KNN), Decision Tree (DTT), Naïve Bayes, Support Vector Machine (SVM), and Neural Network Multi-Layer Perceptron, etc. We then report the most appropriate algorithms (algorithms with the highest prediction accuracy) to build the models to classify fraudulent activities. We also provide the visualization of the data and provide model codes (Github) that others could use as a reference for developing such capabilities. We use undersampling technique to handle the imbalance in the training dataset. We use Python Scikit-Learn (SKLEARN) libraries for developing the classification models. Overall, in the chapter, our objectives are two-fold: discussing the challenges in developing cyber forensic capabilities by automating credit card fraud detections and then proposing machine-learning-based models to handle the same. The discussions would guide and motivate the readers to build similar preventive and forensic capabilities in other financial sectors to deal with the cybersecurity issues.

Our significant contributions are as follows:

1. Proposing a machine learning-driven forensic process for e-commerce transaction
2. Develop and illustrate the classification models using Python machine learning libraries
3. Detailed discussion on solving imbalanced dataset problems during classification model development.

We organize the rest of the chapter as follows. Section 2 provides background information on cyber forensic, machine learning, and data mining, etc. We discuss the main focus of the chapter in Section 3 and present the literature review. In section 4, we propose our financial fraud forensic model and associated processes.

Section 5 presents the case study of developing a machine learning-driven model and capabilities for forensic methods. Finally, section 6 concludes the chapter.

2. BACKGROUND INFORMATION

In this section, we provide some preliminary definitions of financial sectors, cybersecurity, cyber fraud, and cyber forensics. These definitions would help the audience to follow and digest the later sections of this chapter smoothly.

2.1 Financial Sectors

In general, the financial sector is a subset of the economy composed of businesses and institutions that offer financial services to commercial and retail customers. This sector includes a diverse range of businesses, including but not limited to public and private banks, investment firms, insurance firms, and real estate firms. The efficiency of its financial sector largely determines the health of an economy. The stronger the economy, the safer the country or region is. A weak financial sector is frequently indicative of a declining economy. The reason behind this is because a significant portion of the income comes from mortgage and loan incomes. This income increases as the interest rates decline. One example of financial service is the credit cards that we heavily use in our day-to-day purchases. Not only the banks and financial institutions provide the credit card services, but there are other large retailers and distributors who also provide the same services for their customers. Examples are Apple card, and Amazon credit card services, who traditionally do not fall under the financial sector, but we can purchase products and pay the monthly installments using those cards.

2.2 Cybersecurity

The National Institute of Science and Technology (NIST) defines cybersecurity (Force, 2018). The definition treats cybersecurity as a mechanism to prevent, protect, and restore computer devices, electronic communication systems, and information to ensure data availability, integrity, authentication, and confidentiality. In another definition, the NIST Joint Task Force defines (Force & Initiative, 2013) cybersecurity as "the ability to protect or defend the use of cyberspace against cyberattacks." By cybersecurity, we mean the measures and actions necessary to safeguard the information technology infrastructures, networks, devices, and services. This also includes protecting customers' or users' data against unauthorized access or unlawful use, as well as the practice of ensuring information's confidentiality, integrity, and availability. With the advancement of the internet and cloud-based services, there comes the risk of cyberattacks on information technology systems and user data. For instance, there is a significant risk associated with online shopping via credit card services, as the transaction occurs over the internet.

2.3 Cyberfraud and Cybercrime

The terms "cyberfraud" and "cybercrime" are nearly interchangeable. Cyberfraud is a type of deliberate deceptive action undertaken to gain an unfair or unlawful advantage through the use of online facilities such as the internet. Online credit card theft is the most prevalent type of cyberfraud. Additionally, non-delivery of products purchased through online auctions or merchandise are common forms of monetary cyberfraud. Cyberfraud can also refer to data breaches, privacy violations, identity theft, and cyberbullying. On the other hand, in cybercrime, computer devices are used to commit illegal acts. Therefore, cybercrime is another term for computer crime. Cybercrime includes, but is not limited to, online fraud, the trafficking of child pornography and intellectual property, identity theft, and privacy violations. In short, cyberfraud is any deceptive action using online facilities, and cybercrime is the use of computers and online facilities to perform unlawful activities.

2.4 Cyber Forensics

Cyber forensics is synonymous with "computer forensics" or "digital forensics." Although not precisely defined, computer forensics is the practice of collecting and recording evidence from computing devices in a manner suitable for presentation to the law-enforcing agency. Computer forensics conduct a structured investigation process. When we refer to cyber forensics, we refer to the electronic discovery technique used to identify and disclose technical criminal evidence through the use of online or internet-based resources and other computing devices. For instance, in an advanced persistent threat (APT), intruders frequently use internet gateways to infect target computing devices with malware and then steal the necessary login credentials by monitoring traffic and logs.

2.5 Cyber Vulnerabilities

Cyber vulnerability is a flaw in the system hardware or software that a potential attacker could exploit to launch an attack against the system. NIST defines vulnerability (Ross, McEvilley, & Oren, 2016) as "Weakness in an information system, system security procedures, internal controls, or implementation that could be exploited or triggered by a threat source." Although we focus on machine learning-based threat classification in this chapter, the cyber vulnerability definition is necessary to understand any digital forensics analysis.

2.6 Machine Learning

Machine learning is a subfield of artificial intelligence (AI) that enables systems to self-improve. Machine learning is the process of developing computer programs that can access data and self-learn it. The learning process begins with observations or data to identify patterns in data and make more informed decisions using well-defined statistics-based algorithms. The primary objective is to enable computers to automatically learn and adjust their behavior without constant human intervention or assistance. Machine learning comes in various flavors, including supervised learning, unsupervised learning, reinforcement learning, transfer learning, and deep learning. For example, classification tasks can be either supervised or unsupervised, depending on the labeled or unlabeled data available. Numerous resources on machine learning and algorithms are available. Readers are encouraged to familiarize themselves with some of these techniques if they are unfamiliar with machine learning or artificial intelligence.

2.7 Data Mining

Data mining identifies and extracts patterns from large data sets using machine learning, statistics, and database systems. Data mining is an interdisciplinary subfield of computer science and statistics that aims to extract knowledge from large data sets and transform it into a usable framework for subsequent use. Data mining deals with accumulating raw data, pre-processing of data, developing models and inferences, etc. Data mining also deals with performance metrics, the post-processing of discovered structures, visualization, online updating, etc.

The above background information would help the reader dive into the chapter's main focus, which utilizes machine learning-based classifications in financial transaction anomaly identifications.

3. MAIN FOCUS OF THE CHAPTER

Identifying the abnormal pattern in the available datasets is one of the initial steps in any cyber forensic analysis. This chapter focuses on utilizing machine learning-based algorithms to assist in anomaly detection in financial transactions. We apply classification algorithms based on the type of available datasets and develop a machine-learning model that would appropriately classify the illegal or fraudulent transactions among all the transactions. We take the credit card transactions as a sample because of the availability of the data. But provide enough guidance on how to apply the model to other financial datasets. This anomaly detection would help perform any forensic analysis that we may want to solve later by confirming that there is an abnormal pattern of behavior found by analyzing the data using the machine learning model. In the following subsections, we present the issues and controversies that we think necessary in financial data forensics. Also, we discuss similar works that are already available to provide the audience an idea of where we are now and where we are going in financial forensics. Lastly, we would also discuss how machine learning-based models can benefit the financial industry.

3.1 Problems and Controversies in Financial Transaction Predictive Modeling

There could be a couple of challenges that we need to consider while developing any predictive machine learning model on the financial data to build forensic capabilities. Often, the data source anonymizes the original transaction data by applying available algorithms because of customer data's privacy concerns and sensitive nature. One such example is the anonymization of the raw data using the principal component analysis (PCA) technique before applying any training algorithms. However, applying the PCA may directly transform the data into a format that is not understandable or interpretable by human operators. This lack of interpretation may cause some initial problems in data visualization and identify abnormal patterns.

The second problem is that the number of fraudulent transactions is usually significantly less compared to the overall transaction amount. However, even if the number of fraudulent transactions is much less, we cannot ignore the loss incurred through those transactions. Therefore, having a tiny portion of fraudulent or anomalous transactions leads to an imbalanced dataset problem that imposes more challenges in applying machine learning algorithms. An imbalanced dataset is one in which the distribution of examples across known groups or classes is distorted or biased. The distribution can range from a marginal bias to an extreme disparity. For example, there could be a few abnormal classes where the number of regular class data is in the range of hundreds, thousands, or millions of instances. This uneven distribution of the data classes in the training datasets may impose extra challenges in developing the machine learning models. We are interested in classifying those few abnormal cases accurately.

There could be other types of issues in financial forensics. The high cost of research and development and a lack of understanding of the business gain using machine learning models could potentially make the client disinterested. In this chapter, we skip some of those controversies and focus only on the above two problems that we have just discussed. Then, we would provide predictive modeling techniques that could handle the above two issues of anomaly detection in financial transactions.

3.2 Literature Review

There are numerous approaches proposed for financial fraud identification, specifically detecting anomalies in financial transactions and identifying credit card frauds. Therefore, we divide this literature review by focusing on two types of research. The first one deals with applying machine learning in financial statements, and the second one is the application of machine learning in credit card fraud detection.

This section deals with some of the works related to machine learning in the financial sector, specifically anomaly detection in financial statements. Lokanan et al. (Lokanan, Tran, & Vuong, 2019) present methods to detect anomalies in financial statements using machine learning algorithms. Authors have used the listed companies in Vietnam to present their studies. This paper presents some interesting thoughts as they use customer's financial statements to detect anomalies. Jofre and Gerlach (Jofre & Gerlach, 2018) present techniques to fight against accounting fraud using forensic data analysis. Kotsiantis et al. (Kotsiantis, Koumanakos, Tzelepis, & Tampakas, 2006) have applied machine learning techniques to predict fraudulent financial statements.

This section presents some of the works related to credit card fraud detection using machine learning algorithms. Meas et al. (Maes, Tuyls, Vanschoenwinkel, & Manderick, 2002) propose credit card fraud detection using Bayesian and neural networks. Their approach was one of the early contributions in fraud detection in credit card transactions. Yee et al. (Yee, Sagadevan, & Malim, 2018) present the methods to identify credit card fraud detections using machine learning as a data mining technique. Authors have utilized a tool called WEKA (Waikato Environment for Knowledge Analysis) for generating dummy transaction data and then applied Bayesian network classifiers and other machine learning algorithms. Awoyemi et al. (Awoyemi, Adetunmbi, & Oluwadare, 2017) present a comparative study on the credit card fraud classification using naïve Bayes, k-nearest neighbors, and logistic regression classifiers. Randhawa et al. (Randhawa, Loo, Seera, Lim, & Nandi, 2018) use ensemble learning algorithms such as AdaBoost and Majority Voting to perform the credit card fraud detection. Thennakoon et al. (Thennakoon, Bhagyani, Premadasa, Mihiranga, & Kuruwitaarachchi, 2019) use machine learning to detect fraudulent credit card transactions in real-time. This list may go on and on. From the discussion, we understand that researchers have used different machine learning algorithms to identify credit card fraud and financial statements.

Our objective is to present how the machine learning models can assist in financial fraud forensic analysis in this work. For example, there could be a cyberattack on the transaction system for which we need to mine the network logs and events to identify anomalies. Another example is there could be direct anomalies in financial transactions such as credit card transactions. Identifying the fraudulent activities would help in developing a comprehensive forensic analysis. Thus, although we would use the credit card dataset to build machine learning-based classification models, but our goal is to relate the classification result to the financial forensic analysis processes.

4. FINANCIAL FRAUD FORENSIC MODEL

Digital forensic and crime investigation process is not a new thing. There are several works that propose formal guidelines to follow in case of investigating any digital or cyber-crime. NIST (Kent, Chevalier, Grance, & Dang, 2006) proposed a four step digital forensic process. These steps are collection, examination, analysis, and reporting. During the collection process, data of a particular event are identified, recorded, and preserved. The second phase, examination, employs forensic tools and techniques tailored to the collected data to extract relevant information. The assessments may incorporate both automated and manual processes. The next phase, analysis, entails deriving helpful information from the test results to address the questions that motivated the collection and examination. The final step involves reporting the analytical findings. This representation may include describing the actions taken, determining what additional actions are necessary, and making recommendations for improvements to the forensic process. (Brinson, Robinson, & Rogers, 2006) presented a detailed ontology for the cyber crime forensic methods. (Bayuk, 2010) presents detail cyber forensic techniques along with information security forensics.

In this paper, our goal is not to derive a complete novel forensic model. Instead, we aim to utilize available machine learning capabilities to enhance the forensic processes targeting financial sectors. Figure 1 gives the broad outline of our financial forensic model based on the NIST forensic model.

Our proposed model follows the machine learning flows and maps with the NIST forensic model processes. We divide the forensic process into four phases: data collection, data processing, model development, and performance reporting. The data collection process in the model is in line with the NIST collection process. We subdivide this process into data collection, identify correlations, initial analysis, and data visualization subprocesses. The data processing process subdivides into data transformation, feature extraction, sampling methods, and splitting the data into training and testing segments. The model development process subdivides itself into algorithm selection, model development, fitting the model with training data,

Figure 1. Machine learning-based forensic model processes

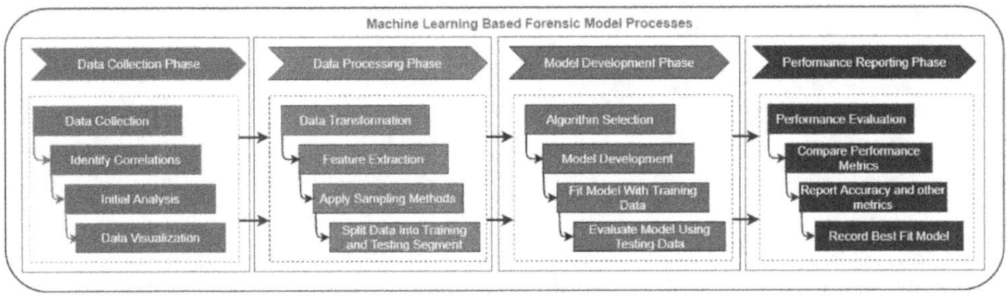

and evaluate the model using testing data. The final step, performance reporting, subdivides into performance evaluation, compares performance metrics, reports the assessment such as accuracy, and records the best fit model for future use. These subprocesses are almost similar in any machine learning model development; here we explicitly map that to the NIST forensic processes to use the machine learning model for forensic purposes. In the next section, we discuss the application and evaluation of our model using credit card transaction datasets.

5. FORENSIC MODEL EVALUATION USING E-COMMERCE CASE STUDY

This section presents the evaluation of the machine learning-based forensic processes using the e-commerce data, more specifically credit card transaction dataset. The reason behind using credit card transaction data is its open-source availability and directly related to the financial sector. We explain the case descriptions, data source, data field descriptions and visualization, model development, and performance evaluation in the following subsections.

5.1 Case Description

In 2009, a datamining contest was organized by the University of California San Diego (UCSD) sponsored by FICO (formerly known as FairIsaac), the credit score reporting institute. An e-commerce dataset was provided which contains both mostly normal credit card transaction and a few abnormal transaction records. The objective of this contest was to maximize the classification accuracy on the test data set given a labelled training dataset. We are using the same dataset to illustrate the forensic model where our objective is to achieve maximum accuracy in the classification

Figure 2. First five rows of training dataset to present the data fields

	amount	hour1	state1	zip1	field1	domain1	field2	hour2	flag1	total	field3	field4	field5	indicator1	indicator2	flag2	flag3	flag4	flag5	label
0	12.95	0	CA	925	3	AOL.COM	1	0	1	12.95	-4276	7	0	1	0	1	1	0	1	0
1	11.01	0	CA	925	3	AOL.COM	1	0	1	11.01	-4276	7	0	1	0	1	1	0	1	0
2	38.85	0	CA	928	3	HOTMAIL.COM	1	0	0	38.85	2602	21	1	0	0	0	0	0	1	0
3	25.90	0	NJ	77	0	AOL.COM	1	0	0	25.90	4139	6	0	0	0	1	1	0	1	0
4	12.95	0	CA	945	3	YAHOO.COM	0	0	1	12.95	3826	9	1	0	0	1	0	0	1	0

of the normal and fraud transactions. The classification result would help in taking next decision-making steps, thus would help in the financial fraud forensic process.

5.2 Data Collection

The original dataset is now available in Kaggle (Kaggle, 2009). Kaggle is an online community of data scientists and machine learning enthusiasts. Kaggle enables users to discover and publish data sets and build models in a web-based environment and collaborate with other data scientists and machine learning engineers.

5.3 Data Description and Dissection

The dataset contains forty text files in total. There are ten files with the training features, ten files with training labels, ten files with testing features, and ten files with testing labels. The input data fields in the training dataset are "amount", "hour1", "state1", "zip1", "field1", "domain1", "field2", "hour2", "flag1", "total", "field3", "field4", "field5", "indicator1", "indicator2", "flag2", "flag3", "flag4", "flag5". Because the e-commerce transaction dataset is sensitive, some features are named field, indicator, and flag. The first task is to combine all the training and testing dataset files into a single training dataset and testing dataset (using Python pandas data frame). A sample first five rows (instances) of training data are given below in Figure 2. As we can see from Figure 2, the transactions are associated with states represented by the field title "state1". The field "domain1" represents the domains used to perform the e-commerce transaction. The "amount" field indicates the dollar amount used in that transaction. The label '0' indicates a normal transaction, and the label '1' indicates a fraud transaction.

5.4 Data Mining, Initial Analysis, and Visualization

In this section, we present some data mining results from the training dataset to understand the distribution of the dataset. We also analyze the correlations among different features. We then visualize some data to present valuable insights from the training dataset.

Figure 3. Features distribution in the training dataset using WEKA tool

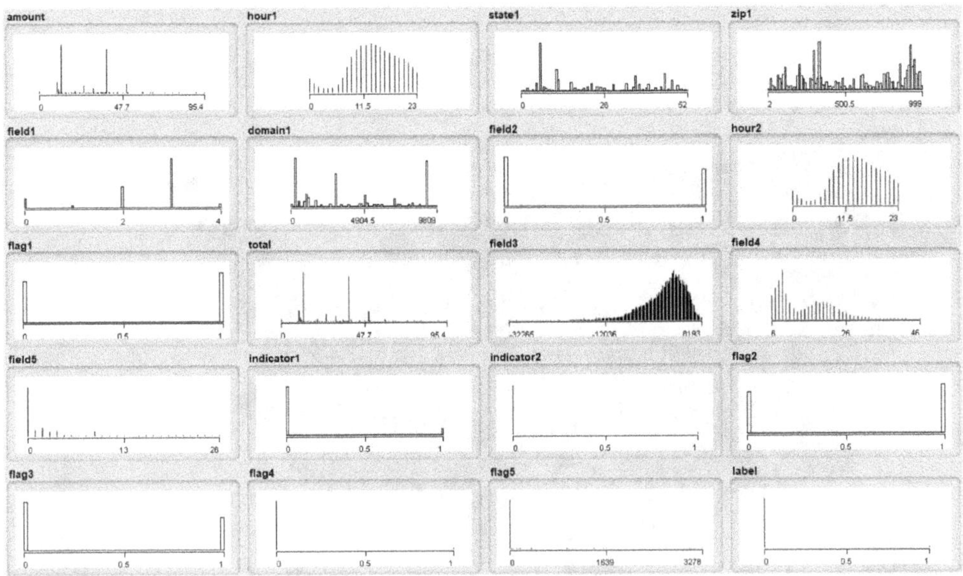

There are 852138 rows (i.e., instances) in the whole training dataset and 94682 cases in the complete testing dataset. In the training dataset, only 2.21% of instances are labeled as a fraud ('label' value '1') among all the transactions, and 97.79% are labeled as normal ('label' value '0') transactions. In number, 833292 transactions are normal transactions, and 18846 transactions are fraud. This disparity in the data distribution makes the dataset a highly imbalanced dataset by having a high skew towards the normal transaction. Thus, this case represents almost a real-life scenario where we have a few fraud cases among all the credit card transactions.

In the data mining process, it is useful to get an initial idea about the distribution of the features in the training data. We use WEKA (Holmes, Donkin, & Witten, 1994) tool to generate the distribution of the training features. WEKA is the short form of Waikato Environment for Knowledge Analysis. The tool was developed at the University of Waikato, New Zealand and is now available as an open-source tool to use in machine learning and data mining analysis. As we can see from Figure 3, each feature is having some distribution that gives an idea about the features. For example, 'label' feature we see that '0' value is much larger compared to '1', which means 'normal' transactions are much larger than the 'fraud' transactions reported in the e-commerce training data.

We also present three interesting visualizations that would help the readers to get an idea about the problem we are dealing before going into the model development. We use Python Plotly module to display the results as given in Figure 4, 5, and

Figure 4. Heatmap representing state-wise total amount of transactions in the training dataset

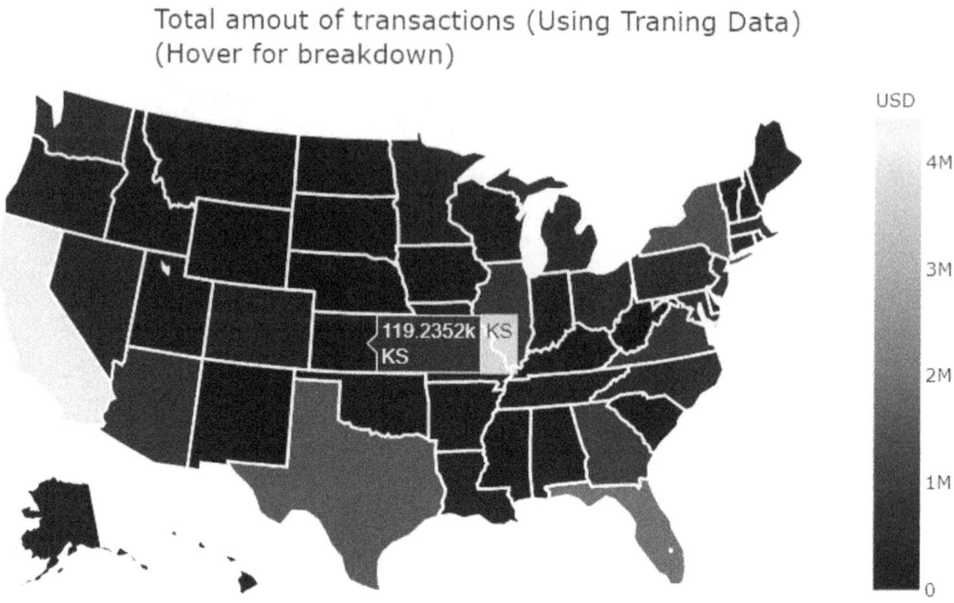

6. From Figure 4, we understand the dollar amount of transactions per state that are reported in the training dataset. This gives an idea which states were having more transactions and which states were having less transactions on that particular e-commerce data.

Similarly, Figure 5 represent the state-wise total number of transactions as reported in the training data. From this geospatial heatmap, we can understand which states were having larger transactions and which states were having small number of transactions compared to others. Figure 6 present the state-wise total fraudulent transactions among all the transactions in that state.

5.5 Data Preprocessing and Transformation

After the initial analysis and visualization of the training data, now is the time for preparing the data to feed into the machine learning model. Some data types may need to be converted into numeric types (for example, the "label" field was considered an object, we transform it into integer type). We use label encoding methods to change the string types into numerical values for some string type inputs. Label Encoding is the process of converting labels to numeric format to make them machine-readable. Then, machine learning algorithms can make a more informed decision about the

Figure 5. Heatmap representing state-wise total number of transactions in the training dataset

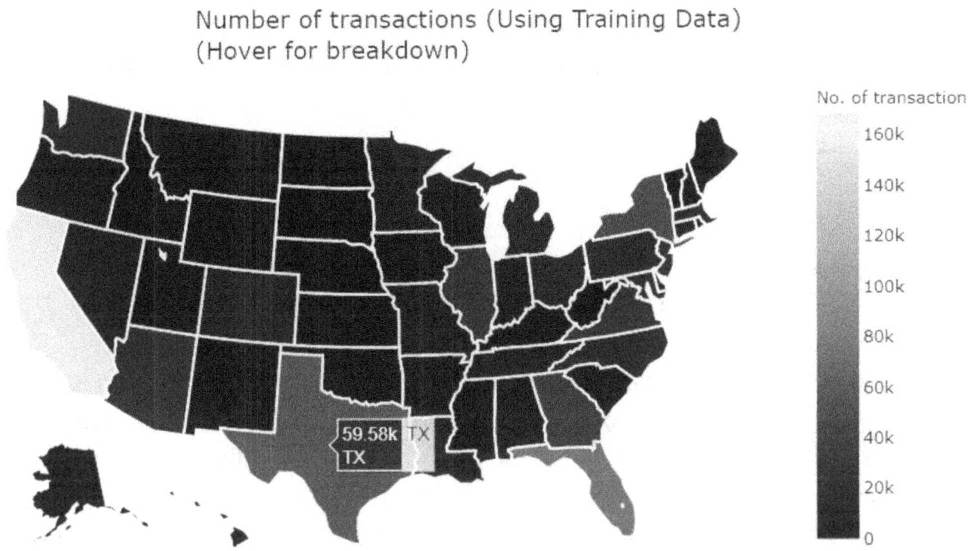

way those labels to handle. For example, the feature "state1" and "domain1" were string types. We transform them to numeric using the label encoding methods. We also transform some other features into numeric values using the python *pd.astype()*

Figure 6. Heatmap of state-wise total fraud transactions among all the transactions in the training dataset

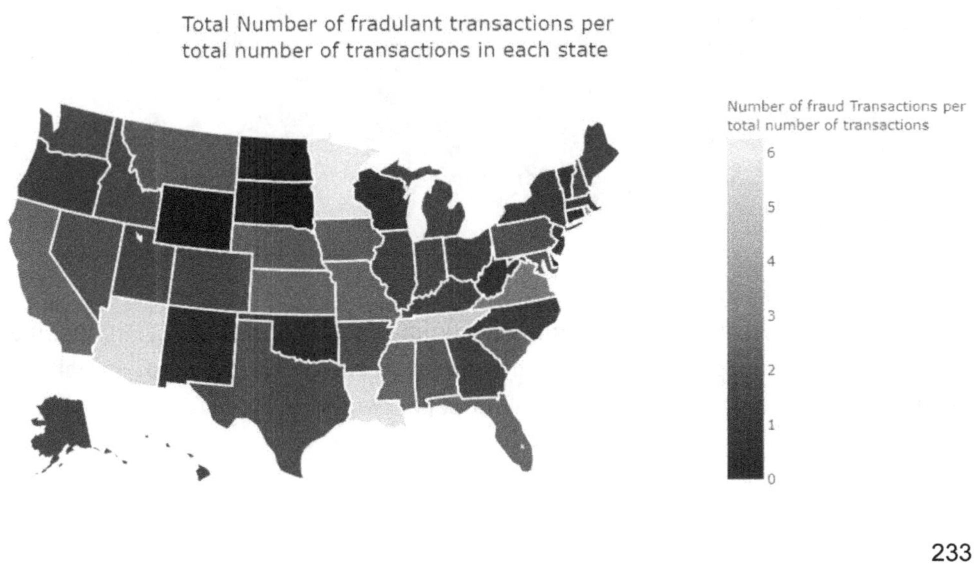

function. After the application of the label encoding technique, the training data (first five rows) looks like Figure 7.

Figure 7. First five instances of the training data after applying label encoding technique

	amount	hour1	state1	zip1	field1	domain1	field2	hour2	flag1	total	field3	field4	field5	indicator1	indicator2	flag2	flag3	flag4	flag5	label
0	12.95	0	6	925	3	277	1	0	1	12.95	-4276	7	0	1	0	1	1	0	1	0
1	11.01	0	6	925	3	277	1	0	1	11.01	-4276	7	0	1	0	1	1	0	1	0
2	38.85	0	6	928	3	3022	1	0	0	38.85	2602	21	1	0	0	0	0	0	1	0
3	25.90	0	33	77	0	277	1	0	0	25.90	4139	6	0	0	0	1	1	0	1	0
4	12.95	0	6	945	3	9147	0	0	1	12.95	3826	9	1	0	0	1	0	0	1	0

Once the numerical transformation is applied on the dataset, the next step is to identify the correlation among the features or fields of the dataset. This will guide to select the most important elements if the number of features is high. For example, in Figure 8, we present the correlation matrix among the features of the training dataset. As we can see, some fields have positive correlation coefficients, and some fields have negative correlation coefficients. This would help in the first step analysis to understand the relations among the features.

Figure 8. Correlation matrix of the training features

	amount	hour1	state1	field1	domain1	field2	hour2	flag1	total	field3	field4	field5	indicator1	indicator2	flag2	flag3
amount	1	0.015	0.00042	-0.081	-0.0017	0.01	0.015	-0.21	1	-0.078	0.069	-0.056	-0.024	-0.014	0.098	0.015
hour1	0.015	1	-0.014	-0.0019	-0.011	0.01	0.99	-0.041	0.015	0.0084	0.012	-0.011	0.00056	-0.0085	-0.015	-0.013
state1	0.00042	-0.014	1	-0.015	0.014	0.0049	-0.014	-0.026	0.00047	0.034	0.019	0.0034	-0.018	0.003	0.0064	-0.00071
field1	-0.081	-0.0019	-0.015	1	0.013	0.041	-0.0019	-0.025	-0.082	0.088	-0.0089	0.025	0.033	0.0015	0.014	0.046
domain1	-0.0017	-0.011	0.014	0.013	1	-0.0077	-0.011	-0.018	-0.0013	0.11	0.069	-0.016	0.052	0.012	-0.019	-0.023
field2	0.01	0.01	0.0049	0.041	-0.0077	1	0.0095	-0.088	0.01	0.027	0.074	-0.024	0.011	0.003	0.013	-0.0035
hour2	0.015	0.99	-0.014	-0.0019	-0.011	0.0095	1	-0.042	0.015	0.0079	0.012	-0.012	0.00078	-0.0079	-0.015	-0.012
flag1	-0.21	-0.041	-0.026	-0.025	-0.018	-0.088	-0.042	1	-0.21	0.052	-0.46	0.18	0.011	0.029	-0.0033	-0.0057
total	1	0.015	0.00047	-0.082	-0.0013	0.01	0.015	-0.21	1	-0.078	0.069	-0.056	-0.024	-0.014	0.098	0.015
field3	-0.078	0.0084	0.034	0.088	0.11	0.027	0.0078	0.052	-0.078	1	0.019	-0.0092	0.13	-0.0062	-0.039	-0.02
field4	0.069	0.012	0.019	-0.0089	0.069	0.074	0.012	-0.46	0.069	0.019	1	-0.17	-0.0095	-0.011	0.045	0.039
field5	-0.056	-0.011	0.0034	0.025	-0.016	-0.024	-0.012	0.18	-0.056	-0.0092	-0.17	1	-0.015	0.0042	0.0033	0.034
indicator1	-0.024	0.00056	-0.018	0.033	0.052	0.011	0.00078	0.011	-0.024	0.13	-0.0095	-0.015	1	-0.048	-0.0019	-0.0014
indicator2	-0.014	-0.0085	0.003	0.0015	0.012	0.003	-0.0079	0.029	-0.014	-0.0062	-0.011	0.0042	-0.048	1	-0.0092	-0.018
flag2	0.098	-0.015	0.0064	0.014	-0.019	0.013	-0.015	-0.0033	0.098	-0.039	0.045	0.0033	-0.0019	-0.0092	1	0.46
flag3	0.015	-0.013	-0.00071	0.046	-0.023	-0.0035	-0.012	-0.0057	0.015	-0.02	0.039	0.034	0.0014	-0.018	0.46	1

The next step is to apply normalization technique on the training dataset. Machine learning algorithms perform better with normalized numerical input variables. Normalization scales each input variable independently to the range 0-1, which is the most precise range for floating-point values. In this case evaluation, we apply *MinMaxScaler()* function from *sklearn.preprocessing* library module.

As we have an imbalanced dataset of the normal and fraud classes, we apply an undersampling technique. Undersampling is the method to handle an imbalanced dataset problem. The undersampling algorithm randomly selects instances from the majority class to remove from the training dataset. This removal results in a decrease in the majority class instances in the transformed version of the training dataset. We can repeat the process until we achieve the desired distribution of classes. In the Python Scikit-Learn library, we apply the *NearMiss* (Mani & Zhang, 2003) method for doing the undersampling.

5.6 Model Development and Algorithm Selection

As we have a labeled training and testing dataset, so the problem falls under the supervised machine learning problem. In supervised machine learning, we are given with labeled datasets where the labels represent different classes. Our goal is to build a model that uses the features in the training dataset and then give associate class properly. This problem is a pure classification type of problem in the context of machine learning. We choose here couple of different algorithms to build our model and compare the performances. These algorithms are Random Forest (RF), Logistic Regression (LR), Naïve Bayes (NB), Support Vector Machine (SVM) Classifier, Decision Tree (DT) Classifier, and Neural Network Multi-Layer Perceptron (MLPC) classifier. We skip the theoretical background of these algorithms because they are widely available in the literature.

5.7 Machine Learning Performance Metrics

The performance of any machine learning model is evaluated based on some performance metrics. In this section, we present the mathematical definitions and significance of some of those metrics in classification type of tasks, so that we can explain them in the context of the case study in the next section.

5.7.1 Confusion Matrix

Confusion matrix is a matrix representation of the performance of the classification task. The matrix compares the predicted class by the model with the actual class (as labeled in the testing dataset). For example, in the case of e-commerce transaction, we have two different classes: normal and fraud. Thus, the confusion matrix for this case would be a 2x2 matrix as given in Figure 9. Here 'normal' class is represented by numeric value '0', and the 'fraud' class is represented by the numeric value '1'.

Figure 9. Confusion Matrix

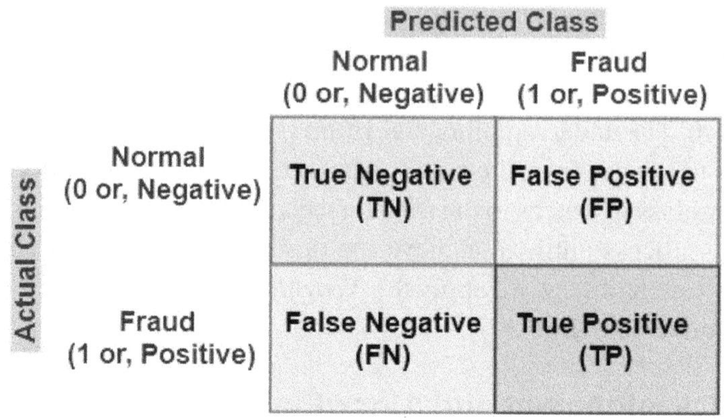

There are four terms associated with the confusion matrix that would help us to derive some of the performance metrics next. These are given below.

- **True Negative (TN):** This is the case when both actual class and the predicted class have the same numeric label value '0' (in our case 'normal' class).
- **True Positive (TP):** This is the case when both the actual class and the predicted class have the same label value of '1' (in our case 'fraud' class)
- **False Negative (FN):** This is the case when the predicted class label is '1' (i.e., fraud) whereas the actual class label is '0' (i.e., normal).
- **False Positive (FP):** This is the case when the predicted class label is '0' (i.e., normal) whereas the actual class label is '1' (i.e., fraud).

There are four performance metrics that we find from the confusion matrix directly. These are given in the below subsection.

5.7.2 Performance Metrics

- **Accuracy:** The accuracy metric represents the corrected predictions among all the samples in the dataset. The prediction accuracy of the model is calculated by the following equation:

$$accuracy = \frac{TP + TN}{TP + TN + FP + FN} \tag{1}$$

- **True Positive Rate (TPR)**: True Positive Rate is also known as sensitivity and defined by eq. (2). True Positive Rate corresponds to the proportion of positive data points that are correctly considered as positive, with respect to all positive data points. In our case, this is the fraction of the 'fraud' class that are correctly predicted as 'fraud' among all the predicted 'fraud' instances.

$$TPR = \frac{TP}{TP + FN} \tag{2}$$

- **False Positive Rate (FPR):** False Positive Rate corresponds to the proportion of negative data points that are mistakenly considered as positive, with respect to all negative data points. In our case, this is the fraction of the 'fraud' instances that are predicted as 'normal' among all the predicted 'normal' instances.

$$FPR = \frac{FP}{FP + TN} \tag{3}$$

- **Precision:** Precision is the number of correct positive results (TP) as a fraction of the number of positive results (TP+ FP) predicted by the classifier.

$$Precision = \frac{TP}{TP + FP} \tag{4}$$

- **Recall:** Recall is the number of correct positive results (TP) divided by the number of predicted positive samples (TP+FN).

$$Recall = \frac{TP}{TP + FN} \tag{5}$$

- **True Negative Rate (TNR):** It is also termed as "specificity". In contrast to recall, specificity is defined as the number of negative results (TN) returned by the model among all the negative samples (TN+FP)

$$Specificity = \frac{TN}{TN + FP} \tag{6}$$

- **Support:** It is defined as the number of samples of the true response that lies in each class of target values.

- **F1-Score:** The F1 Score represents the Harmonic Mean of precision and recall. It indicates the classifier's precision (the number of instances it correctly classifies) and the model's robustness (it does not wrongly predict significant cases). High precision but low recall results in a highly accurate classification, but it also misses many difficult-to-classify instances. The higher the F1 Score, the more accurate our model is. F1 Score tries to balance between precision and recall.

$$F1Score = 2 * \frac{1}{\frac{1}{precision} + \frac{1}{recall}} \tag{7}$$

- **Matthews Correlation Coefficient (MCC):** In machine learning, the Matthews correlation coefficient is a statistical metric to assess the quality of binary and multiclass classifications. It illustrates the degree of agreement between observed and predicted values. The MCC is essentially a value for the correlation coefficient between -1 and +1. A coefficient of +1 indicates a perfect prediction, a coefficient of 0 indicates an average random prediction, and a coefficient of -1 indicates an inverse projection. Additionally, the statistic is also known as the phi coefficient. The expression of MCC is given below:

$$MCC = \frac{TP \times TN - FP \times FN}{\sqrt{(TP+FP)(TP+FN)(TN+FP)(TN+FN)}} \tag{8}$$

- **Receiver Operating Characteristic (ROC) and Area Under Curve (AUC):** A receiver operating characteristic (ROC) curve plots the true positive rate (TPR) against the false positive rate (FPR). AUC is a widely used metric for evaluating binary classification algorithms. A classifier's AUC is equal to the probability that a randomly chosen positive example will be ranked higher than a randomly chosen negative example. The receiver operating characteristic (ROC) curve is a probability curve, while the area under the curve (AUC) indicates separability. AUC-ROC denotes the area under the receiver operating characteristic curve. In simple terms, the AUC-ROC metric indicates a model's ability to distinguish between classes. The AUC value is between [0, 1]. The greater the AUC, the more accurate the model.

Figure 10. Importance of features in Random Forest model in test case 1

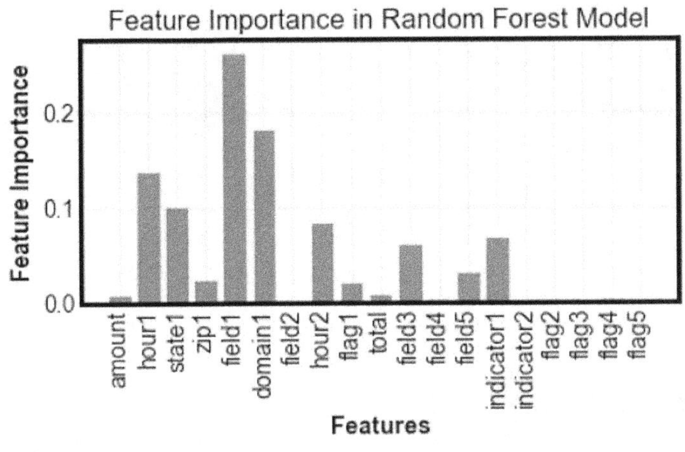

In the next section, we present the performance of our machine learning based forensic model using these metrics that we just present.

5.8 Result Analysis and Performance Evaluation

In this section, we present the results of our machine learning driven forensic model. We separate the evaluation process in two different test cases based on the feature selection process. We present the discussion in below sections.

5.8.1 Test Case 1: Features Exclude Label Data

In this test case, we exclude the 'label' data from the feature selection process. This results comparatively low prediction accuracy because the training data undergoes normalization and under sampling processes before feeding the data into the model. We provide the feature importance for the Random Forest model in Figure 10 as an example to understand the weights of the features in the training model. As found in Figure 10, features named 'field1', 'domain1', 'hour1', 'state1' and 'hour2' are the five most important features in Random Forest model.

We then compare the confusion matrix for the seven algorithms that we have selected for the model development. The algorithms that we have selected are Random Forest (RF), Logistic Regression (LR), K-Nearest Neighbor (KNN), Decision Tree (DT), Naïve Bayes (NB), Multi-Layer Perceptron Classifier (MLPC), and Support Vector Machine Classifier (SVC). We present the confusion matrix in Figure 11. From Figure 11, we found that for the models would have comparatively low prediction accuracy and other performance measures as there are exists a large

Figure 11. Confusion matrix for different models for the test case 1

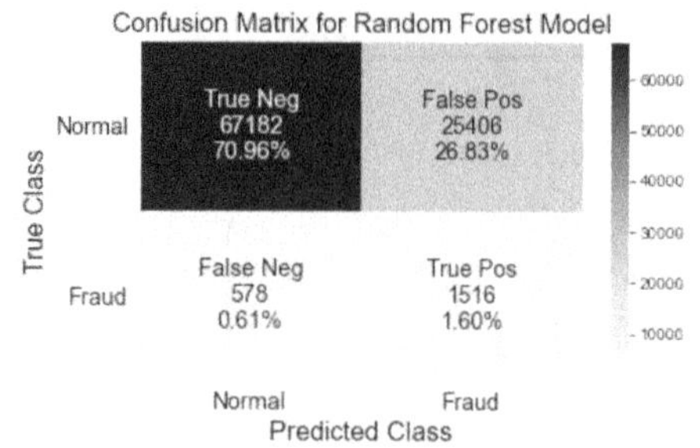

(a) Confusion matrix for RF model

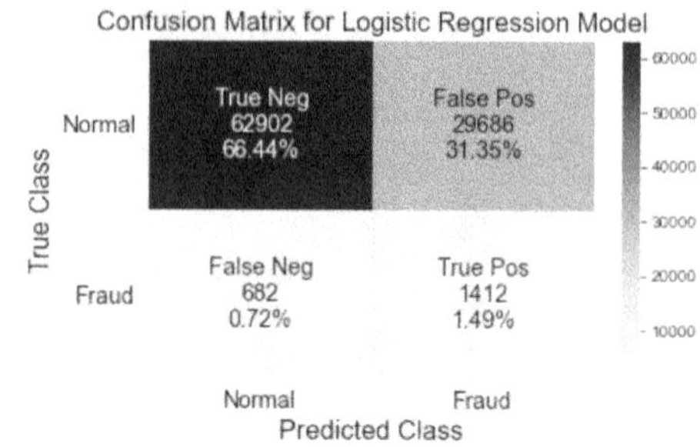

(b) Confusion matrix for LR model

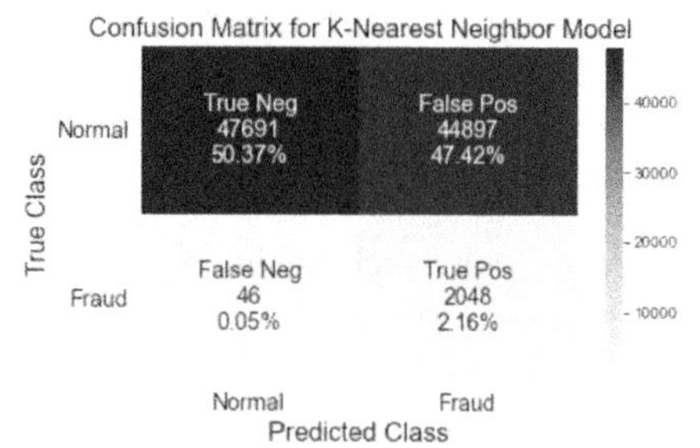

(c) Confusion matrix for KNN model

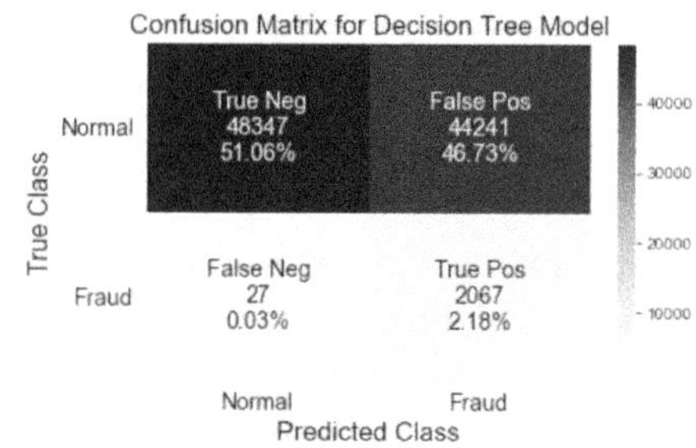

(d) Confusion matrix for DT model

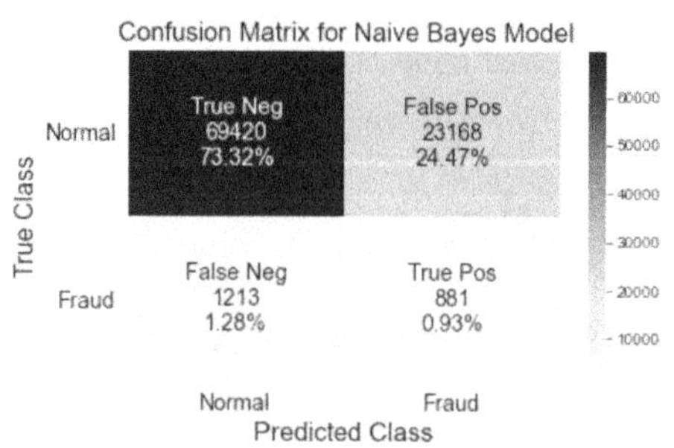

(e) Confusion matrix for NB model

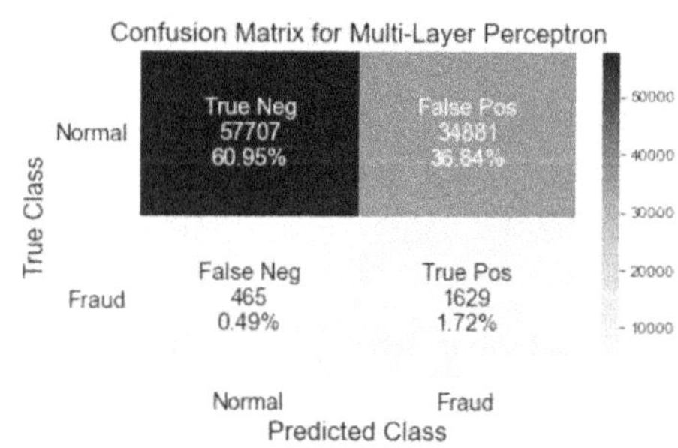

(f) Confusion matrix for MLPC model

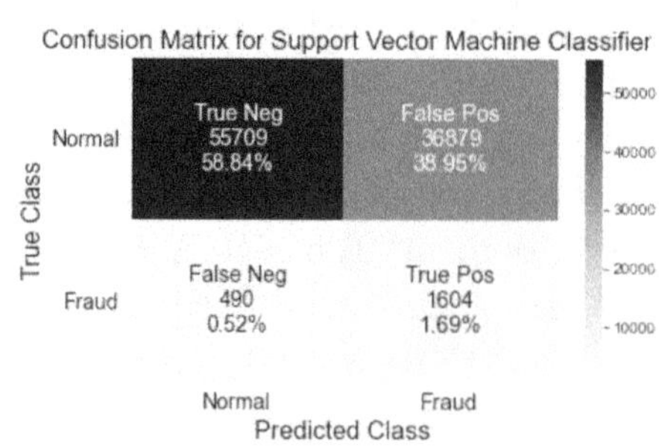

(g) Confusion matrix for SVM model

fraction of false positive and false negative. This is because extracting the 'labels' as a feature parameter make it hard for the model to predict accurately based on the other features.

In Figure 12, we present the performance metrics for the selected algorithms. We choose accuracy, precision, recall, f1 score, MCC, and latency as the performance metrics. The accuracy of the NB (74.25%) and RF (72.56%) models are higher than other models. The precision is low for all the models with an RF (5.63%) model with slightly higher precision. The KNN and DT models have larger recall values of 97.8% and 98.71%, respectively, compared to the other models. In terms of F1 Score RF and NB have values of 0.8217 and 0.8333, respectively, on a scale of 0 to 1. The MCC values are low for all the models having RF (0.1466) as the highest MCC score among the other models. Finally, in terms of latency (the execution time of the model), NB, DT, and LR models are faster, but KNN has the worst execution time of 15576 ms. Comparing these six metrics, we can say that the RF model is the best among the seven algorithms we selected. The next best model candidate is the LR model.

Figure 13 shows the receiver operating characteristic (ROC) curves for all the models along with the area under the ROC curve (AUC-ROC). We find that the AUC of the RF, LR, SVC, MLPC, NB, KNN, and DT models are 0.69, 0.68, 0.68, 0.70, 0.59, 0.75, and 0.75, respectively. The AUC-ROC value of the DT and KNN models are the highest among the other models.

Comparing the performance metrics, we can say that the models are underperforming. The higher accuracy is the NB model, with a value of 74.25%. The RF model is comparatively better among all the other algorithms in overall performance. But still, we need to improve the accuracy and other performance metrics so that we can reduce the false positive rate and false-negative rates. In test case 2, we propose methods to improve the performance metrics and provide the comparative analysis.

5.8.2 Test Case 2: Features Include Label Data

In this test case, we have also included the 'label' value as a feature along with the other features in our training model. Our goal is to improve the performance metrics in this test case. As we already have separate training and testing data, we incorporate the 'label' as an additional feature in the training data. Here, in Figure 14, we present the feature importance values for the Random Forest model. As we can see from Figure 14, the features 'label', 'field1', 'hour1', 'domain1', 'hour2', 'indicatior1' are among the top-performing features. Thus, we see that the 'label' feature contributes a more significant weight in determining the model performance in the Random Forest model.

Figure 12. Performance metrics for the selected algorithms (RF, LR, KNN, DT, NB, MLPC and SVM) for the test case 1.

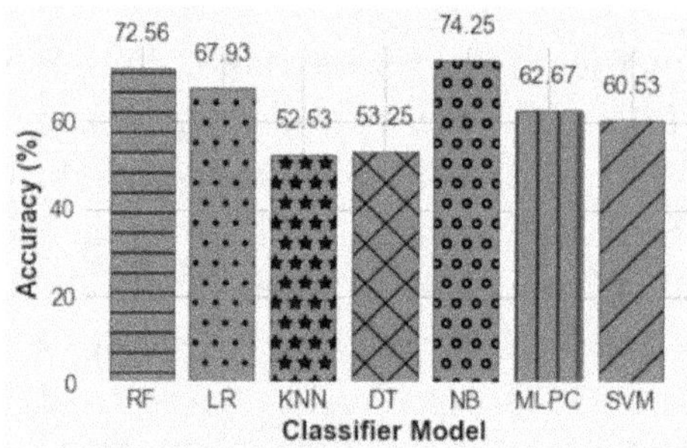

(a) Accuracy

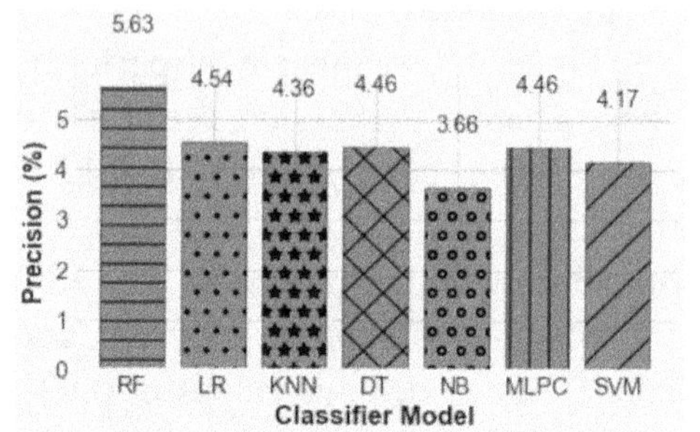

(b) Precision

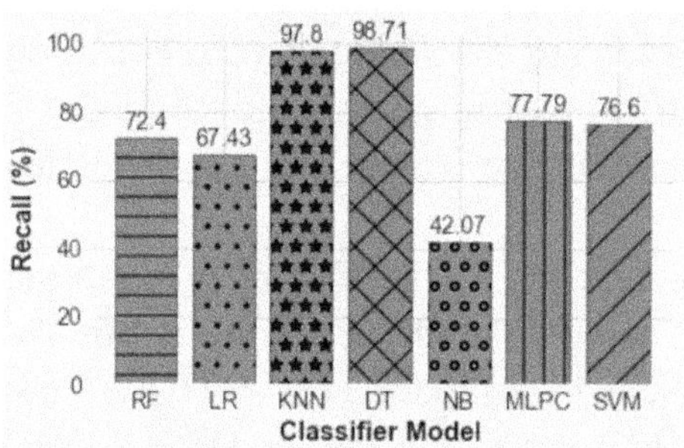

(c) Recall

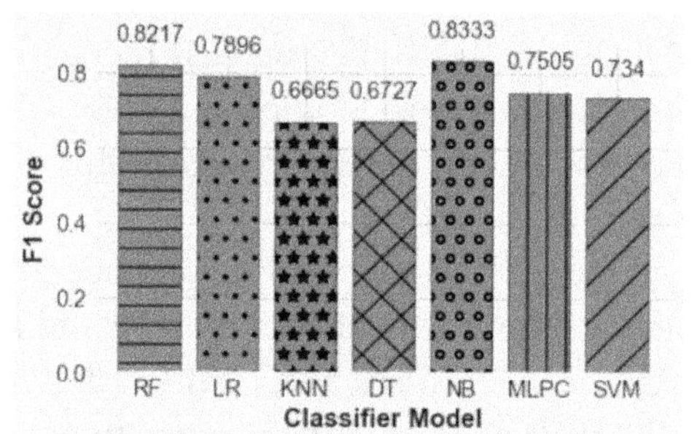

(d) F1 Score

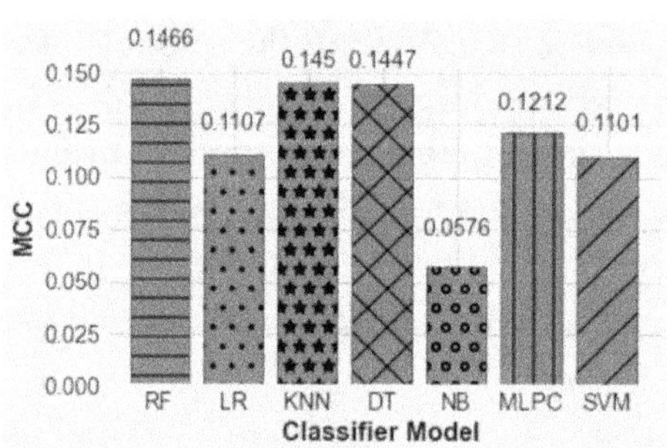

(e) MCC

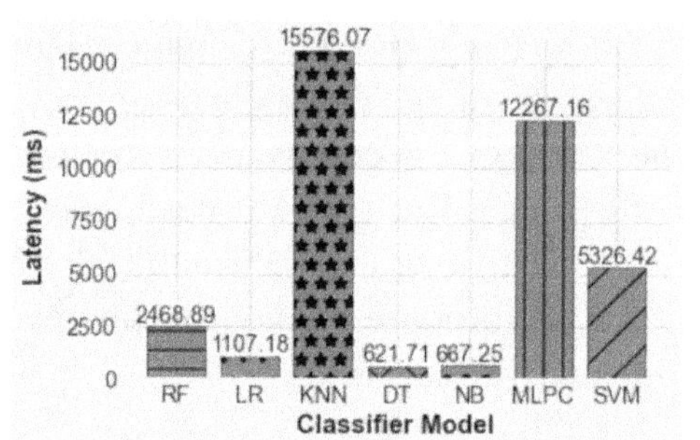

(f) Latency

In Figure 15, we present the confusion matrices of the models that we have developed. In this case, the models are much more accurate in predicting the 'True Positive' (fraud transaction case) and 'True Negative' (normal transaction case) than what we observe in test case 1. For example, only 0.15% were false positives, and the false negative is 0% for the Random Forest model. Similarly, other models' performance also improved significantly.

Figure 16 presents the performance metrics for the seven models like before but with the 'label' feature included. In this case, models' prediction accuracy has

Figure 13. Comparison of the area under receiver operating characteristic curve (AUC-ROC) for test case 1

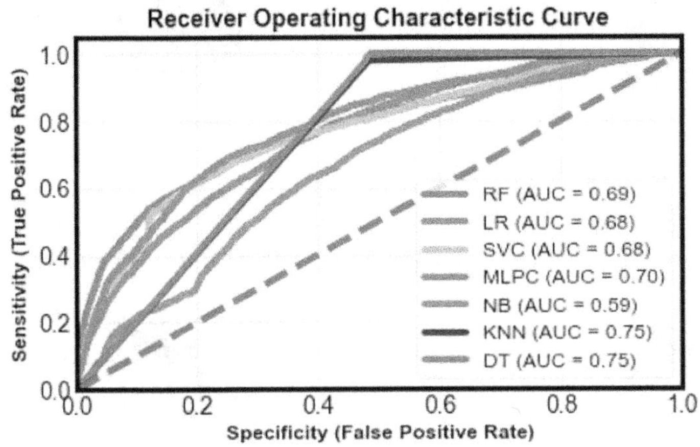

improved significantly. For example, only RF and KNN have an accuracy of 99.85% and 98.18%, respectively. Other models have an accuracy of 100%. Precision is a little lower for KNN (54.85%), but the precision value is 100% for all the other models. The recall is also 100% for all the models; that is why we omit to present the recall chart. F1 Score is also suitable for all the models except KNN (0.7336). For the latency, KNN, MLPC, and SVM are performing poorly. But the other models are pretty fast in execution and fitting the training data. Thus, in this test case, the

Figure 14. Importance of features in Random Forest model in test case 2

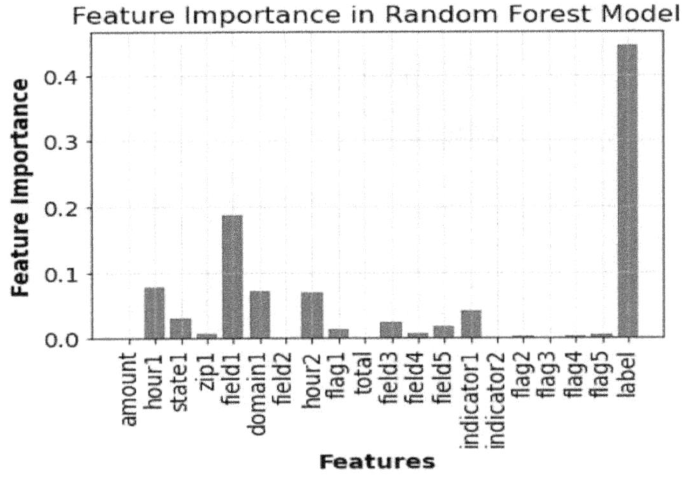

Figure 15. Confusion matrices for different models for the test case 2

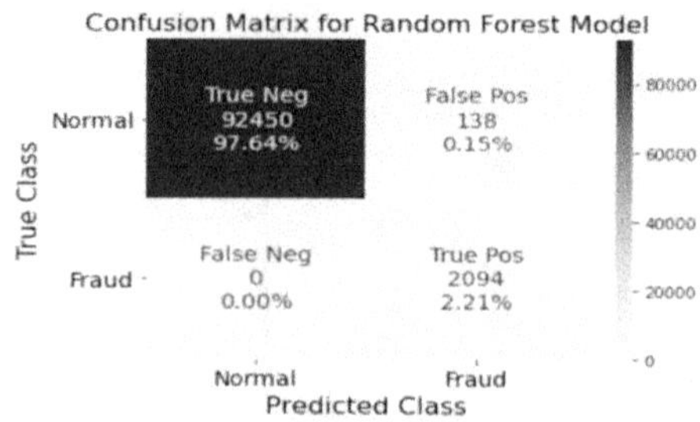

(a) Confusion matrix for RF model

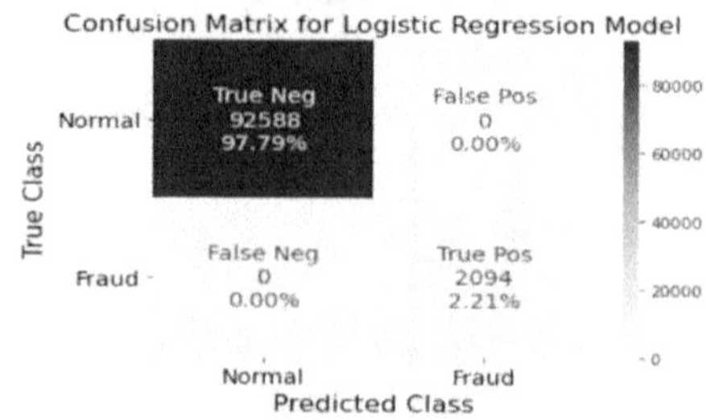

(b) Confusion matrix for LR model

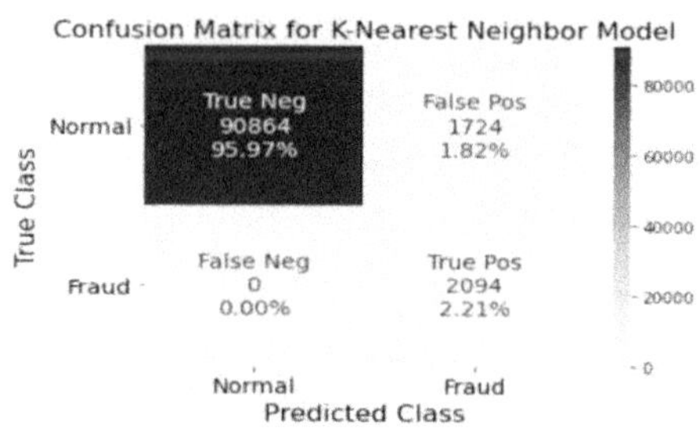

(c) Confusion matrix for KNN model

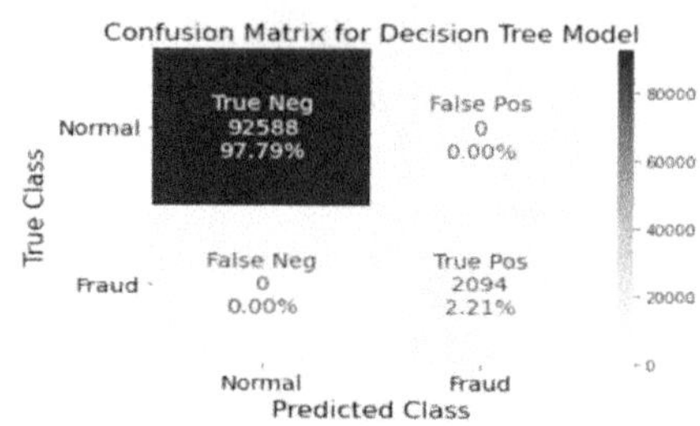

(d) Confusion matrix for DT model

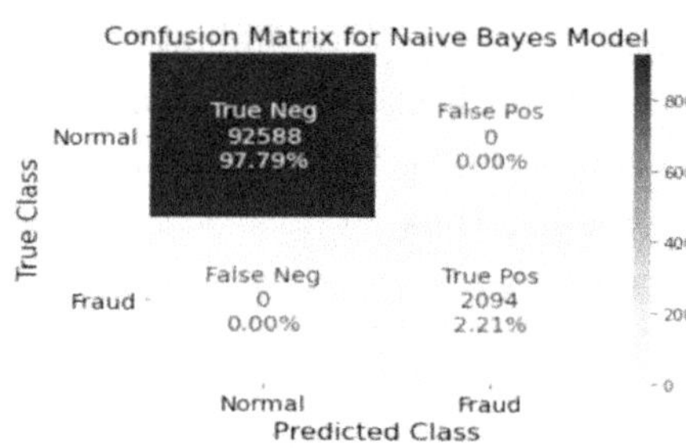

(e) Confusion matrix for NB model

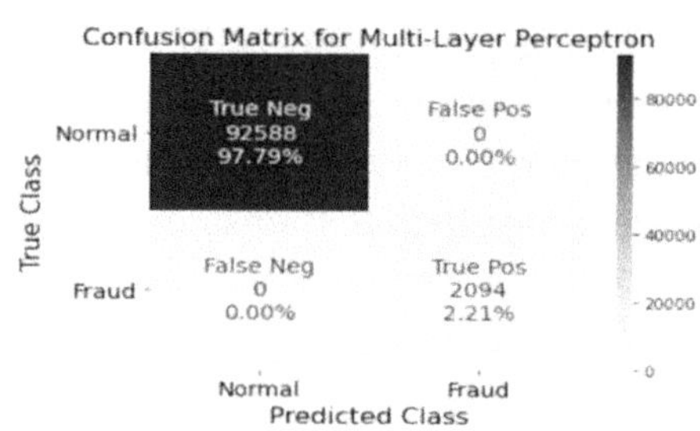

(f) Confusion matrix for MLPC model

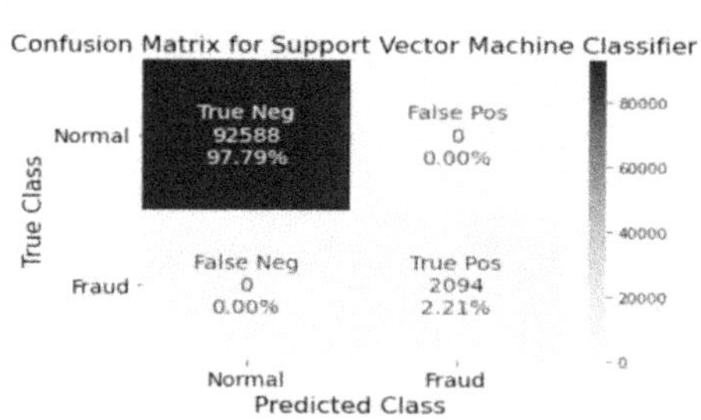

(g) Confusion matrix for SVM model

Figure 16. Performance metrics for the selected algorithms (RF, LR, KNN, DT, NB, MLPC and SVM) for the test case 2

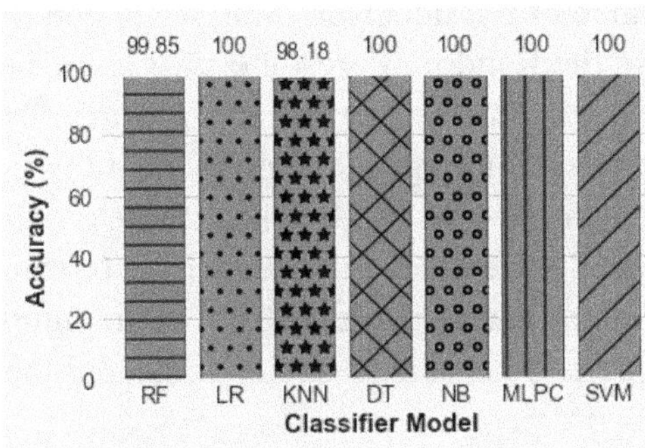

(a) Accuracy

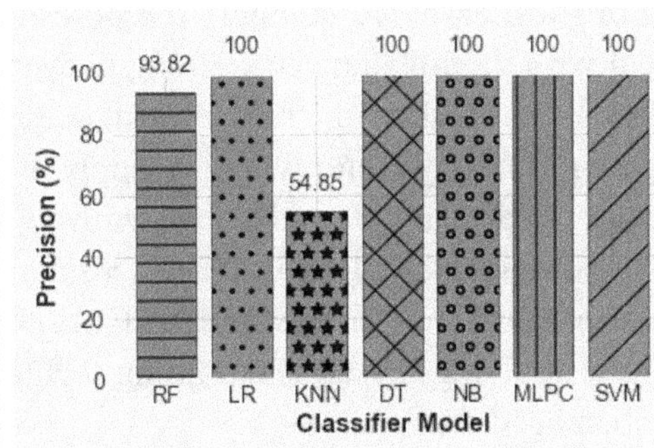

(b) Precision

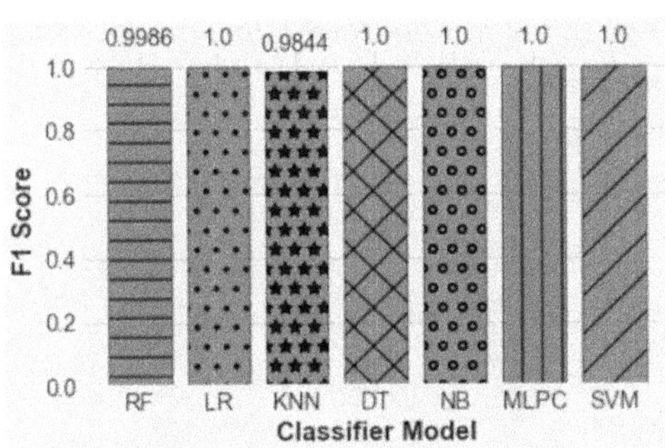

(c) Recall

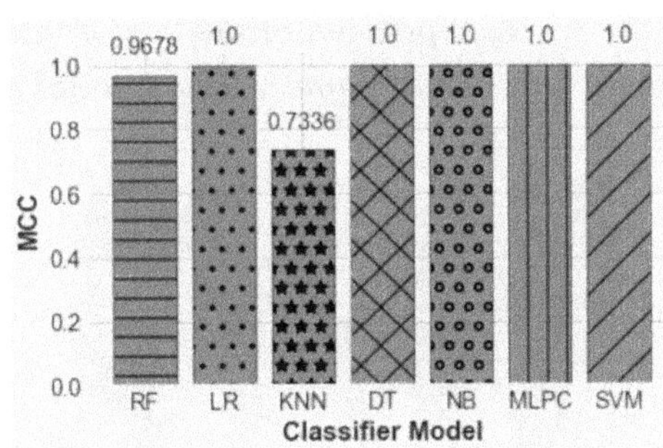

(d) F1 Score

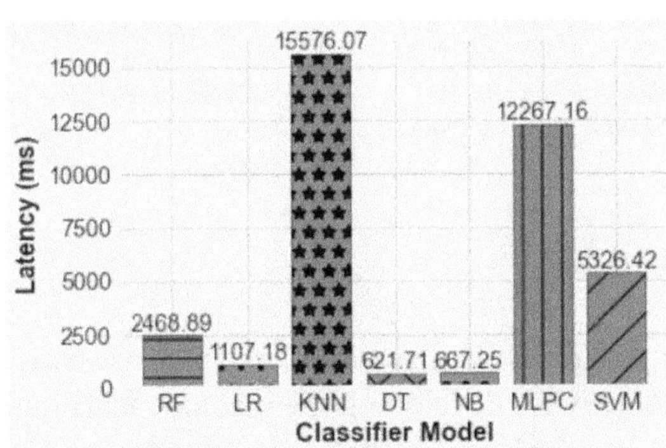

(a) Latency

Logistic Regression classifier and Random Forest classifier are better candidates than other models considering the overall performance.

5.8.2 Test Case 3: Comparison of Sampling Methods

In this test case, we compare the performance of the undersampling technique to deal with the imbalanced training dataset in the Random Forest model. Figure 17 shows the result with the performance metrics. As we see, performance metrics are significantly better when we apply the undersampling technique to the training dataset. For example, the recall with undersampling is 100%, whereas the recall without undersampling is 13.9%. Similar results we get for the sensitivity or true positive rate. The F1 Score and MCC are also significantly better with undersampling. Thus, applying the undersampling technique is an excellent choice to handle the imbalanced dataset.

Figure 17. Comparison of the performance metrics for the Random Forest model with under sampling and without under sampling

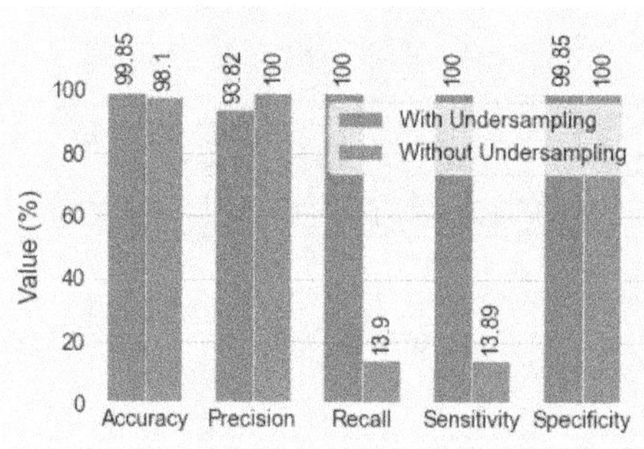

(a) Performance metrics with and without under sampling in Random Forest model

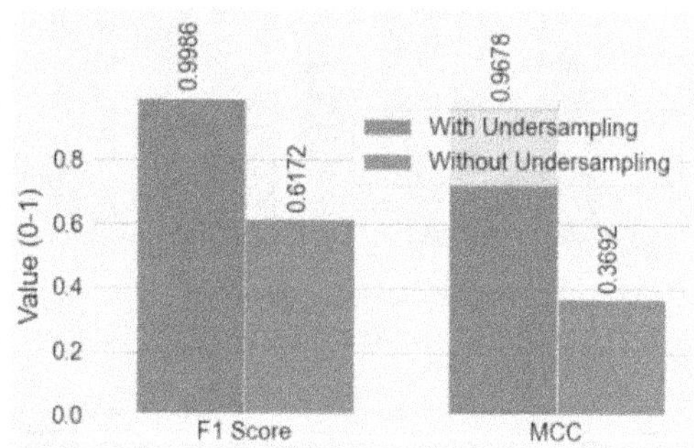

(b) F1 Score and MCC with and without under sampling in Random Forest model

5.9 GitHub Repository

We have used Python programming language for data processing. We have used the Python Scikit-Learn library for developing the machine learning classifier models. For visualization, we have used Python Plotly and Matplotlib. The complete Python codes for data processing, model development, and visualization simulation results are given in the GitHub repository associated with the reference (Haque, 2021).

6. CONCLUSION

Machine learning techniques can help automate forensic scenarios in fighting the sophisticated cyber-attacks on the financial industry. In this work, we first propose a forensic model for the economic sectors. We then focus on developing a machine learning model for the e-commerce transaction data to illustrate how the machine learning models can assist in the financial forensic processes. We have presented the performance metrics of machine learning classifiers using seven different algorithms to classify fraudulent transactions from regular transactions. The algorithm that we have used are Random Forest, Logistic Regression, Naïve Bayes, Decision Tree, K-Nearest Neighbor, Support Vector Machine, and Neural Network Multi-Layer Perceptron. Because the dataset is imbalanced, we need to apply the undersampling technique to process the features before fitting them into the model. We have developed three test cases. The first two test cases handle the feature selection and manipulation, and the third test case deals with comparing the undersampling technique. The performance metrics are satisfactory, and we find that either the Random Forest or the Logistic Regression model has better overall performance than the other models. This chapter describes necessary background information on financial fraud and proposes a detailed forensic technique workflow. Overall, the chapter presents some vital discussion on how the machine learning-based models can boost the automation processes in financial fraud forensics.

ACKNOWLEDGMENT

The Department of Defense's Center of Excellence in AI and Machine Learning (CoE-AIML) funded this work partially through a contract with the US Army Research Laboratory, W911NF-20-2-0277

REFERENCES

Ascent, T. (2020). *Identity Theft and Credit Card Fraud Statistics for 2020*. Retrieved from https://www.fool.com/the-ascent/research/identity-theft-credit-card-fraud-statistics/

Awoyemi, J. O., Adetunmbi, A. O., & Oluwadare, S. A. (2017). *Credit card fraud detection using machine learning techniques: A comparative analysis*. Paper presented at the 2017 International Conference on Computing Networking and Informatics (ICCNI). 10.1109/ICCNI.2017.8123782

Bayuk, J. (2010). *Cyber Forensics*. Springer.

Brinson, A., Robinson, A., & Rogers, M. (2006). A cyber forensics ontology: Creating a new approach to studying cyber forensics. *Digital Investigation, 3*, 37-43.

Force, J. T. (2018). *Risk Management Framework for Information Systems and Organizations: A System Life Cycle Approach for Security and Privacy*. Academic Press.

Force, J. T., & Initiative, T. (2013). Security and privacy controls for federal information systems and organizations. *NIST Special Publication*, *800*(53), 8–13.

Global, P. (2020). *Global Economic Crime and Fraud Survey 2020*. Retrieved from https://www.pwc.com/gx/en/services/forensics/economic-crime-survey.html

Haque, A. (Producer). (2021). *GitHub Code*. Retrieved from https://github.com/arif6008/Credit_Card_Transaction_Classification

Hasham, S., Joshi, S., & Mikkelsen, D. (2019). Financial Crime And Fraud. In *The Age of Cyber Security*. McKinsey & Company.

Holmes, G., Donkin, A., & Witten, I. H. (1994). Weka: A machine learning workbench. *Proceedings of ANZIIS'94-Australian New Zealnd Intelligent Information Systems Conference*. 10.1109/ANZIIS.1994.396988

JofreM.GerlachR. H. (2018). Fighting accounting fraud through forensic data analytics. *Available at* SSRN 3176288. doi:10.2139/ssrn.3176288

Kaggle. (2009). *UCSD-FICO Dataset Task 1*. Author.

Kent, K., Chevalier, S., Grance, T., & Dang, H. (2006). Guide to integrating forensic techniques into incident response. *NIST Special Publication*, *10*(14), 800–886. doi:10.6028/NIST.SP.800-86

Kotsiantis, S., Koumanakos, E., Tzelepis, D., & Tampakas, V. (2006). *Predicting fraudulent financial statements with machine learning techniques*. Paper presented at the Hellenic Conference on Artificial Intelligence. 10.1007/11752912_63

Lokanan, M., Tran, V., & Vuong, N. H. (2019). Detecting anomalies in financial statements using machine learning algorithm. *Asian Journal of Accounting Research*.

Maes, S., Tuyls, K., Vanschoenwinkel, B., & Manderick, B. (2002). Credit card fraud detection using Bayesian and neural networks. *Proceedings of the 1st international naiso congress on neuro fuzzy technologies*.

Mani, I., & Zhang, I. (2003). kNN approach to unbalanced data distributions: a case study involving information extraction. *Proceedings of workshop on learning from imbalanced datasets*.

Randhawa, K., Loo, C. K., Seera, M., Lim, C. P., & Nandi, A. K. (2018). Credit card fraud detection using AdaBoost and majority voting. *IEEE Access: Practical Innovations, Open Solutions*, *6*, 14277–14284. doi:10.1109/ACCESS.2018.2806420

Ross, R., McEvilley, M., & Oren, J. (2016). *Systems security engineering: Considerations for a multidisciplinary approach in the engineering of trustworthy secure systems*. Academic Press.

Thennakoon, A., Bhagyani, C., Premadasa, S., Mihiranga, S., & Kuruwitaarachchi, N. (2019). *Real-time credit card fraud detection using machine learning*. Paper presented at the 2019 9th International Conference on Cloud Computing, Data Science & Engineering (Confluence). doi:10.6028/NIST.SP.800-160

Yee, O. S., Sagadevan, S., & Malim, N. H. A. H. (2018). Credit card fraud detection using machine learning as data mining technique. *Journal of Telecommunication Electronic and Computer Engineering*, *10*(1-4), 23–27.

Zetter, K. (2016). *That Insane, $81 M Bangladesh Bank Heist. Here's What We Know. Wired.*

Conclusion

One century ago, Georg Popp was the first scientist presenting in court a case in which the geological makeup of soils was used to secure a criminal conviction. Ever since, significant advances in the theory and practice of forensic geosciences have taken place. As branch of forensic geosciences, forensic geochemistry has been used in standard contamination characterization with methodology developed in organic geochemistry, a combination we refer to as Forensic Environmental Geochemistry. Other geochemical proxies were also used for forensic purposes including nuclear and isotopic forensic geochemistry.

Compared with other direct methods, in forensics, geophysical techniques currently need to identify, characterize, and distinguish hidden structures in order to obtain useful information and provide accurate and effective results for the research purposes in this field geophysical technology can minimize time and cost factors, and maximize the amount of data, information and knowledge available. It is time to classify environmental polluting activities as crimes. Further, the setting of standards to limit of criminal activities represents the urgency. Traditional crimes may damage a person or a small group while environmental crimes threaten the whole humanity. This chapter discussed advances in forensic geophysics. The focus was on the use of the magnetic susceptibility for the environmental forensic investigations. This tool proves relevant due to its fast and cheap manipulation. In Tunisia, environmental crimes are really neglected and wrongdoer are still damaging the ecosystems. As it is shown in this study pollution of Tunisian coasts is quite evident by using the magnetic susceptibility.

Forensic sedimentology has slowly evolved over time from a core of associated ideas in a very novel way to at least one of the most important forms of crime-fighting investigation. It continues to utilize and revolutionize the fundamental concepts of sedimentology and drive innovation both within the research laboratory and within the research laboratory. Since each case is a completely new problem, inventive solutions must be developed to ensure that the correct conclusions are drawn within the case. However, these solutions are used by sedimentologists in every work area. However, the interpretation and presentation of these conclusions lies somewhere between ancient sedimentology and rhetoric. While ancient sedimentologists give their findings and theories to various scientists, rhetorical sedimentologists give their evidence to the average citizen before excessive judgment and should then leave no room for speculation. Any interpretation that makes rhetoric must be impartial and based primarily on fact, as it has a direct impact on the lives of those involved in the case.

Since maps inception, with the ease of satellite images access and analysis, the use of crime mapping remote sensing technologies has become a relevant tool in criminology. Especially in transnational crimes and terrorist activities, these technologies have decreased the dark figure of crime, shedding light on wrongdoers that otherwise would not be recognized. The accuracy assessments may make remotely sensed crime data more serviceable in courtroom settings and law enforcement planning strategies. In this chapter, high resolution geospatial technology for transnational and terrorism crimes are encouraged. This chapter is meant to investigate the relevance of remote sensing in detecting transnational crimes and terrorist acts. This work should take into the widest definition of transnational crimes and terrorist activities and the link between them including the daily remote sensing monitoring of terrorist mobility and their number evolution, arms quality, and quantity (Bayraktar TB2, S-125 SAM System) within the base and the region, linked illegal activities funding terrorist groups (human trafficking from Africa to Europe, arms trade towards Mali, and smuggling). Further, the countries of the region (including Egypt, Algeria and Tunisia) should initiate anticipatory attacks against terrorist groups to limit their capacity to destabilize regional security. Further, logistic remote sensing preparation to rapidly, safely, and massively react to any terrorist act. The remote sensing become active detection of transnational crimes and terrorism. It is worth noting that first order validation is not always feasible. On the other hand, second order validation techniques that may facilitate validation in difficult situations.

The article enlightens the importance of IoT Forensics as the dependability of human beings on IoT increases exponentially. IoT systems are dynamic and complex and composed of numerous interconnected fragments analogous to the Open System Interconnection model. The proposed Cross-Layer Learning model builds on top of Distributed Logging, Layered IoT Architecture, Cross-Layer Forensics, and Deep Learning Architecture. Applying Deep Learning on a single layer does not give a clear picture as IoT systems are interconnected. The idea behind the model is to fill this gap by merging data of two or more layers and automate the correlation. We chose Bi- LSTM to find long-term dependencies between the data and cross verify obtained data from an investigation point of view. It helps investigators by recommending suspicious portions at the same time preserving privacy. A case study is presented to demonstrate the notion. Development of a Cross-Layer Learning Architecture is in progress and we further aim to develop a tool for extracting data from all the stated layers making application of Deep Learning more feasible and hence examining the specified model in greater detail.

As per the various literatures, analysis of fire scene investigation is one of the most challenging, though not impossible, tasks. This challenge is made more difficult by the reconstruction of a fire scene and the determination of the point of origin of a fire. However, some methods of interacting with onlookers, careful external and internal examination of the fire scene, tracing the fire patterns like V patterns (if the vertical surface is near the point of origin of fire), U patterns (indicates the pool of liquid which could be formed of accidental or intentional spillage of the fuel), and so on may help to arrive at near-accurate conclusions or reasons for the fire. Searching for the protected areas, conditions of the objects especially metallic and glass objects. Careful observation of the window panes can also give hint about the fire propagation and enhancements. Apart from careful observation of the fire scene, an examiner should also notice and try to differentiate the entries made on the scene before (by any culprit, suspects or the residents of that building) and after (firefighters and fire investigators) the termination of the fire. An examiner should first check for any container or the source that could be responsible for the fire and collect all the evidences from the scene and should transport it to the fire laboratory with the proper packaging and chain of custody. In the laboratory, the analysis should be performed unbiasedly and with the proper and appropriate use of the instruments, as discusses in the main text. The interpretation of the results must be performed by an experienced individual who has dealt with these analyses before. These laboratory analyses may prove very useful in the construction of fire scene to determine the point of origin and development of the fire. Hence, it is very important to conduct the systematic fire scene analysis because the opinion formed after the analysis has to stand the court trials and also important for the insurance claims by the victims or affected people.

To detect seam carving-based forgery under JPEG recompression anti-forensics, we first analyze the statistical modification to the DCT coefficients by JPEG down-recompression, and design large-scale feature mining-based approach to the detection. Ensemble classifier is applied to the rich features for the detection. Experimental results show that our approach redefines the state of detecting JPEG down-recompression and it also effectively detects the seam carving forgery under recompression anti-forensics, which was not well explored before. Our following study indicates that by integrating feature mining with deep learning-based approaches together, which may also be expanded to other image forensics problems that were not well exposed before. These combined approaches may be employed to reduce the computational cost and the cost to react for cybercrimes or improve cybersecurity to vulnerable population including older adults and children using digital multimedia data increasingly. Texture analysis will be also considered according to each individual examination image, especially the increasing digital dependence posting images on social media as increasing daily social activities. Furthermore, blind detection will be investigated, and its performance will be evaluated. These techniques are pivotal to understand the implication of forgery detection for digital multimedia data in modern society with increasing digital dependence.

An effective source camera identification and verification method based on PRNU analysis for use in LINE social network had been proposed. The main contribution of this chapter is we consider the effects of LINE social network in possible SCI methods. Experimental results demonstrated our method cannot only provide effectiveness but with the zero FPR and low FNR for use in LINE social network.

An efficient AI solution for LPR on moving scooters is presented. Lightweight MobileNetV2-based SSD is chosen to detect license plates in the scene, and LPRNet is used to recognize the license plates. License plates are detected depending on their character orientations, and are rotation corrected accordingly for better recognition precision. Several data augmentation methods are employed to enhance the robustness of our models while preventing overfitting, and focal CTC loss is adopted to conquer the data imbalance problem. Experimental results on real street scenes show that our model achieves 95.7% precision and 95% recall at high speed during the daytime, and an average precision of 86.1% and an average recall of 90.1% are attained when the speed is lower than 20 km/h during the night. In addition, our system has real-time performance even if there are 7 license plates in the given scene, and with a maximum power consumption of only 91.8W. For future work, our system could have better performance with state-of-the-art object detection methods such as YOLOv4which are both fast and accurate at the same time. Our adopted MobileNetV2-based SSD is quite fast yet less accurate. Furthermore, there are some different ways to solve the data imbalance problem. For instance, use Conditional Generative Adversarial Network (CGAN) to generate different valid license plates for training. CGAN is effective at synthesizing images from given labels, and thus it is quite convenient to control the character distribution of license plate dataset with CGAN owing to its controllability.

Machine learning techniques can help automate forensic scenarios in fighting the sophisticated cyber-attacks on the financial industry. In this work, we first propose a forensic model for the economic sectors. We then focus on developing a machine learning model for the e-commerce transaction data to illustrate how the machine learning models can assist in the financial forensic processes. We have presented the performance metrics of machine learning classifiers using seven different algorithms to classify fraudulent transactions from regular transactions. The algorithm that we have used are Random Forest, Logistic Regression, Naïve Bayes, Decision Tree, K-Nearest Neighbor, Support Vector Machine, and Neural Network Multi-Layer Perceptron. Because the dataset is imbalanced, we need to apply the undersampling technique to process the features before fitting them into the model. We have developed three test cases. The first two test cases handle the feature selection and manipulation, and the third test case deals with the comparison of the undersampling technique. The performance metrics are satisfactory, and we find that either the Random Forest or the Logistic Regression model has better overall performance than the other models. This chapter describes necessary background information on financial fraud and proposes a detailed forensic technique workflow. Overall, the chapter presents some vital discussion on how the machine learning-based models can boost the automation processes in financial fraud forensics.

Compilation of References

Abdul-Ghani, H. A., & Konstantas, D. (2019). A Comprehensive Study of Security and Privacy Guidelines, Threats, and Countermeasures: An IoT Perspective. *Journal of Sensor and Actuator Networks*, *8*(2), 22. doi:10.3390/jsan8020022

Agarwal, A., & Gupta, A. (2020). A maximum relevancy and minimum redundancy feature selection approach for median filtering forensics. *Multimedia Tools and Applications*, *79*(29-30), 21743–21770. doi:10.100711042-020-08994-4

Agarwal, R., Verma, O. P., Saini, A., Shaw, A., & Patel, R. (2021). The Advent of Deep Learning-Based Image Forgery Detection Techniques. In J. S. Raj, A. M. Iliyasu, R. Bestak, & Z. A. Baig (Eds.), *Innovative Data Communication Technologies and Application. Lecture Notes on Data Engineering and Communications Technologies* (Vol. 59). Springer. doi:10.1007/978-981-15-9651-3_56

Alajmi, R., Abdel-Gaber, R., & Haddadi, R. (2020). Molecular identification of forensically important beetles in Saudi Arabia based on mitochondrial 16 srRNA gene. *Entomological Research*, *50*(7), 343–350. doi:10.1111/1748-5967.12438

Al-Garadi, M. A., Mohamed, A., Al-Ali, A. K., Du, X., Ali, I., & Guizani, M. (2020). A Survey of Machine and Deep Learning Methods for Internet of Things (IoT) Security. *IEEE Communications Surveys and Tutorials*, *22*(3), 1646–1685. doi:10.1109/COMST.2020.2988293

Ali, Y., Khan, A., Khan, G., Khan, A. U., & Ahmad, Z. (2021). Evaluation of effectiveness of fence on a country's border: A case study of Pak-Afghan border fence. *Journal of International Migration and Integration*, *22*(2), 769–790. doi:10.100712134-020-00774-1

Al-Masri, E., Bai, Y., & Li, J. (2018). A Fog-Based Digital Forensics Investigation Framework for IoT Systems. *2018 IEEE International Conference on Smart Cloud (SmartCloud)*, 196–201. 10.1109/SmartCloud.2018.00040

Almirall, J. R., & Furton, K. G. (2004). *Analysis and Interpretation of Fire Scene Evidence*. CRC Press.

Amerini, I., Ballan, L., Caldelli, R., Del Bimbo, A., & Serra, G. (2011). A SIFT-based forensic method for copy–move attack detection and transformation recovery. *IEEE Transactions on Information Forensics and Security*, *6*(3), 1099–1110. doi:10.1109/TIFS.2011.2129512

Amran, T. S. T., Amin, M. S. M., Ahmad, M. R., Sani, S., Masenwat, N. A., Abas, A. A., ... Adnan, M. A. K. (2020). Ground Penetrating Radar (GPR) Applications in Forensic. *Jurnal Sains Nuklear Malaysia*, *32*(1), 49–57.

Ascent, T. (2020). *Identity Theft and Credit Card Fraud Statistics for 2020*. Retrieved from https://www.fool.com/the-ascent/research/identity-theft-credit-card-fraud-statistics/

Atlam, H. F., El-Din Hemdan, E., Alenezi, A., Alassafi, M. O., & Wills, G. B. (2020). Internet of Things Forensics: A Review. *Internet of Things*, *11*, 100220. doi:10.1016/j.iot.2020.100220

Atsa'am, D. D., Wario, R., & Alola, U. V. (2021). Why Do Terrorists Attack Other Terrorists? Knowledge Discovery from the Global Terrorism Database. *The International Journal of Intelligence, Security, and Public Affairs*, 1-14.

Attah, C. E. (2020). Opinion Piece–Terrorism, Human Trafficking and Conflict-Related Sexual Violence: Examining the Nexus. Journal of Human Trafficking. *Enslavement and Conflict-Related Sexual Violence*, *1*(2), 117–133. doi:10.7590/266644720X16061196655016

Avidan, S., & Shamir, A. (2007). Seam carving for content-aware image resizing. *ACM Transactions on Graphics*, *26*(3), 10. doi:10.1145/1276377.1276390

Avtar, R., Kouser, A., Kumar, A., Singh, D., Misra, P., Gupta, A., Yunus, A. P., Kumar, P., Johnson, B. A., Dasgupta, R., Sahu, N., & Besse Rimba, A. (2021). Remote Sensing for International Peace and Security: Its Role and Implications. *Remote Sensing*, *13*(3), 439. doi:10.3390/rs13030439

Awoyemi, J. O., Adetunmbi, A. O., & Oluwadare, S. A. (2017). *Credit card fraud detection using machine learning techniques: A comparative analysis.* Paper presented at the 2017 International Conference on Computing Networking and Informatics (ICCNI). 10.1109/ICCNI.2017.8123782

Babun, L., Sikder, A. K., Acar, A., & Uluagac, A. S. (2018). *IoTDots: A Digital Forensics Framework for Smart Environments*. https://arxiv.org/abs/1809.00745

Bachagha, N., Wang, X., Luo, L., Li, L., Khatteli, H., & Lasaponara, R. (2020). Remote sensing and GIS techniques for reconstructing the military fort system on the Roman boundary (Tunisian section) and identifying archaeological sites. *Remote Sensing of Environment*, *236*, 111418. doi:10.1016/j.rse.2019.111418

Baldaino, J., Ommen, D. M., Saunders, C. P., Hietpas, J., & Buscaglia, J. (2021). Characterization and differentiation of aluminum powders used in improvised explosive devices–Part 1: Proof of concept of the utility of particle micromorphometry. *Journal of Forensic Sciences*, *66*(1), 83–95. doi:10.1111/1556-4029.14564 PMID:33006770

Banwarth. (2015). *Fire Service Features of Buildings and Fire Protection Systems*. Academic Press.

Barad, Z., & Goswami, M. (2020). Image Forgery Detection using Deep Learning: A Survey. *Proceedings of 2020 6th International Conference on Advanced Computing and Communication Systems (ICACCS)*, 571-576. doi: 10.1109/ICACCS48705.2020.9074408

Basu, P. (2013). Biomass Characteristics. In *Biomass Gasification, Pyrolysis and Torrefaction* (pp. 72–73). Elsevier Inc., doi:10.1016/B978-0-12-396488-5.00003-4

Bayuk, J. (2010). *Cyber Forensics*. Springer.

Beccaro, A. (2020). ISIS in Libya and beyond, 2014–2016. *Journal of North African Studies*, 1–20. doi:10.1080/13629387.2020.1747445

Berezowski, V., Mallett, X., Ellis, J., & Moffat, I. (2021). Using Ground Penetrating Radar and Resistivity Methods to Locate Unmarked Graves: A Review. *Remote Sensing*, *13*(15), 2880. doi:10.3390/rs13152880

Bergström, F. (2020). *Trafficking as Terrorism: A Theory-testing Case Study on the Islamic State's Human Trafficking of the Yazidi*. Academic Press.

Bhimabhai, K. S. (2018). Comparative Study of Aliphatic and Aromatic Compounds. *The Pharma Innovation Journal, 7*(9), 175–77. www.chemguide.co.uk

Bianchi, T., & Piva, A. (2012). Image forgery localization via block-grained analysis of JPEG artifacts. *IEEE Transactions on Information Forensics and Security*, *7*(3), 1003–1017. doi:10.1109/TIFS.2012.2187516

Bi, F., Zhang, J., Pang, F., Bian, M., & Wang, Y. (2020). Suburban Building Detection from Optical Remote Sensing Images Based on a Deformation Adaptability Model. *Photonirvachak (Dehra Dun)*, *48*(6), 831–839. doi:10.100712524-020-01117-4

Bochkovskiy, A., Wang, C. Y., & Liao, H. Y. M. (2020). *YOLOv4: Optimal Speed and Accuracy of Object Detection*. https://arxiv.org/pdf/2004.10934.pdf

Bodle, E. S., & Hardy, J. K. (2007). Multivariate Pattern Recognition of Petroleum-Based Accelerants by Solid-Phase Microextraction Gas Chromatography with Flame Ionization Detection. *Analytica Chimica Acta*, *589*(2), 247–254. doi:10.1016/j.aca.2007.03.006 PMID:17418188

Bovenkerk, F., & Chakra, B. A. (2005). Terrorism and organized crime. In Forum on Crime and Society (Vol. 4, No. 1 and 2, pp. 3-16). Academic Press.

Bowman, D. M. J. S., Balch, J. K., Artaxo, P., Bond, W. J., Carlson, J. M., Cochrane, M. A., D'Antonio, C. M., DeFries, R. S., Doyle, J. C., Harrison, S. P., Johnston, F. H., Keeley, J. E., Krawchuk, M. A., Kull, C. A., Marston, J. B., Moritz, M. A., Prentice, I. C., Roos, C. I., Scott, A. C., ... Pyne, S. J. (2009). Fire in the Earth System. *Science*, *324*(5926), 481–484. doi:10.1126cience.1163886 PMID:19390038

Brinson, A., Robinson, A., & Rogers, M. (2006). A cyber forensics ontology: Creating a new approach to studying cyber forensics. *Digital Investigation, 3*, 37-43.

Bull, P. A., & Morgan, R. M. (2006). Sediment fingerprints: A forensic technique using quartz sand grains. *Science & Justice*, *46*, 107–124. doi:10.1016/S1355-0306(06)71581

Bumbrah, G. S., Sarin, R. K., & Sharma, R. M. (2018). Derivative Ultraviolet Spectrophotometric Studies on Ignitable Liquids. *Journal of Forensic Chemistry and Toxicology*, *4*(1).

Burhan, M., Rehman, R., Khan, B., & Kim, B.-S. (2018). IoT Elements, Layered Architectures and Security Issues: A Comprehensive Survey. *Sensors (Basel)*, *18*(9), 2796. doi:10.339018092796 PMID:30149582

Burnier, C., Massonnet, G., Coulson, S., DeTata, D., & Pitts, K. (2020). Characterization and classification of water-based compounds in condoms and personal hygiene products using GC-MS. *Forensic Science International*, *317*, 110513. doi:10.1016/j.forsciint.2020.110513 PMID:32987281

Cadwell, Z. (2020). *Pointing the Finger: Applying Extreme Environment Fingermark Development to Forensic Archaeology*. Academic Press.

Canata, R. E., Ferreira, F. J. F., Borges, W. R., & da Silva Salvador, F. A. (2020). Analysis of 2D and 3D GPR responses in the Federal University of Paraná Forensic Geophysics Controlled Site–a Case Study. *Brazilian Journal of Geophysics*, *38*(2). Advance online publication. doi:10.22564/rbgf.v38i2.2045

Caparó Bellido, A., & Rundquist, B. C. (2021). Semi-Automatic Fractional Snow Cover Monitoring from Near-Surface Remote Sensing in Grassland. *Remote Sensing*, *13*(11), 2045. doi:10.3390/rs13112045

Carter, R. (2020). *Investigation of Rare Earth Element Signatures and Surrogate Nuclear Melt Glass for Nuclear Forensic Analysis* (Doctoral dissertation). University Of Notre Dame.

Chang, W., Shih, T. K., & Hsu, H. (2013). Detection of seam carving in JPEG images. *Proceedings of the 2013 International Joint Conference on Awareness Science and Technology and Ubi-media Computing*, 632-638. 10.1109/ICAwST.2013.6765516

Chen, Y. L., & Hsu, C. T. (2011). Detecting recompression of JPEG images via periodicity analysis of compression artifacts for tampering detection. *IEEE Transactions on Information Forensics and Security*, *6*(2), 396–406. doi:10.1109/TIFS.2011.2106121

Chua, C. C., Kwok, H., Yan, J., Cuthbertson, D., Aggelen, G. V., Brunswick, P., & Shang, D. (2021). Development of a tiered analytical method for forensic investigation of mixed lubricating oil samples. *Environmental Forensics*, 1–13. doi:10.1080/15275922.2021.1907821

Colledge, D. (2020). *A review of intrusive search methods and excavation techniques for clandestine grave site recovery in forensic archaeology* (Doctoral dissertation). Murdoch University.

Corbin, S. L. (2021). *Ineffective US immigration policies and deterrence mechanisms at the US-Mexico border: The story of Mexico and the northern triangle*. Academic Press.

Cragin, R. K. (2021). Preventing the next wave of foreign terrorist fighters: Lessons learned from the experiences of Algeria and Tunisia. *Studies in Conflict and Terrorism*, *44*(7), 543–564. doi:10.1080/1057610X.2019.1568005

Cullis, B. (1995). *An Exploratory Analysis of Responses to Geographic Information System Adoption on Tri-Service Military Installations.* Army Engineer Waterways Experiment Station Vicksburg MS Tri-Service Cadd/Gis Technology Center.

Damiata, B. N., Walker, J., Stansell, A., & Steinberg, J. M. (2020, November). On GPR surveying of historical cemeteries and ancient graveyards to aid forensic research. In *18th International Conference on Ground Penetrating Radar* (pp. 220-223). Society of Exploration Geophysicists. 10.1190/gpr2020-059.1

Dargan, S., Kumar, M., Ayyagari, M. R., & Kumar, G. (2020). A Survey of Deep Learning and Its Applications: A New Paradigm to Machine Learning. *Archives of Computational Methods in Engineering, 27*(4), 1071–1092. doi:10.100711831-019-09344-w

Dashnyam, Z. (2020). Joint Statement on Strengthening the Strategic Partnership between Mongolia and India. *Mongolian Journal of International Affairs, 21*, 145–150.

David GossmanL. D. (n.d.). https://www.hgexperts.com/expert-witness-articles/fires-and-explosions-evidence-gathering-and-preservation-47077

Davis, A., Whitehead, C., & Lengke, M. (2021). Subtle early-warning indicators of landfill subsurface thermal events. *Environmental Forensics*, 1–19. doi:10.1080/15275922.2021.1887973

Dawson, L. A., Di Maggio, R. M., McKinley, J., Di Capua, G., Peppoloni, S., & Pringle, J. (2021). A proposal for a White Paper on Geoethics in Forensic Geology. *Geological Society, 508.*

De Angelis, D., Mapelli, G., Mazzullo, F. L., Lorenz, M. T., & Cattaneo, C. (2020). Possible applications of reflected UV photography in forensic odontology: Food for thought. *Legal Medicine, 42*, 101641. doi:10.1016/j.legalmed.2019.101641 PMID:31751798

de Bruin, M., & Schmitz, P. M. U. (2021). The application of forensic geomorphology in rhinoceros poaching investigations in Africa. *The South African Geographical Journal, 103*(1), 43–63. do i:10.1080/03736245.2020.1835700

de Caritat, P., Woods, B., Simpson, T., Nichols, C., Hoogenboom, L., Ilheo, A., ... Hoogewerff, J. (2021). Forensic soil provenancing in an urban/suburban setting: A sequential multivariate approach. *Journal of Forensic Sciences.* PMID:33955554

De Giorgi, L., Fragalá, G., Leucci, G., Mazzaglia, A., Malfitana, D., & Russo, S. (2020). Forensic geophysical survey to investigate illegal looting of a site of historical importance. *Archaeological Prospection*, arp.1796. doi:10.1002/arp.1796

de Oliveira, O. M., Queiroz, A. F. D. S., Cerqueira, J. R., Soares, S. A., Garcia, K. S., Pavani Filho, A., ... Moreira, Í. T. (2020). Environmental disaster in the northeast coast of Brazil: Forensic geochemistry in the identification of the source of the oily material. *Marine Pollution Bulletin, 160*, 111597. doi:10.1016/j.marpolbul.2020.111597 PMID:32896715

DeHann, J. D., & Icove, D. J. (2012). *Kirk's Fire Investigation* (7th ed.). Julie Levin Alexander.

Demirdjian, Z. S., & Mokatsian, Z. (2015). The cost of cyber crimes to business and society. *ASBBS Proceedings, 22*(1), 104.

Desille, A. (2021). On the Use of Visual Methods to Understand Local Immigration Politics. In *Visual Methodology in Migration Studies* (pp. 67–84). Springer. doi:10.1007/978-3-030-67608-7_4

Deveci, C., Atilgan, M., & Demirçin, S. (2021). The forensic aspect of thermal injuries during MRI: A case report with a review of literature. *Revista de Medicina Legala, 29*, 49–52.

Dhabbah, A. M., Al-Jaber, S. S., Al-Ghamdi, A. H., & Aqel, A. (2014). Determination of Gasoline Residues on Carpets by SPME-GC-MS Technique. *Arabian Journal for Science and Engineering, 39*(9), 6749–6756. doi:10.100713369-014-1233-1

Dhanda, S. S., Singh, B., & Jindal, P. (2020). Lightweight Cryptography: A Solution to Secure IoT. *Wireless Personal Communications, 112*(3), 1947–1980. doi:10.100711277-020-07134-3

Diaz, E., Yokosuk, M. O., Myers, T. L., & Johnson, T. J. (2020, September). Preliminary parameter studies to form pellets of hard materials for optical constants derived via single-angle reflectance spectroscopy. In *Counterterrorism, Crime Fighting, Forensics, and Surveillance Technologies IV* (Vol. 11542, p. 115420H). International Society for Optics and Photonics. doi:10.1117/12.2574250

Dong, C. E., Webb, J. B., Bottrell, M. C., Saginor, I., Lee, B. D., & Stern, L. A. (2020). Strengths, limitations, and recommendations for instrumental color measurement in forensic soil characterization. *Journal of Forensic Sciences, 65*(2), 438–449. doi:10.1111/1556-4029.14193 PMID:31524965

Douglas, G. S., Graan, T. P., & Hardenstine, J. H. (2020). Forensic identification and quantification of oil sands-based bitumen released into a complex sediment environment. *Marine Pollution Bulletin, 155*, 111141. doi:10.1016/j.marpolbul.2020.111141 PMID:32469767

Dudley, R. J. (1976). The particle size analysis of soils and its use in forensic science—The determination of particle size distributions within the silt and sand fractions. *Journal - Forensic Science Society, 16*(3), 219–229. doi:10.1016/S0015-7368(76)71064-8 PMID:1022844

Ekim, B., Sertel, E., & Kabadayı, M. E. (2021). Automatic Road Extraction from Historical Maps Using Deep Learning Techniques: A Regional Case Study of Turkey in a German World War II Map. *ISPRS International Journal of Geo-Information, 10*(8), 492. doi:10.3390/ijgi10080492

Ekštajn, H., Kružić, I., & Bašić, Ž. (2021). Forensic investigation of a mass grave at Ovčara, near Vukovar, of victims killed by the Serbian army in 1991. *ST-OPEN, 2*, 1–26. doi:10.48188o.2.3

Elis, V. R., Almeida, E. R., Porsani, J. L., & Stangari, M. C. (2020, November). Ground-penetrating radar, resistivity, and induced polarization applied in forensic research in tropical soils. In *18th International Conference on Ground Penetrating Radar* (pp. 224-227). Society of Exploration Geophysicists.

Eppich, G. R. (2020). Nuclear forensics: How science helps stop the trafficking of nuclear materials. *Bulletin of the Atomic Scientists, 76*(5), 263–270. doi:10.1080/00963402.2020.1806587

Essefi, E. (2013). *Wet aeolian sedimentology and sequence stratigraphy in Eastern Tunisia: implications for wet aeolian sedimentology and sequence stratigraphy on mars* [Ph.D. thesis]. National Engineering School of Sfax.

Essefi, E. (2013). *Wet Aeolian Sedimentology and Sequence Stratigraphy within the Terrestrial Analogues in Eastern Tunisia: Implications for Wet Aeolian Sedimentology and Sequence Stratigraphy on Mars* (Doctoral dissertation). National Engineering School of Sfax, Sfax, Tunisia.

Essefi, E. (2021b). Record of the Anthropocene-Great Acceleration along a core from the coast of Sfax, southeastern Tunisia. *Turkish Journal of Earth Science*, 1-16.

Essefi, E., Gharsalli, N., & Yaich, C. (2015c). Geophysical and geochemical study of the silico-evaporitic sedimentary filling of Boujmal wetland, eastern Tunisia: Inferring the climatic signal within groundwater noise. *J Basic Appl Res Int, 15*(2–1), 14–25.

Essefi, E., Gharsalli, N., Salhi, S., & Yaich, C. (2015b). Cyclicity of grain size parameters along a Core from the Gleysol of Sebkha Ennoual, southeastern Tunisia. *J Basic Appl Res Int, 15*(5–3), 124–134.

Essefi, E. (2020). On the Geochemistry and Mineralogy of the Anthropocene. *Int J Water Wastewater Treat, 6*(2).

Essefi, E. (2020). On the Geochemistry and Mineralogy of the Anthropocene. *International Journal of Water and Wastewater Treatment, 6*(2), 1–14.

Essefi, E. (2021). Record of the Anthropocene-Great Acceleration along a core from the coast of Sfax, southeastern Tunisia. *Turkish Journal of Earth Sciences*.

Essefi, E. (2021a). *Apport des Approches Géochimique et Sédimentologique au Fonctionnement des Systèmes Endoréiques et Exoréiques de l'Holocène à l'Anthropocène: Intérêts Environnemental et Géoéconomique*. Institut Supérieur des Sciences et Techniques des Eaux de Gabès.

Essefi, E., Gharsalli, N., Kalabi, S., Ben Ameur, M., & Yaich, C. (2015d). Spectral analysis of a core from the Sebkha of Sidi Mansour, southernTunisia: The Holocene cyclostratigraphy. *J Geophys Remote Sens, 4*(02), 141. doi:10.4172/2169-0049.1000141

Essefi, E., Gharsalli, N., & Yaich, C. (2015a). Climatic significance aeolian sedimentary filling of sebkha Mhabeul, southeastern Tunisia: Contribution of the genetic approach. *J Basic Appl Res Int, 15*(2–4), 140–152.

Essefi, E., Jmaa, H. B., Touir, J., Tagorti, M. A., & Yaich, C. (2013). Variability to sedimentary dynamics and climatic conditions during the last two millennia at sebkha Souassi in eastern Tunisia. *Papers on Global Change IGBP, 20*(1), 29–45. doi:10.2478/igbp-2013-0001

Essefi, E., Mefteh, S., Medhioub, M., & Yaich, C. (2014). Magnetic study of the heated and unheated sedimentary fillings of Sebkha Mhabeul, Southeast Tunisia: A geophysical method for paleoclimatic investigation and tephrochronological dating. *International Journal of Geophysics, 2014*, 7–7. doi:10.1155/2014/908395

Essefi, E., Tagorti, M. A., Touir, J., & Yaich, C. (2013). Hydrocarbons migration through groundwater convergence toward saline depressions: A case study, Sidi El Hani discharge playa, Tunisian Sahel. *International Scholarly Research Notices.*

Evans-Nguyen, K. (2019). Forensic Analysis of Fire Debris and Explosives. In *An Introduction to Instrumentation Used in Fire Debris and Explosive Analysis*. Academic Press.

Evidence Collection. (n.d.). https://www.iaaievidenceguide.com/

Fabian, K., Reimann, C., Kuzina, D., Kosareva, L., Fattakhova, L., & Nurgaliev, D. (2016, April). GEMAS: Unmixing magnetic properties of European agricultural soil. In EGU General Assembly Conference Abstracts (pp. EPSC2016-12495). Academic Press.

Fadlullah, Z. M., Tang, F., Mao, B., Kato, N., Akashi, O., Inoue, T., & Mizutani, K. (2017). State-of-the-Art Deep Learning: Evolving Machine Intelligence Toward Tomorrow's Intelligent Network Traffic Control Systems. *IEEE Communications Surveys and Tutorials*, *19*(4), 2432–2455. doi:10.1109/COMST.2017.2707140

Faramawy, S., Zaki, T., & Sakr, A. A. E. (2016). Natural Gas Origin, Composition, and Processing: A Review. *Journal of Natural Gas Science and Engineering*, *34*, 34–54. doi:10.1016/j.jngse.2016.06.030

Feng, X., Dawam, E. S., & Amin, S. (2017). A New Digital Forensics Model of Smart City Automated Vehicles. *2017 IEEE International Conference on Internet of Things (IThings) and IEEE Green Computing and Communications (GreenCom) and IEEE Cyber, Physical and Social Computing (CPSCom) and IEEE Smart Data (SmartData)*, 274–279. 10.1109/iThings-GreenCom-CPSCom-SmartData.2017.47

Fernandes, E., Rahmati, A., Eykholt, K., & Prakash, A. (2017). Internet of Things Security Research: A Rehash of Old Ideas or New Intellectual Challenges? *IEEE Security and Privacy*, *15*(4), 79–84. doi:10.1109/MSP.2017.3151346

Ferrell, M. (2020). *Applications of Close-range Photogrammetry for Documenting Human Skeletal Remains in Obstructed Wooded Environments*. Academic Press.

Fillion, C., & Sharma, G. (2010). Detecting content adaptive scaling of images for forensic applications. *Proceedings of SPIE, Media Forensics and Security II*, *7541*, 888–896. doi:. 2010.10.1117/12.838647

Finkelstein, N. S., Rosengarten, H., & Levy, O. (2020). Photographic Image Comparison of Vegetation. *Journal of Forensic Sciences*, *65*(6), 2165–2169. doi:10.1111/1556-4029.14505 PMID:33460121

Fires Classification. (2019). http://global-intrima.com/media/news/FIRE_CLASSIFICATION.pdf

Fisabillah, A., & Pujiyono, U. R. (2021). *Policy to Eradicate Crime Funding of Terrorism as Transnational Organized Crime*. Academic Press.

Fitzpatrick, R. W. (2009). Soil: forensic analysis. *Wiley encyclopedia of forensic science*, 1-14.

Fitzpatrick, R. W., & Baker, A. (2006). Magnetic susceptibility and low altitude aerial photography using a helium balloon: applications to forensic soil examinations. In Book of Abstracts (p. 19). Academic Press.

Fleuriet, K. J., & Castellano, M. (2020). Media, place-making, and concept-metaphors: The US-Mexico border during the rise of Donald Trump. *Media Culture & Society*, *42*(6), 880–897. doi:10.1177/0163443719890539

Fogel, I., & Sagi, D. (1989). Gabor filters as texture discriminator. *Biological Cybernetics*, *61*(2). doi:10.1007/BF00204594

Force, J. T. (2018). *Risk Management Framework for Information Systems and Organizations: A System Life Cycle Approach for Security and Privacy*. Academic Press.

Force, J. T., & Initiative, T. (2013). Security and privacy controls for federal information systems and organizations. *NIST Special Publication*, *800*(53), 8–13.

Franceschetti, L., Galante, N., Del Sordo, S., Casali, M. B., & Genovese, U. (2021). Forensic Considerations on the Two Major Civilian Terrorist Events occurred in Milan: A Retrospective Autopsy-Based Study. *Forensic Science International*, *326*, 110929. doi:10.1016/j.forsciint.2021.110929 PMID:34329954

Frankl, A., Evrard, O., Verleyen, E., Cammeraat, E., & Stokes, A. (2021, March). How forensic science can lead the way in identifying culprit soil fingerprints in European mountains. In *MCAA Annual Conference* (p. 38). Academic Press.

Fridrich, J., & Kodovsky, J. (2012). Rich models for steganalysis of digital images. *IEEE Transactions on Information Forensics and Security*, *7*(3), 868–882. doi:10.1109/TIFS.2012.2190402

Furton, K. G. (2000). Application of Solid-Phase Microextraction to the Recovery of Explosives and Ignitable Liquid Residues from Forensic Specimens. *Journal of Chromatography A*, *885*, 419–432. papers2://publication/uuid/F55C29AC-08CD-4F01-A104-6F9164B92A72

Furton, K. G., Bruna, J., & Almirall, J. R. (1995). A Simple, Inexpensive, Rapid, Sensitive and Solventless Technique for the Analysis of Accelerants in Fire Debris Based on SPME. *Journal of High Resolution Chromatography*, *18*(10), 625–629.

García, M. G., Pérez-Cárceles, M. D., Osuna, E., & Legaz, I. (2020). Impact of the human microbiome in forensic sciences: A systematic review. *Applied and Environmental Microbiology*, *86*(22). Advance online publication. doi:10.1128/AEM.01451-20 PMID:32887714

Ghannem, N., Gargouri, D., Sarbeji, M. M., Yaich, C., & Azri, C. (2014). Metal contamination of surface sediments of the Sfax–Chebba coastal line, Tunisia. *Environmental Earth Sciences*, *72*(9), 3419–3427. doi:10.100712665-014-3248-z

Gharsalli, N., Essefi, E., Baydoun, R., & Yaich, C. (2020). The Anthropocene and great acceleration as controversial epoch of human-induced activities: Case study of the Halk el Menjel Wetland, Eastern Tunisia. *Applied Ecology and Environmental Research*, *18*(3), 4137–4166. doi:10.15666/aeer/1803_41374166

Global, P. (2020). *Global Economic Crime and Fraud Survey 2020*. Retrieved from https://www.pwc.com/gx/en/services/forensics/economic-crime-survey.html

Gonçalves, da Silva, Menotti, & Schwartz. (2016). Benchmark for License Plate Character Segmentation. *Journal of Electronic Imaging, 25*(5).

Gorbett, G. E., Hicks, W., Hopkins, R., & Kennedy, P. M. (2010). Fire Patterns Analysis with Low Heat Release Rate Initial Fuels. *International Symposium on Fire Investigation Science and Technology*: 269–94.

Graves, A., Fernandez, S., Gomez, F., & Schmidhuber, J. (2006). Connectionist Temporal Classification: Labelling Unsegmented Sequence Data with Recurrent Neural Networks. *Proceedings of International Conference on Machine Learning*, 369-376. 10.1145/1143844.1143891

Grigorescu, S. E., Petkov, N., & Kruizinga, P. (2002). Comparison of texture features based on Gabor filters. *IEEE Transactions on Image Processing, 11*(10), 1160–1167. doi:10.1109/TIP.2002.804262 PMID:18249688

Gruber, K. (2021). *Remote Sensing Techniques for Middle Eastern Archaeology-K.* Gruber.

Gupta, G., Lalita Jawale, M., & Bhattacharya, B. (2015). Various Methods for the Determination of the Burning Rates of Solid Propellants - an Overview. *Central European Journal of Energetic Materials, 12*(3), 593–620.

Hachem, M., Sharma, B. K., El Naggar, A., Pilankar, I., & Anwar, N. (2020). Systematic Approaches For Soil Analysis In Forensic Investigation. In *Advances in Science and Engineering Technology International Conferences (ASET)* (pp. 1-5). IEEE.

Hall, J. R. (2006). *Home Cooking Fire Patterns and Trends*. https://iflss.net/CookingFires-NFPA report2006.pdf

Han, R., Ke, Y., Du, L., Qin, F., & Guo, J. (2018). Exploring the location of object deleted by seam-carving. *Expert Systems with Applications, 95*, 162–171. doi:10.1016/j.eswa.2017.11.023

Han, Y., Ma, S., Zhang, F., & Li, C. (2020, July). Object Detection of Remote Sensing Airport Image Based on Improved Faster R-CNN. *Journal of Physics: Conference Series, 1601*(3).

Haque, A. (Producer). (2021). *GitHub Code*. Retrieved from https://github.com/arif6008/Credit_Card_Transaction_Classification

Harbawi, M., & Varol, A. (2017). An improved digital evidence acquisition model for the Internet of Things forensic I: A theoretical framework. *2017 5th International Symposium on Digital Forensic and Security (ISDFS)*, 1–6. 10.1109/ISDFS.2017.7916508

Harvey, A. J., & Sekulla, A. (2021). Evidence of alcohol induced weapon focus in eyewitness memory. *Applied Cognitive Psychology, 35*(5), 1263–1272. doi:10.1002/acp.3858

Hasham, S., Joshi, S., & Mikkelsen, D. (2019). Financial Crime And Fraud. In *The Age of Cyber Security*. McKinsey & Company.

Hashmi, T. A., Qureshi, R., Tipre, D., & Menon, S. (2020). Investigation of pesticide residues in water, sediments and fish samples from Tapi River, India as a case study and its forensic significance. *Environmental Forensics*, *21*(1), 1–10. doi:10.1080/15275922.2019.1693441

Hassija, V., Chamola, V., Saxena, V., Jain, D., Goyal, P., & Sikdar, B. (2019). A Survey on IoT Security: Application Areas, Security Threats, and Solution Architectures. *IEEE Access: Practical Innovations, Open Solutions*, *7*, 82721–82743. doi:10.1109/ACCESS.2019.2924045

Hine, G. (2004). Fire scene investigation: an introduction for chemists. In J. Almirall & K. Furton (Eds.), *Analysis and interpretation of fire scene evidence* (pp. 71–72). CRC Press.

Holleyhead, R. (1996). Ignition of Flammable Gases and Liquids by Cigarettes: A Review. *Science and Justice - Journal - Forensic Science Society*, *36*(4), 257–266.

Holmes, G., Donkin, A., & Witten, I. H. (1994). Weka: A machine learning workbench. *Proceedings of ANZIIS'94-Australian New Zealnd Intelligent Information Systems Conference*. 10.1109/ANZIIS.1994.396988

Hossain, M., Hasan, R., & Zawoad, S. (2017). Trust-IoV: A trustworthy forensic investigation framework for the internet of vehicles (IoV). *Proceedings - 2017 IEEE 2nd International Congress on Internet of Things, ICIOT 2017*, 25–32. 10.1109/IEEE.ICIOT.2017.13

Howard, R., & Wilson, I. B. (1948). The effect of surface on melting point. *Journal of Colloid Science*, *3*(6), 551–561.

Huang, F. J., Huang, J., & Shi, Y. (2010). Detecting double JPEG compression with the same quantization matrix. *IEEE Transactions on Information Forensics and Security*, *5*(4), 848–856. doi:10.1109/TIFS.2010.2072921

Hussain, F., Hussain, R., Hassan, S. A., & Hossain, E. (2020). Machine Learning in IoT Security: Current Solutions and Future Challenges. *IEEE Communications Surveys and Tutorials*, *22*(3), 1686–1721. doi:10.1109/COMST.2020.2986444

Inkret, J., Podovšovnik, E., Zupanc, T., Haring, G., & Pajnič, I. Z. (2021). Intra-bone nuclear DNA variability in Second World War metatarsal and metacarpal bones. *International Journal of Legal Medicine*, 1–12. PMID:33624158

Intel. (2020). *Intel Distribution of OpenVINO Toolkit*. https://software.intel.com/en us/openvino

International Association of Arson Investigators. (1999). *A pocket guide to accelerant evidence collection (Massachusetts chapter)* (2nd ed.). IAAI.

ISO - ISO/IEC 29100:2011 - Information technology — Security techniques — Privacy framework. (n.d.). Retrieved October 2, 2020, from https://www.iso.org/standard/45123.html

Isola, P., Zhu, J. Y., Zhou, T., & Efros, A. A. (2017). Image to Image Translation with Conditional Adversarial Networks. *Proceedings of IEEE Conference on Computer Vision and Pattern Recognition*, 1125-1134. 10.1109/CVPR.2017.632

Jakubec, P., Qubaiová, J., Novák, M., & Růžička, J. (2021). Developmental biology of forensically important beetle, Necrophila (Calosilpha) brunnicollis (Coleoptera: Silphidae). *Journal of Medical Entomology, 58*(1), 64–70. PMID:32829403

Joffé, G. (2020). Libya: The new geopolitical arena. *Journal of North African Studies, 25*(5), 681–688. doi:10.1080/13629387.2020.1790110

JofreM.GerlachR. H. (2018). Fighting accounting fraud through forensic data analytics. *Available at* SSRN 3176288. doi:10.2139/ssrn.3176288

John, S. O. O., Usman, I. T., Akpa, T. C., & Ibrahim, U. (2021, February). Rare earth elements in Uranium ore for nuclear forensic application. *IOP Conference Series. Earth and Environmental Science, 655*(1), 012075. doi:10.1088/1755-1315/655/1/012075

Kadir, N. K. (2020). Cyberterrorism's Dilemma: Renewal of Conventional Terrorism. *International Journal of Global Community, 3*(2), 97-110.

Kadir, N. K., Judhariksawan, J., & Maskun, M. (2019). Terrorism and cyberspace: A phenomenon of cyber-terrorism as transnational crimes. *FIAT JUSTISIA: Jurnal Ilmu Hukum, 13*(4), 333–344. doi:10.25041/fiatjustisia.v13no4.1735

Kaggle. (2009). *UCSD-FICO Dataset Task 1*. Author.

Karie, N. M., Kebande, V. R., & Venter, H. S. (2019). Diverging deep learning cognitive computing techniques into cyber forensics. *Forensic Science International: Synergy, 1*, 61–67. doi:10.1016/j.fsisyn.2019.03.006 PMID:32411955

Kaur, H. (2021). Machine learning and border security. *PalArch's Journal of Archaeology of Egypt/Egyptology, 18*(1), 3467-3475.

Kebande, V. R., & Ray, I. (2016). A Generic Digital Forensic Investigation Framework for Internet of Things (IoT). *2016 IEEE 4th International Conference on Future Internet of Things and Cloud (FiCloud)*, 356–362. 10.1109/FiCloud.2016.57

Kebande, V. R., Mudau, P. P., Ikuesan, R. A., Venter, H. S., & Choo, K.-K. R. (2020). Holistic digital forensic readiness framework for IoT-enabled organizations. *Forensic Science International: Reports, 2*, 100117. doi:10.1016/j.fsir.2020.100117

Kelly, T. B., Angel, M. N., O'Connor, D. E., Huff, C. C., Morris, L. E., & Wach, G. D. (2021). A novel approach to 3D modelling ground-penetrating radar (GPR) data–A case study of a cemetery and applications for criminal investigation. *Forensic Science International, 325*, 110882. doi:10.1016/j.forsciint.2021.110882 PMID:34182205

Kent, K., Chevalier, S., Grance, T., & Dang, H. (2006). Guide to integrating forensic techniques into incident response. *NIST Special Publication, 10*(14), 800–886. doi:10.6028/NIST.SP.800-86

Khalifa, A., Bashir, B., Çakir, Z., Kaya, Ş., Alsalman, A., & Henaish, A. (2021). Paradigm of Geological Mapping of the Adıyaman Fault Zone of Eastern Turkey Using Landsat 8 Remotely Sensed Data Coupled with PCA, ICA, and MNFA Techniques. *ISPRS International Journal of Geo-Information, 10*(6), 368. doi:10.3390/ijgi10060368

Kiperwasser, E., & Goldberg, Y. (2016). Simple and Accurate Dependency Parsing Using Bidirectional LSTM Feature Representations. *Transactions of the Association for Computational Linguistics, 4*, 313–327. doi:10.1162/tacl_a_00101

Kodovsky, J., Fridrich, J., & Holub, V. (2012). Ensemble classifiers for steganalysis of digital media. *IEEE Transactions on Information Forensics and Security, 7*(2), 432–444. doi:10.1109/TIFS.2011.2175919

Kootker, L. M., Plomp, E., Ammer, S. T., Hoogland, V., & Davies, G. R. (2020). Spatial patterns in 87Sr/86Sr ratios in modern human dental enamel and tap water from the Netherlands: Implications for forensic provenancing. *The Science of the Total Environment, 729*, 138992. doi:10.1016/j.scitotenv.2020.138992 PMID:32361454

Koroniotis, N., Moustafa, N., & Sitnikova, E. (2019). Forensics and Deep Learning Mechanisms for Botnets in Internet of Things: A Survey of Challenges and Solutions. *IEEE Access: Practical Innovations, Open Solutions, 7*, 61764–61785. doi:10.1109/ACCESS.2019.2916717

Koroniotis, N., Moustafa, N., & Sitnikova, E. (2020). A new network forensic framework based on deep learning for Internet of Things networks: A particle deep framework. *Future Generation Computer Systems, 110*, 91–106. doi:10.1016/j.future.2020.03.042

Koşaroğlu, S., Kamacı, Z., Erdoğan, S., Bektaş, Ö., & Büyüksaraç, A. (2021). Determination of historical graves by ground penetrating radar method: Sakarya Field Battle (August 23–September 13, 1921, Turkey). *Australian Journal of Forensic Sciences*, 1-19.

Kotsiantis, S., Koumanakos, E., Tzelepis, D., & Tampakas, V. (2006). *Predicting fraudulent financial statements with machine learning techniques.* Paper presented at the Hellenic Conference on Artificial Intelligence. 10.1007/11752912_63

Kristo, M. J. (2020). Nuclear forensics. In *Handbook of Radioactivity Analysis* (Vol. 2, pp. 921–951). Academic Press. doi:10.1016/B978-0-12-814395-7.00013-1

Kubai, A. (2021). The "Africa Rising" Paradox, Human Trafficking, and Perilous Migration Across the Sahara and the Mediterranean to Europe. In The Governance, Security and Development Nexus (pp. 355-370). Palgrave Macmillan.

Kupi, T. G., Uushona, V., Mathuthu, M., Coetzee, M., & van Tonder, D. (2020). Using lead isotope ratios to distinguish between samples of diffrent uranium mines. *Journal of Radioanalytical and Nuclear Chemistry, 324*(1), 1–5. doi:10.100710967-020-07048-1

Kurosawa, K., Kuroki, K., & Saitoh, N. (1999). CCD Fingerprint Method-Identification of a Video Camera from videotaped Images. *Proceedings of IEEE International Conference on Image Processing*. 10.1109/ICIP.1999.817172

Lanier, A. (2017). *Applications of forensic sedimentology*. Academic Press.

Laroca, R., Zanlorensi, L. A., Gonçalves, G. R., Todt, E., Schwartz, W. R., & Menotti, D. (2019). *An Efficient and Layout Independent Automatic License Plate Recognition System Based on the YOLO Detector*. https://arxiv.org/pdf/1909.01754.pdf

Lawrence, A. H. (1983). Chromatography Discussion Group Offers Reduced Subscriptions. *Trends in Analytical Chemistry*, *2*(12), V.

Lee, C., & de Vries, W. T. (2021). Testing and Validating the Suitability of Geospatially Informed Proxies on Land Tenure in North Korea for Korean (Re-) Unification. *Remote Sensing*, *13*(7), 1301. doi:10.3390/rs13071301

Legleiter, C. J., & Kinzel, P. J. (2020). Inferring surface flow velocities in sediment-Laden Alaskan rivers from optical image sequences acquired from a helicopter. *Remote Sensing*, *12*(8), 1282. doi:10.3390/rs12081282

Lentini, J. J. (2013). Evidence Collection at Fire Scenes. *Encyclopedia of Forensic Sciences: Second Edition, 3*(December), 387–91. doi:10.1016/B978-0-12-382165-2.00211-7

Leucci, G. (2020). *Advances in geophysical methods applied to forensic investigations: New developments in acquisition and data analysis methodologies*. Springer Nature. doi:10.1007/978-3-030-46242-0

Lewis, J. A. (2018). *Economic impact of cybercrime-no slowing down*. Available: http://csis-website-prod.s3.amazonaws.com/s3fs-public/publication/economic-impact-cybercrime.pdf

Li, H., Ota, K., & Dong, M. (2018). Learning IoT in Edge: Deep Learning for the Internet of Things with Edge Computing. *IEEE Network*, *32*(1), 96–101. doi:10.1109/MNET.2018.1700202

Li, H., Wang, P., & Shen, C. (2018). Toward End—to—End Car License Plate Detection and Recognition with Deep Neural Networks. *IEEE Transactions on Intelligent Transportation Systems*, *20*(3), 1126–1136. doi:10.1109/TITS.2018.2847291

Lilley, D. G. (1997). Structural Fire Development. *35th Aerospace Sciences Meeting and Exhibit*. https://arc.aiaa.org/doi/abs/10.2514/6.1997-265

Lilley, D. G. (1996). Fire Causes and Ignition. *Proceedings of the Intersociety Energy Conversion Engineering Conference, 1*, 85–90.

Lim, P. W., Ahmad, F. L. A., & Chang, K. H. (2018). Forensic Gas Chromatography Analysis of Time Elapsed Gasoline in Fire Scene Investigation. *The Malaysian Journal of Analytical Sciences*, *22*(1), 72–79.

Lin, T. Y., Goyal, P., Girshick, R., He, K., & Dollár, P. (2017). Focal Loss for Dense Object Detection. *Proceedings of IEEE International Conference on Computer Vision*, 2980-2988.

Listos, P., Gryzinska, M., Panasiuk-Flak, K., Ciesielka, M., & Teresinski, G. (2021). Assessment of temperature changes in carcasses in the early post-mortem period using the spectrum of a thermal imaging camera. *Medycyna Weterynaryjna*, *77*(5), 253–257. doi:10.21521/mw.6535

Liu, C., Zhu, T., Zhang, J., & Zhou, W. (2020). *Privacy Intelligence: A Survey on Image Sharing on Online Social Networks.* arXiv preprint arXiv:2008.12199.

Liu, W., Anguelov, D., Erhan, D., Szegedy, C., Reed, S., Fu, C. Y., & Berg, A. C. (2016). SSD: Single Shot MultiBox Detector. *Proceedings of the European Conference on Computer Vision*, 21-37. 10.1007/978-3-319-46448-0_2

Liu, Q. (2011). Detection of misaligned cropping and recompression with the same quantization matrix and relevant forgery. *Proc. 3rd International ACM workshop on Multimedia in Forensics and Intelligence*, 25-30. 10.1145/2072521.2072528

Liu, Q. (2017). An approach to detecting JPEG down-recompression and seam carving forgery under recompression anti-forensics. *Pattern Recognition*, *65*, 35–46. doi:10.1016/j.patcog.2016.12.010

Liu, Q. (2019). An Improved Approach to Exposing JPEG Seam Carving under Recompression. *IEEE Transactions on Circuits and Systems for Video Technology*, *29*(7), 1907–1918. doi:10.1109/TCSVT.2018.2859633

Liu, Q., & Chen, Z. (2015). Improved approaches with calibrated neighboring joint density to steganalysis and seam-carved forgery detection in JPEG images. *ACM Transactions on Intelligent Systems and Technology*, *5*(4), 63. doi:10.1145/2560365

Liu, Q., Sung, A. H., Zhou, B., & Qiao, M. (2016). Exposing inpainting forgery in JPEG images under recompression attacks. *Proc. 15th IEEE International Conference on Machine Learning and Applications*, 164-169. 10.1109/ICMLA.2016.0035

Lockwood, E., & Masters, P. (2021). Application of viewshed analysis and probability mapping for search area determination based on the Moors murders on Saddleworth Moor, the United Kingdom. *Journal of Forensic Sciences*, *66*(5), 1770–1787. doi:10.1111/1556-4029.14799 PMID:34296762

Lokanan, M., Tran, V., & Vuong, N. H. (2019). Detecting anomalies in financial statements using machine learning algorithm. *Asian Journal of Accounting Research.*

Ludwig, E. E. (Ed.). (1999). Process Safety and Pressure-Relieving Devices. In *Applied Process Design for Chemical and Petrochemical Plants.* Gulf Professional Publishing. https://www.sciencedirect.com/science/article/pii/S1874863599800089

Lukas, J., Fridrich, J., & Goljan, M. (2005a) Determining Digital Image Origin Using Sensor Imperfections. In *Proceedings SPIE Electronic Imaging* (pp. 249-260). 10.1117/12.587105

Lukas, J., Fridrich, J., & Goljan, M. (2005b) Digital Bullet Scratches for Images. *Proceedings of IEEE International Conference on Image Processing.*

Lukas, J., Fridrich, J., & Goljan, M. (2006). Digital Camera Identification from Sensor Noise. *IEEE Transactions on Information Forensics and Security*, *1*(2), 205–214. doi:10.1109/TIFS.2006.873602

Lu, Y., Chen, P., & Harrington, P. B. (2009). Comparison of Differential Mobility Spectrometry and Mass Spectrometry for Gas Chromatographic Detection of Ignitable Liquids from Fire Debris Using Projected Difference Resolution. *Analytical and Bioanalytical Chemistry*, *394*(8), 2061–2067.

Lu, Y., & Harrington, P. B. (2007). Forensic Application of Gas Chromatography-Differential Mobility Spectrometry with Two-Way Classification of Ignitable Liquids from Fire Debris. *Analytical Chemistry*, *79*(17), 6752–6759.

Madureira-Carvalho, Á., Ribeiro, H., Newman, G., & Brewer, J.,, M., Guedes, A., Abreu, I., ... Dawson, L. (2020). Geochemical analysis of sediment samples for forensic purposes: Characterisation of two river beaches from the Douro River, Portugal. *The Australian Journal of Forensic Sciences*, *52*(2), 222–234. doi:10.1080/00450618.2018.1533037

Madzunya, D., Uushona, V., Mathuthu, M., & Heike, W. (2021). Rare earth elements in uranium ore deposits from Namibia: A nuclear forensics tool. *Journal of Environmental Radioactivity*, *237*, 106668. doi:10.1016/j.jenvrad.2021.106668 PMID:34116457

Maes, S., Tuyls, K., Vanschoenwinkel, B., & Manderick, B. (2002). Credit card fraud detection using Bayesian and neural networks. *Proceedings of the 1st international naiso congress on neuro fuzzy technologies.*

Mahaney, W. C. (2020). The Hannibal enigma of 218 BC: A forensic exercise of importance to historical archaeology. *Sci. Cult*, *6*, 7–24.

Mahdavi, A., & Siegel, J. A. (2020). Quantitative filter forensics: Size distribution and particulate matter concentrations in residential buildings. *Indoor Air.*

Mahmoudi, L., Doumari, S. A., Safarianzengir, V., Maleki, R., Kianinejad, S., & Kianian, M. (2021). Monitoring and Prediction of Dust and Investigating its Environmental Impacts in the Western Half of Iran and the Eastern Borders of Turkey and Iraq, Using Remote Sensing and GIS. *Photonirvachak (Dehra Dun)*, *49*(4), 713–724. doi:10.100712524-020-01224-2

Majumdar, S. (2021). The Role of Remote Sensing and GIS in Military Strategy to Prevent Terror Attacks. *Intelligent Data Analytics for Terror Threat Prediction: Architectures, Methodologies, Techniques and Applications*, 79-94.

Makarenko, T. (2004). The crime-terror continuum: Tracing the interplay between transnational organised crime and terrorism. *Global Crime*, *6*(1), 129–145. doi:10.1080/1744057042000297025

Malejko, J., Deoniziak, K., Tomczuk, M., Długokencka, J., & Godlewska-Żyłkiewicz, B. (2020). Puparial cases as toxicological indicators: Bioaccumulation of cadmium and thallium in the forensically important blowfly Lucilia sericata. *Frontiers in Chemistry*, *8*, 8. doi:10.3389/fchem.2020.586067 PMID:33330373

Malmborg, J., Kooistra, K., Kraus, U. R., & Kienhuis, P. (2020). Evaluation of light petroleum biomarkers for the 3rd edition of the European Committee for Standardization methodology for oil spill identification (EN15522-2). Environmental Forensics, 1-15.

Malmgren, A., & Riley, G. (2012). 5 Comprehensive Renewable Energy. In *Biomass Power Generation*. Elsevier Ltd. doi:10.1016/B978-0-08-087872-0.00505-9

Mani, I., & Zhang, I. (2003). kNN approach to unbalanced data distributions: a case study involving information extraction. *Proceedings of workshop on learning from imbalanced datasets.*

Mankoff, K. D., van As, D., Lines, A., Bording, T., Elliott, J., Kraghede, R., Cantalloube, H., Oriot, H., Dubois-Fernandez, P., Ruault du Plessis, O., Christiansen, A. V., Auken, E., Hansen, K., Colgan, W., & Karlsson, N. B. (2020). Search and recovery of aircraft parts in ice-sheet crevasse fields using airborne and in situ geophysical sensors. *Journal of Glaciology*, *66*(257), 496–508. doi:10.1017/jog.2020.26

Manrong, C., Lizhong, Y., Xiangfeng, N., & Bin, C. (2009). Application of environmental magnetism on crime detection in a highway traffic accident from Yangzhou to Guazhou, Jiangsu Province, China. *Forensic Science International*, *187*(1-3), 29–33. doi:10.1016/j.forsciint.2009.02.012 PMID:19303726

Mansegosa, D. A., Giannotti, P. S., Marchiori, J. I., Jofré, F. N., Aballay, F. H., & Aisa, C. F. (2021). The story of a homicide: The location, exhumation, and multidisciplinary analysis of a clandestine burial. *Forensic Science International: Reports*, *3*, 100165.

Marshall, J. R., Bull, P. A., & Morgan, R. M. (2012). Energy regimes for aeolian sand grain surface textures. *Sedimentary Geology*, *253-254*, 17–24. doi:10.1016/j.sedgeo.2012.01.001

Matczak, G., & Mazurek, P. (2021). Comparative Monte Carlo Analysis of Background Estimation Algorithms for Unmanned Aerial Vehicle Detection. *Remote Sensing*, *13*(5), 870. doi:10.3390/rs13050870

Mclain, D., Canaday, J., Chamberlain, D., Deangeles, K., Naik, S., Quigley, N., ... El-Jaby, A. (2020). *Approaches for Developing and Sustaining the Sealed Radioactive Source and Material Component of a National Nuclear Forensics Library* (No. IAEA-TECDOC--1896 (SUPPLEMENTARY FILES)). Academic Press.

Meffert, C., Clark, D., Baggili, I., & Breitinger, F. (2017). Forensic state acquisition from internet of things (FSAIoT): A general framework and practical approach for IoT forensics through IoT device state acquisition. *ACM International Conference Proceeding Series, Part F1305*. 10.1145/3098954.3104053

Melo, V. F., Testoni, S. A., Dawson, L. A., & Salvador, F. A. D. S. (2020). Sand fraction is not suitable for forensic investigations in subtropical soils. *Revista Brasileira de Ciência do Solo*, 44.

Mia, R., Panchal, V., & Guglot, Y. (2021). *The Role of 'Digital Forensic Photography'.* The Indian Criminal Justice System.

Migeon, V., Fitoussi, C., Pili, E., & Bourdon, B. (2020). Molybdenum isotope fractionation in uranium oxides and during key processes of the nuclear fuel cycle: Towards a new nuclear forensic tool. *Geochimica et Cosmochimica Acta*, *279*, 238–257. doi:10.1016/j.gca.2020.03.046

Mihcak, M. K., Kozintsev, I., & Ramchandran, K. (1999) Spatially adaptive statistical modeling of wavelet image coefficients and its application to denoising. In Proceedings IEEE International Conference Acoustics, Speech, Signal Processing (vol. 6, pp. 3253-3256). IEEE.

Mirai (malware) - Wikipedia. (n.d.). Retrieved October 17, 2020, from https://en.wikipedia.org/wiki/Mirai_(malware)

Miraz, M., Ali, M., Excell, P., & Picking, R. (2018). Internet of Nano-Things, Things and Everything: Future Growth Trends. *Future Internet*, *10*(8), 68. doi:10.3390/fi10080068

Mirdad, M. A. (2020). Taliban insurgency and transnational organized crime nexus. *Masyarakat, Kebudayaan dan Politik, 33*(3), 266-277.

Mobariz, M. A., & Kaplan, G. (2021). Moving Borders: Mapping and Monitoring Amu Darya River Dynamics Using Remote Sensing Data and Techniques. *Geodetski list, 75*(1), 29-44.

Mohamad Noor, M., & Hassan, W. H. (2019). Current research on Internet of Things (IoT) security: A survey. *Computer Networks*, *148*, 283–294. doi:10.1016/j.comnet.2018.11.025

Molina, C. M., Pringle, J. K., Saumett, M., & Hernández, O. (2015). Preliminary results of sequential monitoring of simulated clandestine graves in Colombia, South America, using ground penetrating radar and botany. *Forensic Science International*, *248*, 61–70. doi:10.1016/j.forsciint.2014.12.011 PMID:25596556

Molina, C. M., Wisniewski, K. D., Drake, J., Baena, A., Guatame, A., & Pringle, J. K. (2020). Testing application of geographical information systems, forensic geomorphology and electrical resistivity tomography to investigate clandestine grave sites in Colombia, South America. *Journal of Forensic Sciences*, *65*(1), 266–273. doi:10.1111/1556-4029.14168 PMID:31424577

Morgan, R. M., & Bull, P. A. (2007). The use of grain size distribution analysis of sediments and soils in forensic enquiry. *Science & Justice*, *47*, 125–135. doi:10.1016/j.scijus.2007.02.001

Morgan, R. M., Wiltshire, P., Parker, A., & Bull, P. A. (2006). The role of forensic geoscience in wildlife crime detection. *Forensic Science International*, *162*, 152–162. doi:10.1016/j.forsciint.2006.06.045

Morrison, B. A., Coventry, L., & Briggs, P. (2020). Technological Change in the Retirement Transition and the Implications for Cybersecurity Vulnerability in Older Adults. *Frontiers in Psychology*, *11*, 623. doi:10.3389/fpsyg.2020.00623 PMID:32425841

Morrison, R. D. (2020). Forensic techniques for establishing the origin and timing of a contaminant release. In *Expert Witnessing* (pp. 145–172). Routledge. doi:10.1201/9781003067764-10

Nde, S. C., Manjoro, M., & Mathuthu, M. (2019). Farm dam siltation and sediment source tracing in the Zeerust Swartruggens area in the north-west of South Africa. *International Journal of Hydrology Science and Technology*, *9*(3), 324–337. doi:10.1504/IJHST.2019.102322

NFPA 921- Guide for Fire and Explosion Investigations, 2017 Edition, Chapter 17.

Nieto, A., Rios, R., & Lopez, J. (2017). A Methodology for Privacy-Aware IoT-Forensics. *2017 IEEE Trustcom/BigDataSE/ICESS*, 626–633. doi:10.1109/Trustcom/BigDataSE/ICESS.2017.293

NIJ (National Institute of Justice). (2009). A Guide for Investigating Fire and Arson | National Institute of Justice. *Office of Justice Programs*. https://www.nij.gov/topics/law-enforcement/investigations/crime-scene/guides/fire-arson/pages/welcome.aspx

Nitrean, A. (2020). *Development of interagency cooperation for nuclear forensics examinations in Moldova* (No. IAEA-TECDOC--1896 (SUPPLEMENTARY FILES)). Academic Press.

Noura, H. N., Salman, O., Chehab, A., & Couturier, R. (2020). DistLog: A distributed logging scheme for IoT forensics. *Ad Hoc Networks*, *98*, 102061. doi:10.1016/j.adhoc.2019.102061

Nuzzo, M., & Gehlen, M. (2021). *Forensic use of diagnostic biomarker ratios in environmental oil spill geochemistry*. Academic Press.

Okubo, S., & Shelley, L. (Eds.). (2011). Human security, transnational crime and human trafficking: Asian and Western perspectives (Vol. 6). Routledge.

Omelicheva, M. Y., & Markowitz, L. P. (2021). Rethinking Intersections of Crime and Terrorism: Insights from Political Economies of Violence. *International Studies Review*, viab025. doi:10.1093/isr/viab025

Osbourne, M. G., & Christensen, L. S. (2020). Pathways to Child Sexual Offending: Applying Ward and Siegert's Pathways Model on Offenders Convicted of Penetrative Offenses on Children. *Sexuality & Culture*, *24*(6), 1756–1773. doi:10.100712119-020-09718-4

Palomino, W., Barreto, R., & Cerón, A. (2020). Drug Trafficking and Terrorism as Transnational Crimes in Latin America. In Maritime Security: Counter-Terrorism Lessons from Maritime Piracy and Narcotics Interdiction (pp. 29-46). IOS Press.

Panigrahi, N., & Tripathy, S. (2021). Design Criteria of a UAV for ISTAR and Remote Sensing Applications. *Photonirvachak (Dehra Dun)*, *49*(3), 665–669. doi:10.100712524-020-01249-7

Parise, G., Mohla, D., Parise, L., & Lombardi, M. (2020, October). Forensic Implications in the Continuous Discontinuity of IoT Innovations. In *2020 IEEE Industry Applications Society Annual Meeting* (pp. 1-6). IEEE.

Pastoor, K. J., Kemp, R. S., Jensen, M. P., & Shafer, J. C. (2021). Progress in Uranium Chemistry: Driving Advances in Front-End Nuclear Fuel Cycle Forensics. *Inorganic Chemistry*, *60*(12), 8347–8367. doi:10.1021/acs.inorgchem.0c03390 PMID:33619961

Pawliszyn & Pawliszyn. (1997). Solid-Phase Microextraction. The Chemical Educator, 1–7.

Peng, F. K. (2015). *The Spatial and Temporal Analysis of the Impacts of Video Surveillance System on the Theft Crime Rates*. Master Chapter, Department of Crime Prevention and Corrections, Central Police University. https://hdl.handle.net/11296/4swq89

Pensieri, M. G., Garau, M., & Barone, P. M. (2020). Drones as an integral part of remote sensing technologies to help missing people. *Drones (Basel)*, *4*(2), 15. doi:10.3390/drones4020015

Perry, B. (2020). "Button-Down Terror": The Metamorphosis of the Hate Movement. In Experiencing Social Research (pp. 281-302). Routledge.

Perumal, S., Norwawi, N. M., & Raman, V. (2015). Internet of Things(IoT) digital forensic investigation model: Top-down forensic approach methodology. *2015 Fifth International Conference on Digital Information Processing and Communications (ICDIPC)*, 19–23. 10.1109/ICDIPC.2015.7323000

Petrich, K. (2021). The Crime–Terror Nexus. In Oxford Research Encyclopedia of International Studies. OUP.

Pierini, M. (2020). New power struggles in the Mediterranean. *Carnegie Europe*, (July), 30.

Pirrie, D., Power, M. R., Rollinson, G. K., Wiltshire, P. E., Newberry, J., & Campbell, H. E. (2009). Automated SEM-EDS (QEMSCAN®) mineral analysis in forensic soil investigations: testing instrumental reproducibility. In *Criminal and environmental soil forensics* (pp. 411–430). Springer.

Pitts, K. M., & Clarke, R. M. (2020). The forensic discrimination of quartz sands from the Swan Coastal Plain, Western Australia. *Forensic Science International: Reports*, *2*, 100130.

Podhade, D., Lal, S. B., Singh, S., Mehera, B., Khare, N., & James, A. (2020). Evaluating the Impact of Wetland Health on Wildlife Health by Soil and Water Quality Analysis. *International Journal of Current Microbiology and Applied Sciences*, *9*(9), 839–849.

Poole, C. F. (2005). Gas Chromatography - Detectors. Encyclopedia of Analytical Science: Second Edition, 95–105.

Poulson, S. R., & Drever, J. I. (2020). Organic contaminant stable isotope geochemistry, and application to contaminant characterization and remediation. In *Groundwater 2000* (pp. 335–336). CRC Press. doi:10.1201/9781003078593-165

Pringle, J. K., Giubertoni, M., Cassidy, N. J., Wisniewski, K. D., Hansen, J. D., Linford, N. T., & Daniels, R. M. (2015). The use of magnetic susceptibility as a forensic search tool. *Forensic Science International*, *246*, 31–42. doi:10.1016/j.forsciint.2014.10.046 PMID:25460105

Pringle, J. K., Stimpson, I. G., Wisniewski, K. D., Heaton, V., Davenward, B., Mirosch, N., Spencer, F., & Jervis, J. R. (2021). Geophysical monitoring of simulated clandestine burials of murder victims to aid forensic investigators. *Geology Today*, *37*(2), 63–65. doi:10.1111/gto.12344

Profumo, A., Gorroni, A., Guarnieri, S. A., Mellerio, G. G., Cucca, L., & Merli, D. (2020). GC-MS qualitative analysis of the volatile, semivolatile and volatilizable fractions of soil evidence for forensic application: A chemical fingerprinting. *Talanta*, *219*, 121304.

Psomiadis, E., Tomanis, L., Kavvadias, A., Soulis, K. X., Charizopoulos, N., & Michas, S. (2021). Potential dam breach analysis and flood wave risk assessment using HEC-RAS and remote sensing data: A multicriteria approach. *Water (Basel), 13*(3), 364. doi:10.3390/w13030364

Pun, C., Yuan, X., & Bi, X. (2015). Image forgery detection using adaptive oversegmentation and feature point matching. *IEEE Transactions on Information Forensics and Security, 10*(8), 1705–1716. doi:10.1109/TIFS.2015.2423261

Pundir, S. K., & Garg, R. D. (2021). Development of rule based approach for assessment of off-road trafficability using remote sensing and ancillary data. *Quaternary International, 575*, 308–316. doi:10.1016/j.quaint.2020.07.017

Puttonen, R., & Romiti, F. (2020). *The Linkages between Organized Crime and Terrorism.* Academic Press.

Pye, K., & Blott, S.J. (2004). Particle size analysis of sediments, soils and related particulate materials for forensic purposes using laser granulometry. *Forensic Science International, 144*, 19-27. doi:10.1016/j.forsciint.2004.02.028

Qasim, M., Xiao, H., He, K., Noman, A., Liu, F., Chen, M. Y., Hussain, D., Jamal, Z. A., & Li, F. (2020). Impact of landfill garbage on insect ecology and human health. *Acta Tropica, 211*, 105630. doi:10.1016/j.actatropica.2020.105630 PMID:32673623

Radil, S. M., Irmischer, I., & Walther, O. J. (2021). Contextualizing the Relationship Between Borderlands and Political Violence: A Dynamic Space-Time Analysis in North and West Africa. *Journal of Borderlands Studies*, 1–19. doi:10.1080/08865655.2021.1968926

Randhawa, K., Loo, C. K., Seera, M., Lim, C. P., & Nandi, A. K. (2018). Credit card fraud detection using AdaBoost and majority voting. *IEEE Access: Practical Innovations, Open Solutions, 6*, 14277–14284. doi:10.1109/ACCESS.2018.2806420

Rashid, M. M., Musa, A., Rahman, M. A., Farahana, N., & Farhana, A. (2012). Automatic Parking Management System and Parking Fee Collection Based on Number Plate Recognition. *International Journal of Machine Learning and Computing, 2*(2), 93–98. doi:10.7763/IJMLC.2012.V2.95

Rawlinson, K. (2014). *HP Study Reveals 70 Percent of Internet of Things Devices Vulnerable to Attack.* HP Inc. https://www8.hp.com/us/en/hp-news/press-release.html?id=1744676#.X4prekIzbBJ

Redmon, J., & Farhadi, A. (2017). YOLO9000: Better, Faster, Stronger. *Proceedings of IEEE Conference on Computer Vision and Pattern Recognition*, 6517-6525.

Redowan, M., Phinn, S. R., Roelfsema, C. M., & Aziz, A. A. (2020). CLASlite unmixing of Landsat images to estimate REDD+ activity data for deforestation in a Bangladesh forest. *Journal of Applied Remote Sensing, 14*(2), 024505. doi:10.1117/1.JRS.14.024505

Redsicker, D. R., & O'Connor, J. J. (1996). *Practical Fire and Arson Investigation (2nd ed.).* CRC Press.

Reinhard, A., Inglis, J., Steiner, R., LaMont, S., & Palacz, Z. (2021). Isotopic analysis of sub-nanogram neodymium loads using new ATONA™ amplifiers. *Rapid Communications in Mass Spectrometry*, *35*(7), e9032. doi:10.1002/rcm.9032 PMID:33340159

Reno, J., Marcus, D., Leary, M. L., & Samuels, J. E. (2000). *Fire and Arson Scene Evidence: A Guide for Public Safety Personnel.* https://www.crime-scene-investigator.net/fireandarsonsceneevidenceNIJ.pdf

Rocke, B., Ruffell, A., & Donnelly, L. (2021). Drone aerial imagery for the simulation of a neonate burial based on the geoforensic search strategy (GSS). *Journal of Forensic Sciences*, *66*(4), 1506–1519. doi:10.1111/1556-4029.14690 PMID:33576508

Roffo, G., Melzi, S., & Cristani, M. (2015). Infinite feature selection. *Proceedings of the 2015 IEEE International Conference on Computer Vision*, 4202-4210.

Rondeau, C. M., Temple, M. A., & Lopez, J. (2019). Industrial IoT cross-layer forensic investigation. *Wiley Interdisciplinary Reviews: Forensic Science*, *1*(1), e1322. doi:10.1002/wfs2.1322

Rosen, J. D. (2021). The Drug War in Colombia. In The US War on Drugs at Home and Abroad (pp. 17-43). Palgrave Macmillan.

Ross, R., McEvilley, M., & Oren, J. (2016). *Systems security engineering: Considerations for a multidisciplinary approach in the engineering of trustworthy secure systems*. Academic Press.

Różycki, S., Zapłata, R., Karczewski, J., Ossowski, A., & Tomczyk, J. (2020). Integrated Archaeological Research: Archival Resources, Surveys, Geophysical Prospection and Excavation Approach at an Execution and Burial Site: The German Nazi Labour Camp in Treblinka. *Geosciences*, *10*(9), 336. doi:10.3390/geosciences10090336

Ruffell, A., & McKinley, J. (2005). Forensic geoscience: Applications of geology, geomorphology and geophysics to criminal investigations. *Earth-Science Reviews*, *69*, 235–247. doi:10.1016/j.earscirev.2004.08.002

Ryan, P., & Watson, R. (2017). Research Challenges for the Internet of Things: What Role Can OR Play? *Systems*, *5*(1), 24. doi:10.3390ystems5010024

Ryu, S., Lee, H., & Lee, H.-K. (2014). Detecting trace of seam carving for forensics analysis. *IEICE Transactions on Information and Systems*, *97*(5), 1304–1311. doi:10.1587/transinf.E97.D.1304

Sacchi, E., & Nicosia, U. (2013). Forensic paleontology: A tool for "intelligence" and investigation. *Journal of Forensic Sciences*, *58*, 651–657. doi:10.1111/1556-4029.12084

Sadineni, L., Pilli, E., & Battula, R. B. (2019). *A Holistic Forensic Model for the Internet of Things*., doi:10.1007/978-3-030-28752-8_1

Sandler, M., Howard, A., Zhu, M., Zhmoginov, A., & Chen, L. C. (2018). MobileNetV2: Inverted Residuals and Linear Bottlenecks. *Proceedings of IEEE Conference on Computer Vision and Pattern Recognition*, 4510-4520.

Sangwan, P., Nain, T., Singal, K., Hooda, N., & Sharma, N. (2020). Soil as a tool of revelation in forensic science: A review. *Analytical Methods, 12*(43), 5150–5159. doi:10.1039/D0AY01634A PMID:33135702

Sant'Ovaia, H., Marques, G., Santos, A., Gomes, C., & Rocha, A. (2015). Magnetic susceptibility and isothermal remanent magnetization in human tissues: A study case. *Biometals, 28*(6), 951–958. doi:10.100710534-015-9879-z PMID:26373856

Sarkar, A., Nataraj, L., & Manjunath, B. (2009). Detection of seam carving and localization of seam insertions in digital images. *Proceedings of the 11th ACM workshop on Multimedia and security*, 107–116. 10.1145/1597817.1597837

Sathwara, S., Dutta, N., & Pricop, E. (2018). IoT Forensic A digital investigation framework for IoT systems. *2018 10th International Conference on Electronics, Computers and Artificial Intelligence (ECAI)*, 1–4. 10.1109/ECAI.2018.8679017

Saul, B. (2017). The legal relationship between terrorism and transnational crime. *International Criminal Law Review, 17*(3), 417–452. doi:10.1163/15718123-01703001

Schmidt-Rohr, K. (2015). Why Combustions Are Always Exothermic, Yielding about 418 KJ per Mole of O2. *Journal of Chemical Education, 92*(12), 2094–2099.

Schneider, B., Knippel, E., & Miller, R. (2020). Utilizing ground-penetrating radar to assist in a forensic cold case investigation. *SEG Technical Program Expanded Abstracts, 2020*, 3324–3330. doi:10.1190egam2020-3420999.1

Scientific Working Group on Digital Evidence. (2018). *SWGDE Best Practices for Image Authentication.* https://drive.google.com/file/d/1Z0DsJMa6aDZlFJ9kRfOL8cow5VjhVT0t/view

Sharifi, K., & Leon-Garcia, A. (1995). Estimation of shape parameter for generalized Gaussian distributions in subband decompositions of video. *IEEE Transactions on Circuits and Systems for Video Technology, 5*(1), 52–56. doi:10.1109/76.350779

Sharma, R. C., Kumar, S., Kumar, S., Mann, M., & Sharma, M. (2020). Photoacoustic remote sensing of suspicious objects for defence and forensic applications. *Spectrochimica Acta. Part A: Molecular and Biomolecular Spectroscopy, 224*, 117445. doi:10.1016/j.saa.2019.117445 PMID:31382229

Shelley, L. I., & Picarelli, J. T. (2005). Methods and motives: Exploring links between transnational organized crime and international terrorism. *Trends in Organized Crime, 9*(2), 52–67. doi:10.100712117-005-1024-x

Sheng, G., & Gao, T. (2016). Detection of seam-carving image based on benford's law for forensic applications. *International Journal of Digital Crime and Forensics, 8*(1), 51–61. doi:10.4018/IJDCF.2016010104

Sherman. (1933). *Radiation From Luminous and Non-Luminous Natural Gas Flames.* The American Society of Mechanical Engineers.

Shina, H. G., Leeb, J., Yunc, Y. H., Yood, S. H., Jange, J., Ohf, S. H., ... Leea, J. (2021). *χ-separation: Magnetic susceptibility source separation toward iron and myelin mapping in the brain.* Academic Press.

Shvedchikova, T., Barone, P. M., & Moghaddam, N. (2021). Crimes in the Past: Archaeological and Anthropological Evidence. *Crimes in the Past.*

Siegel, J., Knupfer, G., & Saukko, P. (2000). Methods. *Encyclopedia of Forensic Science.*

Silva, S. M., & Jung, C. R. (2018). License Plate Detection and Recognition in Unconstrained Scenarios. *Proceedings of the European Conference on Computer Vision*, 593-609. 10.1007/978-3-030-01258-8_36

Silva, J., Rodrigues, J., Al-Muhtadi, J., Rabêlo, R., & Furtado, V. (2019). Management Platforms and Protocols for Internet of Things: A Survey. *Sensors (Basel)*, *19*(3), 676. doi:10.339019030676 PMID:30736424

Silva, S. M., & Jung, C. R. (2017). Real—Time Brazilian License Plate Detection and Recognition Using Deep Convolutional Neural Networks. *Proceedings of Conference on Graphics, Patterns, and Images (SIBGRAPI)*, 55—62.

Simeoni, A., Reszka P., Colella F., & Torero J.L. (2011). *Flaming Ignition of Solid Fuels.* Academic Press.

Simonyan, K., & Zisserman, A. (2014). *Very Deep Convolutional Networks for Large—Scale Image Recognition.* https://arxiv.org/pdf/1409.1556.pdf

Singh, A. (2021). *An Interlink between White Collar Crimes and Terrorism.* Neolexvision Blogs. https://www. aequivic. in/post/an-interlink-between-white-collar-crimes-and-terrorism

Smith, C. D., & Dietz, M. L. (2021). Support loading effects on the performance of an extraction chromatographic resin: Toward improved separation of trivalent lanthanides. *Talanta*, *222*, 121541. doi:10.1016/j.talanta.2020.121541 PMID:33167249

Song, X., Liu, F., Zhang, Z., Yang, C., Luo, X., & Chen, L. (2017). 2D Gabor filters-based steganalysis of content-adaptive JPEG steganography. *Multimedia Tools and Applications*, *76*(24), 26391–26419. doi:10.100711042-016-4157-9

Spy Blog - Watching Them Watching Us. (2009). *Operation Algebra child rape convictions in Scotland: open WiFi tracking, digital camera image forensics.* https://spyblog.org.uk/ssl/spyblog/2009/05/09/operation-algebra-child-rape-convictions-in-scotland-open-wifi-tracking-digi.html

Stamm, M. C., Wu, M., & Liu, K. J. R. (2013). Information Forensics: An Overview of the First Decade. *IEEE Access: Practical Innovations, Open Solutions*, *1*, 167–200. doi:10.1109/ACCESS.2013.2260814

Stauffer, E., Dolan, J. A., & Newman, R. (2008). Chemistry and Physics of Fire and Liquid Fuels. In *Fire Debris Analysis* (pp. 85–129). Elsevier.

Sutliff-Johansson, S. (2021). *Tracing anthropogenic Technology Critical Elements using Environmental Forensics: Case studies at Kallholmsfjärden and the Bothnian Bay, Northern Sweden* (Doctoral dissertation). Luleå University of Technology.

Swain, A. (2020). Big Data Challenges and Hype Digital Forensic: A Review in Health Care Management. *Big Data Analytics and Computing for Digital Forensic Investigations*, 43-63.

Sydow, L., Johns, D. A., Zappitello, S. J., & Maurer, T. (2020). Environmental forensic investigation of mystery sediment plumes at Barton Springs. Academic Press.

Taspinar, S., Mohanty, M., & Memon, N. (2017). PRNU-based camera attribution from multiple seam-carved images. *IEEE Transactions on Information Forensics and Security*, *12*(12), 3065–3080. doi:10.1109/TIFS.2017.2737961

Thennakoon, A., Bhagyani, C., Premadasa, S., Mihiranga, S., & Kuruwitaarachchi, N. (2019). *Real-time credit card fraud detection using machine learning*. Paper presented at the 2019 9th International Conference on Cloud Computing, Data Science & Engineering (Confluence). doi:10.6028/NIST.SP.800-160

Thienpont, J. R., Yang, Z., Hall, R. I., Wolfe, B. B., Hollebone, B. P., & Blais, J. M. (2021). Tracking petrogenic hydrocarbons in lakes of the Peace-Athabasca Delta in Alberta, Canada using petroleum biomarkers. *Environmental Pollution*, *286*, 117286. doi:10.1016/j.envpol.2021.117286 PMID:33990046

Ubelaker, D. H., & Francescutti, C. (2020). The role of stable isotope analysis in forensic anthropology. *Forensic Science and Humanitarian Action: Interacting with the Dead and the Living*, 273-284.

Valjarevic, A., & Venter, H. S. (2015). A Comprehensive and Harmonized Digital Forensic Investigation Process Model. *Journal of Forensic Sciences*, *60*(6), 1467–1483. doi:10.1111/1556-4029.12823 PMID:26258644

Vanderheyden, N., Verhoeven, E., Vermeulen, S., & Bekaert, B. (2020). Survival of forensic trace evidence on improvised explosive devices: Perspectives on individualisation. *Scientific Reports*, *10*(1), 1–11. doi:10.103841598-020-69385-1 PMID:32732923

Varga, Z., Wallenius, M., Nicholl, A., & Mayer, K. (2020). Assessment of uranium inhomogeneity and isotope imaging for nuclear forensics. *Spectrochimica Acta. Part B, Atomic Spectroscopy*, *171*, 105920. doi:10.1016/j.sab.2020.105920

Verma, R., Govindaraj, J., & Gupta, G. (2018). DF 2.0: Designing an automated, privacy preserving, and efficient digital forensic framework. *Annual ADFSL Conference on Digital Forensics, Security and Law*. https://commons.erau.edu/adfsl/2018/presentations/12

Wallace, J. R., Scala, J. R., Gultepe, I., & LeBlanc, H. N. (2021). Bridging Disciplines: The Application of Forensic Science to Meteorological Evidence. *Pure and Applied Geophysics*, *178*(3), 693–704. doi:10.100700024-020-02646-2

Wang, W., & Dong, J. (2010). Image tampering detection based on stationary distribution of Markov chain. *Proceedings of the 17th International Conference on Image Processing*, 2101-2104. 10.1109/ICIP.2010.5652660

Wattanachote, K., Shih, T. K., Chang, W.-L., & Chang, H.-H. (2015). Tamper detection of jpeg image due to seam modifications. *IEEE Transactions on Information Forensics and Security*, *10*(12), 2477–2491. doi:10.1109/TIFS.2015.2464776

Wei, J., Lin, Y., & Wu, Y. (2014). A patch analysis method to detect seam carved images. *Pattern Recognition Letters*, *36*, 100–106. doi:10.1016/j.patrec.2013.09.026

Weisburd, D., Savona, E. U., Hasisi, B., & Calderoni, F. (Eds.). (2020). *Understanding recruitment to organized crime and terrorism*. Springer. doi:10.1007/978-3-030-36639-1

Weiss, C., & Roy, B. (2020). President's Page: Near-surface geophysics for the future. *The Leading Edge*, *39*(5), 304–306. doi:10.1190/tle39050304.1

Welch, S. A. (2016). Human trafficking and terrorism: Utilizing national security resources to prevent human trafficking in the Islamic state. *Duke J. Gender L. & Pol'y*, *24*, 165.

Whyte, C. J., Vengosh, A., Warner, N. R., Jackson, R. B., Muehlenbachs, K., Schwartz, F. W., & Darrah, T. H. (2021). Geochemical evidence for fugitive gas contamination and associated water quality changes in drinking-water wells from Parker County, Texas. *The Science of the Total Environment*, *780*, 146555. doi:10.1016/j.scitotenv.2021.146555 PMID:34030322

Williams, M. R., & Sigman, M. (2007, May). Performance testing of commercial containers for collection and storage of fire debris evidence. *Journal of Forensic Sciences*, *52*(3), 579–585.

Xu, J., Xiao, W., He, T., Deng, X., & Chen, W. (2021). Extraction of built-up area using multi-sensor data—A case study based on Google earth engine in Zhejiang Province, China. *International Journal of Remote Sensing*, *42*(2), 389–404. doi:10.1080/01431161.2020.1809027

Xu, X., Du, C., Ma, F., Shen, Y., & Zhou, J. (2020). Forensic soil analysis using laser-induced breakdown spectroscopy (LIBS) and Fourier transform infrared total attenuated reflectance spectroscopy (FTIR-ATR): Principles and case studies. *Forensic Science International*, *310*, 110222. doi:10.1016/j.forsciint.2020.110222 PMID:32193128

Yakubu, A., & Idehen, R. (2021). *Border Patrol And Surveillance In The Control Of Transnational Terrorism In Nigeria And Niger*. Academic Press.

Yang, J., Wang, C., Jiang, B., Song, H., & Meng, Q. (2020). Visual perception enabled industry intelligence: State of the art, challenges and prospects. *IEEE Transactions on Industrial Informatics*, *17*(3), 2204–2219. doi:10.1109/TII.2020.2998818

Yang, W. C., Jiang, J., & Chen, C. H. (2021). A Fast Source Camera Identification and Verification Method Based on PRNU Analysis for Use in Video Forensic Investigations. *Multimedia Tools and Applications*, *80*(5), 6617–6638. doi:10.100711042-020-09763-z

Yaqoob, I., Hashem, I. A. T., Ahmed, A., Kazmi, S. M. A., & Hong, C. S. (2019). Internet of things forensics: Recent advances, taxonomy, requirements, and open challenges. *Future Generation Computer Systems*, *92*, 265–275. doi:10.1016/j.future.2018.09.058

Yee, O. S., Sagadevan, S., & Malim, N. H. A. H. (2018). Credit card fraud detection using machine learning as data mining technique. *Journal of Telecommunication Electronic and Computer Engineering*, *10*(1-4), 23–27.

Yendaw, E. (2021). Cross-Border Migration of Itinerant Immigrant Retailers in Ghana. *Journal of International Migration and Integration*, 1–21. PMID:34025218

Yin, T., Yang, G., Li, L., Zhang, D., & Sun, X. (2015). Detecting seam carving based image resizing using local binary patterns. *Computers & Security*, *55*, 130–141. doi:10.1016/j.cose.2015.09.003

Yu, C. D., Wang, K. X., Liu, X. D., Cuney, M., Pan, J. Y., Wang, G., Zhang, L., & Zhang, J. (2020). Uranium mineralogical and chemical features of the Na-metasomatic type uranium deposit in the Longshoushan metallogenic belt, northwestern China. *Minerals (Basel)*, *10*(4), 335. doi:10.3390/min10040335

Yu, Y., Barthaud, D., Price, B. A., Bandara, A. K., Zisman, A., & Nuseibeh, B. (2019). LiveBox: A Self-Adaptive Forensic-Ready Service for Drones. *IEEE Access: Practical Innovations, Open Solutions*, *7*, 148401–148412. doi:10.1109/ACCESS.2019.2942033

Zacharias, D. C., Gama, C. M., & Fornaro, A. (2021). Mysterious oil spill on Brazilian coast: Analysis and estimates. *Marine Pollution Bulletin*, *165*, 112125. doi:10.1016/j.marpolbul.2021.112125 PMID:33582423

Zakiev, E. S., Pankov, S. V., & Kalabay, K. B. (2020). Application of geoinformation systems in the armed forces and other military formations in the Republic of Kazakhstan. *Vojnotehnički glasnik*, *68*(2), 356-382.

Zawadzki, J., Fabijańczyk, P., & Przeździecki, K. (2020). *Geostatistical Methods as a Tool Supporting Revitalization of Industrially Degraded and Post-Mining Areas*. Academic Press.

Zetter, K. (2016). *That Insane, $81 M Bangladesh Bank Heist. Here's What We Know. Wired.*

Zhang, D., Yin, T., Yang, G., Xia, M., Li, L., & Sun, X. (2017). Detecting image seam carving with low scaling ratio using multi-scale spatial and spectral entropies. *Journal of Visual Communication and Image Representation*, *48*(C), 281–291. doi:10.1016/j.jvcir.2017.07.006

Zhang, K., Cao, J., Liu, H., Maharjan, S., & Zhang, Y. (2020). Deep Reinforcement Learning for Social-Aware Edge Computing and Caching in Urban Informatics. *IEEE Transactions on Industrial Informatics*, *16*(8), 5467–5477. doi:10.1109/TII.2019.2953189

Zhang, Q., Yang, L. T., Chen, Z., & Li, P. (2018). A survey on deep learning for big data. *Information Fusion*, *42*, 146–157. doi:10.1016/j.inffus.2017.10.006

Zhang, W., Guo, W., Liu, X., Liu, Y., Zhou, J., Li, B., Lu, Q., & Yang, S. (2018). LSTM-Based Analysis of Industrial IoT Equipment. *IEEE Access: Practical Innovations, Open Solutions*, *6*, 23551–23560. doi:10.1109/ACCESS.2018.2825538

Zherzdev, S., & Gruzdev, A. (2018). *LPRNet: License Plate Recognition via Deep Neural Networks*. https://arxiv.org/pdf/1806.10447.pdf

Zhou, C., Yu, S., Huang, W., Zhang, H., Xiao, Z., Zeng, L., Huang, W., & Pan, C. (2021). Oil maturities, mixing and charging episodes in the cratonic regions of the Tarim Basin, NW China: Insight from biomarker and diamondoid concentrations and oil bulk properties. *Marine and Petroleum Geology*, *126*, 104903. doi:10.1016/j.marpetgeo.2021.104903

Zia, T., Liu, P., & Han, W. (2017). Application-Specific Digital Forensics Investigative Model in Internet of Things (IoT). *Proceedings of the 12th International Conference on Availability, Reliability and Security*, 1–7. 10.1145/3098954.3104052

About the Contributors

Chung-Hao Chen received his Ph.D. in Electrical Engineering at the University of Tennessee, Knoxville in August 2009, and his B.S. and M.S. in Computer Science and Information Engineering from Fu-Jen Catholic University, Taiwan in 1997 and 1999, respectively. In April 2001, he joined the Panasonic Taiwan Laboratory Company, Ltd. as a research and development engineer where he remained until August 2003. In August 2009, he joined in the department of Math and Computer Science at North Carolina Central University as an assistant professor. Since August 2011, he joins in the department of Electrical and Computer Engineering at Old Dominion University and he is currently an associate professor. His research interests include robotics, automated surveillance systems, pattern recognition, image processing, artificial intelligent systems, data analysis and mining, and network performance analysis. He has published more than 70 peer reviewed articles.

Wen-Chao Yang received the B.S. degree from Central Police University in 1997, M.B.A. degree from National Central University in 2002 and Ph.D. degree from National Chiao Tung University in 2014. He worked for the Forensic Laboratory at Central Police University from 1997 through 2005 and became an information system manager in Computer Center at Central Police University from 2005 through 2014. In 2015, he joined the Department of Forensic Science at Central Police University to be an Assistant Professor. His major research interests include image processing, information hiding, and forensic sciences. Dr. Yang is a founder of the Multimedia Processing Laboratory at Central Police University and a frequent forensic appraiser to government in the area of image, video, and digital evidence. He has published numerous articles related to steganography, multimedia forensic, and computer aided forensic examination. Dr. Yang is a senior member of Taiwan Academy of Forensic Sciences and the Chinese Cryptology and Information Security Association. He is now a Secretary General of Taiwan Academy of Forensic Sciences.

Lijia Chen is an Associate Professor of Communication Engineering in the School of Physics and Electronics, Henan University, Henan Province, China. At present, Dr. Chen is the leader of the first-class specialty and excellent engineer pilot specialty of communication engineering in Henan Province, the leader of the excellent experimental class of communication engineering, the member of the innovation team of Henan Province, and the deputy director of the experimental teaching center of electronic and communication engineering in Henan Province. Director of Kaifeng Intelligent Manufacturing Engineering Technology Research Center, member of Henan Communication Society and member of Henan intelligent manufacturing Committee. In addition, he is serving as Director of Graduate Studies in Detection Technology and Automation, head of Communication Engineering, and member of the research group in the Laboratory of advanced computation methods and intelligent applications. His research interests are in Evolutionary Computation, Communication Network, Digital Signal Processing and Evolutionary Design Methods on Electronic Circuits and Systems. He has published over 20 journal papers, over 10 conference papers and 7 granted patents.

* * *

Chiou-Shann Fuh received the BS degree in computer science and information engineering from National Taiwan University, Taipei, Taiwan, in 1983, the MS degree in computer science from the Pennsylvania State University, University Park, PA, in 1987, and the PhD degree in computer science from Harvard University, Cambridge, MA, in 1992. He was with AT&T Bell Laboratories and engaged in performance monitoring of switching networks from 1992 to 1993. He was an associate professor in Department of Computer Science and Information Engineering, National Taiwan University, Taipei, Taiwan from 1993 to 2000 and then promoted to a full professor. His current research interests include digital image processing, computer vision, pattern recognition, mathematical morphology, artificial intelligence, deep learning, chatbot, big data analysis, cloud computing, and their applications to defect inspection, automatic optical inspection, industrial automation, digital still camera, digital video camcorder, surveillance camera, and camera module such as color interpolation, auto exposure, auto focus, auto white balance, color calibration, and color management.

Md Ariful Haque is a PhD candidate in the Department of Computational Modeling and Simulation Engineering at Old Dominion University. He is currently working as a graduate research assistant in the Virginia Modeling Analysis and Simulation Center. He has earned Master of Science (M.S.) in Modeling Simulation and Visualization Engineering from Old Dominion University (ODU) in 2018. He has received Master of Business Administration (M.B.A) degree from the Institute of Business Administration (IBA), University of Dhaka in 2016. He holds a Bachelor of Science degree in Electrical and Electronic Engineering from Bangladesh University of Engineering and Technology in 2006. Before joining Old Dominion University as a graduate student, he has worked in the Telecommunication industry for seven years. His research interests include but not limited to cyber-physical system security, power grid security analytics, cloud computing, machine learning, and big data.

Tze-Li Hsu received his doctoral degree in Sociology at Florida State University in 2014, and he is a tenure track faculty in the department of Sociology at Sam Houston State University since 2015. As an assistant profession of Sociology, he teaches various courses at the undergraduate/graduate level, including Social Statistics, Introduction to Ethnic Studies, and Sociology of Aging. He is interested in the intersection of mental health, aging, gender difference, and marital status in aging and the life course perspective. He has advanced statistical training including structural equation modeling, latent growth curve analysis and managed multiple large-scale projects throughout his time in graduate school. His scholarly work can be seen in Sociological Spectrum and Asian Population Studies. His research examines the association between marital status and mental health, with special attention to variations across age groups and to dimensions of psychological well-being among the elderly, using data from the Midlife and Development in the United States (MIDUS). His research on these topics has won Sociological Spectrum 2013 Outstanding Paper Award.

Chi-Hsuan Huang got his master degree from Graduate Institute of Networking and Multimedia, National Taiwan University in 2020.

Sachil Kumar received his Ph.D. in Forensic Chemistry from the King Georges Medical University, India in 2015. He received his Postgraduate degree in Forensic Science (2010) from Dr. B. R. Ambedkar University, Agra, India. Kumar holds a bachelor's degree in Chemistry (Hons) from the University of Delhi, India, 2008. He has worked as Guest Lecturer at the Department of Bio-Technology, Babasaheb Bhimrao Ambedkar University, India, from 2015 to 2016 and has taught many classes in Forensic Sciences to Masters students. Currently, he works as an Assistant Professor in the Department of Forensic Sciences, College of Criminal Justice, Naif Arab University for Security Sciences, Riyadh, Saudi Arabia. His research interests lie in the field of forensic chemistry, which focuses on the temperature-dependent postmortem degradation of cardiac Troponin-T and its association with postmortem interval. He has worked extensively with researchers in a variety of other forensic science disciplines, including personal identification using advanced imaging techniques, hair analysis for time mapping of drug use, and evidentiary aspects of bite-mark analysis. Dr. Kumar has been qualified the UGC-National Eligibility Test for Lectureship, and UGC-National Eligibility Test for Junior Research Fellowship (J.R.F.) in the subject Forensic Science. He has been awarded competitive grants from national funding bodies, including the International Travel Support Scheme, Science and Engineering Research Board (SERB), Department of Science and Technology (DST), Government of India, 2013. Kumar has published more than 30 peer reviewed scholarly research articles, as well as authored several book chapters published by IGI Global, Springer and Walter De Gruyter. He presented many posters and oral lectures at various national and international conferences. He is on numerous Editorial Boards and acts as a peer reviewer for a number of Forensic and Medicine journals.

Sachin Shetty is an Executive Director for Center of Secure and Intelligent Critical Systems in the Virginia Modeling, Analysis and Simulation Center at Old Dominion University. He holds a joint appointment as a Professor with the Department of Computational, Modeling and Simulation Engineering. Sachin Shetty received his PhD in Modeling and Simulation from the Old Dominion University in 2007 under the supervision of Prof. Min Song. Prior to joining Old Dominion University, he was an Associate Professor with the Electrical and Computer Engineering Department at Tennessee State University. He was also the associate director of the Tennessee Interdisciplinary Graduate Engineering Research Institute and directed the Cyber Security laboratory at Tennessee State University. He also holds a dual appointment as an Engineer at the Naval Surface Warfare Center, Crane Indiana. His research interests lie at the intersection of computer networking, network security and machine learning. His laboratory conducts cloud and mobile security research and has received over $12 million in funding from National Science Foundation, Air Office of Scientific Research, Air Force Research Lab, Office of Naval Research, Department of Homeland Security, and Boeing. He is the site lead on the DoD Cyber Security Center of Excellence, the Department of Homeland Security National Center of Excellence, the Critical Infrastructure Resilience Institute (CIRI), and Department of Energy, Cyber Resilient Energy Delivery Consortium (CREDC). He has authored and coauthored over 140 research articles in journals and conference proceedings and two books. He is the recipient of Fulbright Specialist award, EPRI Cybersecurity Research Challenge award, CCI Fellow, DHS Scientific Leadership Award and has been inducted in Tennessee State University's million dollar club. He has served on the technical program committee for ACM CCS, IEEE INFOCOM, IEEE ICDCN, and IEEE ICCCN. He is a Senior Member of IEEE.

Anu Singla has 19 years of total work experience in teaching and research in the field of Forensic Science. She was also the founder faculty member of Institute of Forensic Science & Criminology, Bundelkhand University, Jhansi and continuously involved in the development and growth of the department since its inception. Presently, she is working as Associate Professor at Institute of Forensic Science & Criminology, Bundelkhand University, Jhansi, India.

Yu Sun is a master student in Graduate Institute of Networking and Multimedia, National Taiwan University.

Index

A

Accidents 91
analysis 1-2, 5, 11-14, 18, 30-34, 36-41, 43, 45-47, 50, 55-57, 59-60, 64-65, 67, 69-70, 72-74, 76, 84, 89, 91, 94-99, 101-103, 112-118, 120, 124, 126, 133-134, 144-149, 153, 156, 159-160, 218, 225-228, 230-232, 234, 239, 241, 247
artificial intelligence 66, 89, 161, 225, 248

C

classification 29, 69, 77, 107, 115, 117, 163, 174-175, 218, 222, 225, 227-230, 235, 238, 248
Computer Vision 120, 126, 143, 146, 162, 179, 218-219
Convolutional Neural Networks 219
Credit card fraud 221-222, 227, 247-249
crimes 1, 9, 15-17, 29, 36, 38, 48-53, 55-56, 58, 60-61, 145, 148
cyber forensics 72, 77, 87, 220, 223-224, 247

D

data mining 222, 225, 227, 230-231, 249
deep learning 58, 62-64, 66-69, 72, 79, 84-87, 89, 143-144, 161-163, 179, 225
DIFFICULTIES 39, 42
digital forensics 62-63, 65-66, 84-86, 89-90, 143, 224-225
double compression 119, 121-122

E

e-commerce 220-222, 229-232, 235, 247
ensemble 119, 121, 137, 139, 143, 145, 227
environmental crimes 1, 9, 15-17, 29
Environmental Forensic Geophysics 15

F

fires 92, 94, 107-108, 115
forensic 1-3, 5, 9-13, 15-18, 22, 24-27, 29-48, 50, 62-63, 65, 68-69, 71-74, 76-77, 83, 86-89, 99-100, 102-103, 112, 115-118, 144-146, 148-149, 153, 156, 160, 220-223, 225-230, 239, 246-248
forgery detection 119, 121, 134, 143-144, 146

G

geochemistry 1-3, 6, 9-13, 31
grain size 28, 31, 37, 40-42, 46

I

Ignitable Liquid Residue 91-92, 105
image forensics 119-121, 125, 143, 160
imbalanced dataset 222, 226, 231, 235, 246
inorganic 1-2, 6, 13
Internet of Things 62, 64-65, 77, 85-90
IoT forensics 62-65, 68-69, 71-73, 75-76, 84-85, 88
IoT security 63-64, 67-68, 74, 86-87
isotopic 1-2, 9-11, 13

J

JPEG 119-126, 131, 134-135, 137-140, 142-147, 149

L

laboratories 5-6, 38, 94, 101
large feature mining 119, 121, 126, 140
license plate recognition 161, 165-166, 180, 184-185, 196, 199-200, 218-219

M

machine learning 58, 63, 66-67, 70-71, 86-87, 146, 163, 218, 220, 222, 225-232, 234-235, 238-239, 246-249
magnetic susceptibility 15, 17-18, 20, 22-29, 32, 35-36
methods 2, 5, 10, 16-17, 29-30, 33, 35-40, 44, 47, 57, 61, 67, 85-86, 98, 100, 112, 114, 116, 118-119, 121-123, 143, 149, 159, 162, 165, 180, 191-192, 195, 204-205, 223, 227-228, 232-233, 241, 246
Morphoscopy 42
Multimedia Forensics 148

N

nuclear 1, 9-13, 33, 46

O

organic 1-8, 11, 13, 22, 26, 112-113

P

pollution 1, 3, 6-7, 9, 11, 13-15, 17, 24, 26-27, 29, 45

R

Remote Sensing 16, 29, 35-36, 48, 50-52, 55-61

S

seam-carving 119-121, 134, 145-146
sedimentology 11, 37-39, 42, 44-46
smoke 91-93, 96-97, 104
social network 143, 153, 156, 159
source camera identification (SCI) 149

T

Terrorism 48-52, 56-61
terrorist groups 48, 52-56
transnational 48-53, 55-56, 58-61
Tunisia 1, 6, 9, 11-12, 15, 18, 29, 31-32, 37, 45, 48, 52-53, 55-57
Tunisian-Libyan borders 48

W

water 1, 3, 10, 12, 14-16, 22, 31, 33, 39, 45, 47, 60, 98, 105-106, 109

www.ingramcontent.com/pod-product-compliance
Lightning Source LLC
LaVergne TN
LVHW081331021026
849161LV00008BA/1937